Spire Study System

AFOQT Study Guide

AFOQT Prep & Study Book for the Air Force Officer Qualifying Test

STUDY SYSTEM + TEST PREP GUIDE + PRACTICE TEST QUESTIONS

Thank you for purchasing the *AFOQT Study Guide* from Spire Study System!

Your opinion matters!

As a dedicated educational publisher, we depend on you, our customer, to provide feedback (email: MyBookFeedback@outlook.com). We want to know what we did well and what areas we could further improve.

Likewise, your comments can help other shoppers make informed decisions about whether our system is right for them. If you are happy with your purchase, please take a few minutes to leave a review on Amazon!

Thank you for trusting Spire Study System for your test prep needs. We are always here for you.

Andrew T. Patton
Chief Editor
Spire Study System
Email: MyBookFeedback@outlook.com

Email: MyBookFeedback@outlook.com.

ISBN: 978-1-950159-61-1

Printed in the United States of America

Contents

INTRODUCTION

ABOUT THE AFOQT

Congratulations on embarking on the journey to become an officer in the United States Air Force! This decision marks a significant milestone in your career path, requiring commitment, discipline, and an unwavering sense of purpose. One of the first challenges on this journey is the Air Force Officer Qualifying Test (AFOQT). Success on this exam not only opens the door to countless opportunities within the Air Force but also affirms your readiness for the demands of leadership and service.

The AFOQT is a specialized assessment designed exclusively for individuals aspiring to become officers in the Air Force. While it shares some similarities with the Armed Services Vocational Aptitude Battery (ASVAB), the two tests differ significantly in their focus and structure. The ASVAB serves as a general enlistment test for all branches of the U.S. Armed Forces and evaluates candidates for enlisted roles. In contrast, the AFOQT is tailored specifically for officer candidates, placing a stronger emphasis on advanced cognitive abilities and specialized knowledge. For example, the AFOQT includes unique subtests such as Aviation Information and Instrument Comprehension, which are critical for Air Force operations and leadership roles. These distinctions make the AFOQT more comprehensive and challenging, underscoring the level of preparation required to succeed.

The Structure of the AFOQT

The AFOQT is a rigorous and comprehensive test designed to evaluate candidates across a broad spectrum of skills and knowledge areas. It consists of 12 subtests, each individually timed with varying time limits. These subtests assess critical skills, including word knowledge, mathematics, reading comprehension, aviation concepts, and instrument comprehension. This diverse subject range ensures that candidates are evaluated comprehensively, reflecting the wide range of responsibilities officers face in the Air Force.

The exam includes a total of 550 questions, divided into two primary categories: the Self-Description Inventory and the remaining 11 subtests. The Self-Description Inventory is a personality assessment comprising 240 questions that require no preparation. Instead, this section focuses on evaluating your traits and tendencies to assess suitability for leadership roles. The remaining 310 questions are spread unevenly across the other subtests, testing technical, academic, and cognitive skills essential for officer roles. The broad scope of these subtests demonstrates the importance of a balanced study plan, emphasizing both subject-matter knowledge and problem-solving skills.

Completing the AFOQT requires around 3.5 hours of active test time. However, when factoring in breaks and instructions, candidates should plan for a total commitment of approximately 5 hours. This extended duration mirrors the real-life endurance and focus required of Air Force officers, making it as much a test of perseverance as of knowledge and ability.

Preparation for the AFOQT is critical. Candidates should assess their strengths and focus on areas for improvement, whether it's mastering verbal analogies, refining mathematical skills, or understanding aviation concepts. Success depends on more than just knowledge; it requires time management, critical thinking, and composure under pressure—skills that are vital in the dynamic environment of Air Force operations.

Exam Areas of the AFOQT

Here are the details of the AFOQT's 12 subtests:

1. **Verbal Analogies** evaluates your English language skills. Based on your mastery of the precise and nuanced meaning of words and their usage, you must be able to use logic and context to determine the right answer.

2. **Arithmetic Reasoning** tests your understanding of basic math terms and problem-solving methods: addition, subtraction, multiplication, division, ratio, rate, proportion, percentage, time, distance, and so on.

3. **Word Knowledge** evaluates your capability to understand the written English language. You need a decent vocabulary to do well on this subtest as you will be asked to find synonyms for given words.

4. **Math Knowledge** evaluates your mastery of arithmetic, algebra, and geometry. The questions test your knowledge of absolute values, algebraic expressions, equations, exponents, factors, geometric properties, and so on.

5. **Reading Comprehension** measures your ability to read and understand the written English language. You will read written passages and then answer multiple-choice questions about the content. You need to be

able to read the materials at a reasonable speed and have good comprehension as a lot of reading must be completed before you can answer the questions.

6. **Situational Judgment** measures your ability to effectively lead by asking you to determine the MOST effective and LEAST effective means to resolve an issue or problem. These problems are intended to resemble what you may encounter in real life as an officer.

7. **Self-Description Inventory** evaluates your personality. Answers to questions within this subtest are not graded and do not count toward the outcome of the test (i.e., your comprehensive score).

8. **Physical Science** covers a broad spectrum of middle to high school-level science concepts. These concepts include Chemistry, Mechanics, Electronics, and Earth Science.

9. **Table Reading** evaluates your ability to locate a specific number on large grid using X & Y coordinates, without a straight edge, and under relatively tight time pressure.

10. **Instrument Comprehension** evaluates your ability to determine the position of an aircraft in flight by instrument readings of altitude, compass heading, rate of climb or dive, degree of bank, and so on.

11. **Block Counting** tests your ability to visualize in three-dimensional space. You will see the three-dimensional rending of a group of blocks and then count how many "other" blocks from all sides are touching one individual block.

12. **Aviation Information** tests your knowledge of general flight and airplane mechanics, procedures, and functions.

AFOQT Minimum Score Requirements

Even though there are 12 subtests, your AFOQT results will only be scored across 6 composites, 5 of which have minimum required scores.

Your scores are reported as a 1st–99th percentile ranking that indicates indicating how your performance compares relative to other test takers. AFOQT scores are not comprised of the percentage of questions you answered correctly. For instance, if your score is 75, it does NOT mean you got 75% of the test questions correct. Instead, a score of 75 means you did better than 75% of other test-takers.

AFOQT composite scores are calculated from different combinations of the AFOQT subtests.

However, an important nuance of the AFOQT composite scoring methodology is that the same subtest can be included in multiple composite scores. For instance, the results from the Word Knowledge subtest are used in 3 different composite scores: Verbal, Academic Aptitude, and Air Battle Manager. Hence, excelling in one subtest can influence multiple composite scores.

Here are the six composite scores and the corresponding subtests from which they are derived:

- **Air Battle Manager (ABM).** This composite score assesses the knowledge and abilities considered necessary for candidates to successfully complete ABM training.
    - Included subtests: Verbal Analogies, Word Knowledge, Table Reading, Instrument Comprehension, Block Counting, and Aviation Information.
- **Academic Aptitude.** This composite score assesses verbal and quantitative knowledge and abilities. The Academic Aptitude composite combines all subtests included in the Verbal and Quantitative composites (as described later in this list).
    - Included subtests: Verbal Analogies, Word Knowledge, Arithmetic Reasoning, and Math Knowledge.
- **Combat Systems Officer (CSO).** This composite assesses the knowledge and abilities considered necessary for candidates to successfully complete CSO training.
    - Included subtests: Verbal Analogies, Arithmetic Reasoning, Table Reading, Math Knowledge, Block Counting, and Physical Science.
- **Pilot.** This composite assesses the knowledge and abilities considered necessary for candidates to successfully complete manned and unmanned pilot training.
    - Included subtests: the Instrument Comprehension, Table Reading, Aviation Information, and Math Knowledge.

- **Quantitative.** This composite assesses candidates' quantitative knowledge and abilities.
    - Included subtests: Arithmetic Reasoning and Math Knowledge.
- **Verbal.** This composite assesses candidates' verbal communication knowledge and abilities.
    - Included subtests: Verbal Analogies, Word Knowledge, and Reading Comprehension.

Rated Composites

The term "rated composites" refers to specific scores tied to career paths that require specialized training and certification within the Air Force. These include the Pilot, Combat Systems Officer (CSO), and Air Battle Manager (ABM) composites. Individuals pursuing these roles must meet rigorous minimum score requirements, as these positions involve responsibilities such as flying aircraft, managing complex systems, and overseeing air operations.

For example, the Pilot composite evaluates aptitude in areas essential for manned and unmanned flight training, including Instrument Comprehension and Aviation Information. Similarly, the CSO composite assesses skills crucial for navigation and mission management, such as Arithmetic Reasoning and Table Reading. The ABM composite focuses on abilities needed for air operations coordination, leveraging subtests like Block Counting and Verbal Analogies.

These rated positions are critical to Air Force missions and demand highly specialized skills. As such, achieving strong scores in the corresponding composites is not only a requirement but also an indication of a candidate's potential success in these demanding career paths.

AFOQT Composite Score Composition

Subtest	Items	**Pilot**	**CSO**	**ABM**	Academic	Verbal	Quantitative
Verbal Analogies	25			✓	✓	✓	
Arithmetic Reasoning	25				✓		✓
Word Knowledge	25		✓		✓	✓	
Math Knowledge	25	✓	✓	✓	✓		✓
Reading Comprehension	25				✓	✓	
Situational Judgment Test	50						
Self-Description Inventory	240						
Physical Science	20						
Table Reading	40	✓	✓	✓			
Instrument Comprehension	25	✓		✓			
Block Counting	30		✓	✓			
Aviation Information	20	✓		✓			

Note. Rated composites are in **bold**.

The minimum score you must achieve depends on the U.S. Airforce career path you would like to pursue, as illustrated in the following lists. Note that the Academic Aptitude composite does not have a minimum score requirement for any of these career paths.

General Commissioning (Non-Rated Positions):

- Verbal Composite: Minimum score of 15
- Quantitative Composite: Minimum score of 10

Pilot Candidates (Including Remotely Piloted Aircraft Operators):

- Pilot Composite: Minimum score of 25
- Combat Systems Officer (CSO) Composite: Minimum score of 10
- Verbal Composite: Minimum score of 15
- Quantitative Composite: Minimum score of 10

Combat Systems Officer (CSO) Candidates:

- CSO Composite: Minimum score of 25
- Pilot Composite: Minimum score of 10
- Verbal Composite: Minimum score of 15
- Quantitative Composite: Minimum score of 10

Air Battle Manager (ABM) Candidates:

- ABM Composite: Minimum score of 25
- Verbal Composite: Minimum score of 15
- Quantitative Composite: Minimum score of 10

AFOQT Super-Scoring

Here is a piece of good news if you plan to take the AFOQT. According to the Secretary of the Air Force Public Affairs, AFOQT composite scores in the past were counted only once, for the specific test you took. Since late 2021, though, the highest AFOQT composite scores from any AFOQT attempt are used instead of the most recent score. In other words, a candidate's best composite score across all test attempts will be used as the score of record. Plus, the minimum time allowed between testing is now 90 days instead of 180 days.

Schedule and Register for the AFOQT

After deciding to take the AFOQT and pursue a U.S. Air Force career, you will want to reach out to your recruiter or your Air Force ROTC POC as soon as possible to get help with scheduling and registration. This step will also help you plan your preparation for the test.

There is no fee to take the test. However, the U.S. Air Force generally only gives you two chances. If you do not get satisfactory scores the first time, you must wait for at least 90 days to attempt the AFOQT a second time. (Prior to 2021, this waiting period was 180 days.) Under rare circumstances, the Air Force may allow you a third attempt, but a waiver is required. Our recommendation is therefore to prepare early, prepare well, and pass the test on your first try!

Ace the AFOQT: Strategies for Success

You know you will have to take the AFOQT. No matter how good a student you have been in the past, those butterflies in your stomach flutter anyway. Luckily, you still have a few weeks or months to prepare. Now what?

Your performance on the test will be largely determined by 1) how well you've studied the relevant materials in the past and 2) how well you familiarize and prepare for the test in the weeks and months prior to the test.

Based on our own test-taking experiences and scientific studies of test preparation, memorization, and cognitive psychology, we summarized the following strategies to facilitate and optimize your test preparation.

Choose the Right Test Preparation Tool

Test preparation tools can include courses, books, tutors, flash cards, and so on. Depending on your individual needs and current readiness, the best tool or combination of tools can be different. You are the most qualified person to assess what's best for yourself. When chosen properly, the right tool(s) can save you considerable time and enhance your test performance.

This book strives to strike the right balance between comprehensiveness and conciseness. We provide sufficient coverage to enable you to pass the test with a comfortable safety margin or earn a high score, but keep things relatively concise to help you get ready for the test as quickly as feasible.

Study the Preparation Materials with Systematic Repetition

We ourselves have been tested countless times in academic and professional settings. Based on those experiences and scientific research, we know that learning is more effective when repeated in spaced-out sessions.

Information repeatedly learned over a spaced-out period allows a learner to better remember and better recall the information being learned. In fact, strategically spaced repetition has been scientifically proven to better encode information into long-term memory.

When applying this learning principle, you will want to be systematic about test preparation: First, start early so that you do not have to cram right before the test. Second, while studying new materials, methodically review the older materials you studied a few days ago, well before your memory begins to fade. Using this strategy, you can go through the exam preparation materials at an aggressive rate and still maintain satisfactory retention because you repeatedly reinforce your fading memories.

Another highly beneficial practice can also be helpful: as you take practice tests, take note of those questions you answered incorrectly, or you answered correctly but want to review again anyway. Then write down the page number and question number for each. In the days and weeks that follow, you can systematically revisit the questions multiple times so that you can commit the current information to memory.

Apply Memorization Techniques

Rule number one for memorization is comprehension. A good understanding will go a long way in ensuring long-term memory and effective recall. However, rote memorization is still invariably needed in almost all studies. In such situations, the application of various memorization techniques can help tremendously with retaining information effectively.

Many books have been written on this subject. But most advice boils down to two techniques: imagination and association. These techniques are beyond the scope of this book, but a quick Google search will generate ample resources. You will be amazed how handy these techniques can be in test preparation and beyond.

Take Practice Tests

Taking practice tests is beneficial on multiple fronts:

- Helps familiarize you with the format, coverage, and difficulty level of the test.
- Helps you apply the knowledge you have learned in a different context, reinforcing memory retention.
- Helps you learn from having to figure out the correct answer since you'll probably run into questions where you don't know the answer, or you answer incorrectly.
- Boosts your confidence and eases your nervousness when taking the actual exam thanks to performing increasingly better on practice exams.

Consequently, taking practice tests is an integral part of preparing for any test.

Devote Yourself

For tests with set dates that you have no choice over, plan early and leave yourself enough time to prepare. For those tests that you can choose a test date, you have your choice of two diametrically different approaches to prepare:

- Approach #1. Devote every hour you can to preparing for the exam and then take it as soon as you feel ready.
- Approach #2. Fall prey to all the distractions in life, and study the materials at your leisure, halfheartedly in a drawn-out process. Procrastinate until you have your back against a wall and must take the test.

The more drawn-out the process, the more time you will have to spend reinforcing fading memories. In the end, you spend more total hours on exam preparation if you adopt the second approach.

Thus, in test preparation, if you would like to reduce the cumulative amount of "pain" you must endure and increase your odds of scoring well, then the best approach is to completely commit yourself to a single-minded, intensive preparation period. You can then take the exam when you feel confident and ready.

AFOQT 60-Day Test Prep Study Plan

The study plan at the end of this chapter is designed to help you prepare effectively for the AFOQT exam in 60 days. It is realistic and works for a typical high school student in his/her junior or senior year or someone with equivalent academic preparedness. Here's how you can optimize your study sessions by following the plan:

1. Daily Focus and Goal Setting

Each day of the plan is intentionally focused on one or two core areas. By dedicating specific days to each area, you avoid overloading yourself and maintain a steady, focused progression. Ensure you set realistic goals for each session, such as completing a certain number of pages, a chapter, or finishing a set of practice questions.

2. Efficient Time Management

- Study Blocks: Allocate at least a couple of hours per day to your studies, breaking them into shorter, focused blocks (e.g., 30–45 minutes per block). Use techniques like the Pomodoro method (25 minutes study, 5 minutes break) to stay focused and avoid burnout.
- Prioritize Weak Areas: If you know certain sections are more challenging for you, e.g., math, then plan to spend more time on these areas.
- The amount of work designed for each day is not uniform. Days of lighter workload are sprinkled in the schedule so you can have more time to relax and recuperate on those days.

3. Balancing Study and Review

- Initial Study Days: We designed the study plan so that you start with studying new materials, which builds your foundational skills and familiarity with test concepts.
- Review Days: We include dedicated review days dispersed throughout the test preparation. Reviews days are critical for reinforcing knowledge and identifying areas where you may need additional practice. Use these days to go over key concepts, formulas, and incorrect answers from practice tests, etc.

4. Practice Tests and Test Simulation

- Practice Under Real Conditions: On practice test days, take full-length practice tests in one sitting, replicating test-day conditions. This will help you gauge your readiness and stamina for the actual exam.
- Analyze Your Performance: Use the review days following each practice test to analyze your mistakes and review the correct answers. Look for patterns in your errors—whether it's certain types of math problems or recurring grammar mistakes—and focus on improving those areas.

5. Rest and Recharge

Remember to schedule short breaks during your study sessions to maintain focus and retain information more effectively. Even a 5-minute break can help reset your mind. If you make good progress and end up ahead of the study plan, feel free to take a day off from time to time.

6. Building Confidence for Test Day

- Final Reviews: The last couple of weeks are dedicated to reviewing and taking practice tests. These days are designed to boost your confidence by reinforcing what you've learned.
- Positive Mindset: Use these final days to cultivate a positive mindset. Remind yourself of how much progress you've made, and trust in the preparation you've done.

By sticking to this study plan, managing your time effectively, and focusing on both learning and review, you'll be well-prepared to tackle the AFOQT exam with confidence. It is always advisable to start preparing for the test early though. If you have more than 60 days to prepare, the best time to get started is right away!

Success Skills for Test Day

Time Management

Each AFOQT subtest has strict time limits. Hence, managing your time effectively is essential. Practice pacing yourself by taking timed practice tests. During the test, avoid spending too much time on difficult questions—make your best guess and move on, as you won't have time to dwell on every question.

Staying Calm and Focused

Test anxiety can hinder performance, so practice relaxation techniques like deep breathing or visualization to stay calm during the exam. Maintaining focus for the entire test is crucial, as it's a lengthy process. Breaks are built into the test, so use them to refresh your mind.

Reading and Following Instructions Carefully

Ensure you read and understand all instructions before beginning each subtest. Rushing through instructions or missing key details can lead to avoidable mistakes.

Answer Every Question

The Air Force Officer Qualifying Test does not penalize for incorrect answers. Therefore, it's advantageous to answer every question, even if it requires guessing. With each question offering multiple-choice answers, educated guessing becomes a valuable skill. By eliminating obviously incorrect options, you enhance your chances of selecting the correct answer.

It's important to note that the AFOQT is usually administered in a paper-and-pencil (P&P) format, not as a computer-adaptive test (CAT). In the P&P format, you can review and change your answers within a subtest's time limit. However, once you complete a subtest and move on, you cannot return to it. Therefore, ensure you're confident in your answers before proceeding to the next subtest.

Rest and Hydration

Ensure you get a good night's sleep before the test. Being well-rested helps with concentration and memory. Staying hydrated also supports cognitive function, but avoid excessive caffeine, as it can increase anxiety. By practicing these strategies and remaining calm, focused, and prepared, you will improve your chances of performing well on the test day.

Conquer Test Anxiety

Dealing with test anxiety is a crucial skill that can significantly impact one's academic and professional life. Test anxiety is a type of performance anxiety that occurs when an individual feels an intense fear or panic before, during, or after an examination. It can manifest through various symptoms, including nervousness, difficulty concentrating, negative thoughts, physical symptoms such as headaches or nausea, and even panic attacks. Fortunately, there are effective strategies to manage and overcome test anxiety, ensuring that it doesn't hinder one's ability to perform to the best of their abilities.

Test anxiety can stem from fear of failure, lack of preparation, previous negative experiences, or high pressure to perform well. Identifying the root cause is essential in developing a targeted approach to manage anxiety.

One of the most effective ways to reduce test anxiety is thorough preparation. Begin studying well in advance of the test date. This allows ample time to understand the material, reducing the likelihood of feeling overwhelmed as the test approaches. Create a study plan that breaks down the material into manageable sections. Use organizers, such as outlines, flashcards, or mind maps, to make the study process more efficient and less daunting.

Simulating test conditions can help alleviate anxiety. Practice with timed quizzes or tests in a quiet environment. This not only helps with time management but also makes the actual test environment feel more familiar.

Mindset and Attitude Adjustments

Transform negative thoughts into positive affirmations. Instead of thinking, "I'm going to fail," tell yourself, "I'm prepared and will do my best." Positive thinking can enhance self-confidence and reduce nervousness. Furthermore, accept that it's okay to be nervous and that feeling anxious doesn't mean you will perform poorly. Recognize that anxiety can sometimes motivate you to prepare better.

Relaxation Techniques

Deep breathing exercises can be effective in managing physical symptoms of anxiety. Techniques such as the 4-7-8 method, where you inhale for four seconds, hold your breath for seven seconds, and exhale for eight seconds, can help calm the nervous system.

The Progressive Muscle Relaxation technique can also help. This involves tensing and then slowly relaxing each muscle group in the body to reduce the physical symptoms of stress and anxiety.

During the Test

If you find yourself becoming anxious, pause for a moment, take a few deep breaths, and refocus on the question in front of you. Avoid dwelling on what you might have gotten wrong or what's coming next.

Test anxiety is a common challenge that many face, but it doesn't have to be a barrier to success. By understanding its causes, implementing effective study and preparation strategies, adopting relaxation techniques, and seeking support when needed, individuals can overcome test anxiety. Remember, the goal is not to eliminate anxiety completely but to manage it effectively so that it doesn't interfere with performance.

AFOQT 60-Day Study Plan

Day 1 Chapter 1	Day 2 Chapter 2 & Practice Set	Day 3 Chapter 3 & Practice Set	Day 4 Chapter 4 Section 1	Day 5 Chapter 4 Section 1	Day 6 Chapter 4 Section 1
Day 7 Chapter 4 Section 2	Day 8 Chapter 4 Section 2	Day 9 Chapter 4 Section 2	Day 10 Chapter 4 Section 2	Day 11 Review: Chapter 1	Day 12 Chapter 5
Day 13 Review: Chapter 2 & 3	Day 14 Chapter 6 §1	Day 15 Chapter 6 §1	Day 16 Chapter 6 §2	Day 17 Chapter 6 §2	Day 18 Chapter 6 §3
Day 19 Chapter 6 §4 & Practice Set 1	Day 20 Chapter 6 §4 Practice Set 2	Day 21 Review: Chapter 4	Day 22 Review: Chapter 4	Day 23 Chapter 7	Day 24 Chapter 7 Practice Set
Day 25 Review: Chapter 5	Day 26 Chapter 8	Day 27 Chapter 8 Practice Set	Day 28 Chapter 9 & Practice Set	Day 29 Chapter 10 Section 1 & 2	Day 30 Chapter 10 Section 3
Day 31 Chapter 10 Section 4	Day 32 Review: Chapter 6 §1 & §2	Day 33 Review: Chapter 6 §3 & §4	Day 34 Review: Chapter 6 Practice Sets	Day 35 Chapter 10 Section 5	Day 36 Chapter 10 Section 6 & 7
Day 37 Chapter 10 Section 8 & 9	Day 38 Chapter 10 Section 10	Day 39 Review: Chapter 7	Day 40 Chapter 10 Section 11 & 12	Day 41 Chapter 10 Section 13 & 14	Day 42 Chapter 10 Practice Set 1
Day 43 Chapter 10 Practice Set 2	Day 44 Chapter 11 & 12	Day 45 Review: Chapter 8	Day 46 Review: Chapter 9	Day 47 Practice Test 1	Day 48 Practice Test 2
Day 49 Practice Test 3	Day 50 Review: Chapter 4	Day 51 Practice Test 4	Day 52 Practice Test 5	Day 53 Practice Test 6	Day 54 Practice Test 7
Day 55 Practice Test 8	Day 56 Practice Test 9	Day 57 Review: Chapter 10 Secction 1-5	Day 58 Review: Chapter 10 Secction 6-14	Day 59 Practice Test 10	Day 60 Practice Test 10

I'm Ready!

CHAPTER 1. INSTRUMENT COMPREHENSION

COMPASS

The Instrument Comprehension section requires you to determine which image of an aircraft corresponds to a given compass heading and artificial horizon. This section is quite easy once you get familiarized with the concept, but you first must know how the artificial horizon works and how the compass corresponds to the images.

The compass is quite simple. If the compass is showing a heading of "East," the correct image of the aircraft will be pointing "right" on the page. If the compass heading is "South," the aircraft image will appear to be heading directly at you. See the following examples:

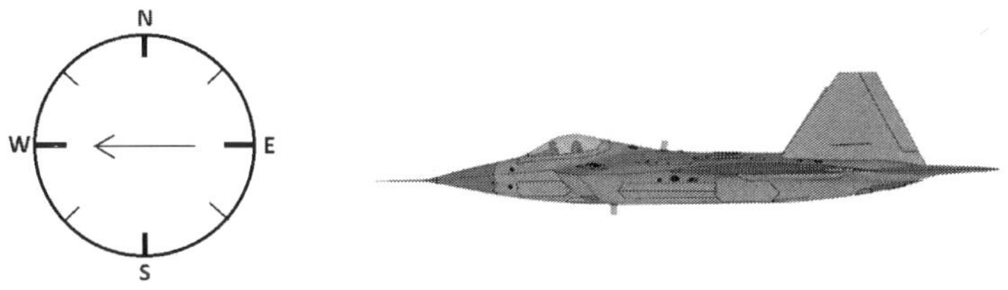

ARTIFICIAL HORIZON

The artificial horizon is slightly more difficult to understand for those who have not seen it before, but only takes a few minutes to grasp the concept. In simple terms, the artificial horizon represents both the aircraft's pitch (i.e., whether it is climbing or descending in altitude) and its roll angle relative to the horizon (i.e., it shows whether the aircraft is banking left or right).

The following image is an artificial horizon showing an aircraft flying level and not banking in either direction. This serves as a baseline for your reference. The flat line you see going from left to right across the center is the horizon. On the actual instrument, the horizon will appear to move while the plus sign suspended above a carrot stays stationary.

The following image is an artificial horizon that represents an aircraft in a dive. As you see, the carrot is now below the horizon.

A banking aircraft is the trickier part for many people to visualize because it often feels reversed. The tip here is to pretend you are in the aircraft and view what the horizon would look like from that vantage point. The following image shows an artificial horizon of an aircraft that is flying level with the horizon, but banking to the right.

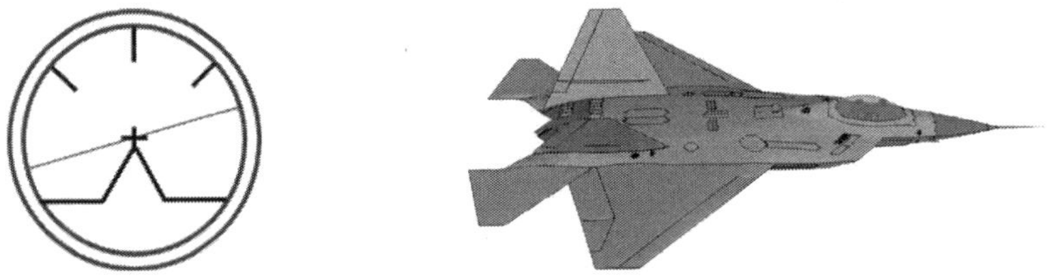

The following image shows an artificial horizon of an aircraft that is flying level with the horizon, but banking to the left.

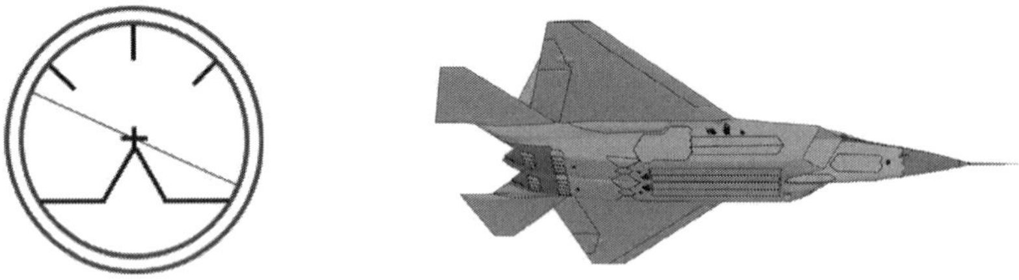

COMPASS AND ARTIFICIAL HORIZON COMBINED

Finally, to put it all together, look at the following examples showing a compass, artificial horizon, and the corresponding image of the aircraft.

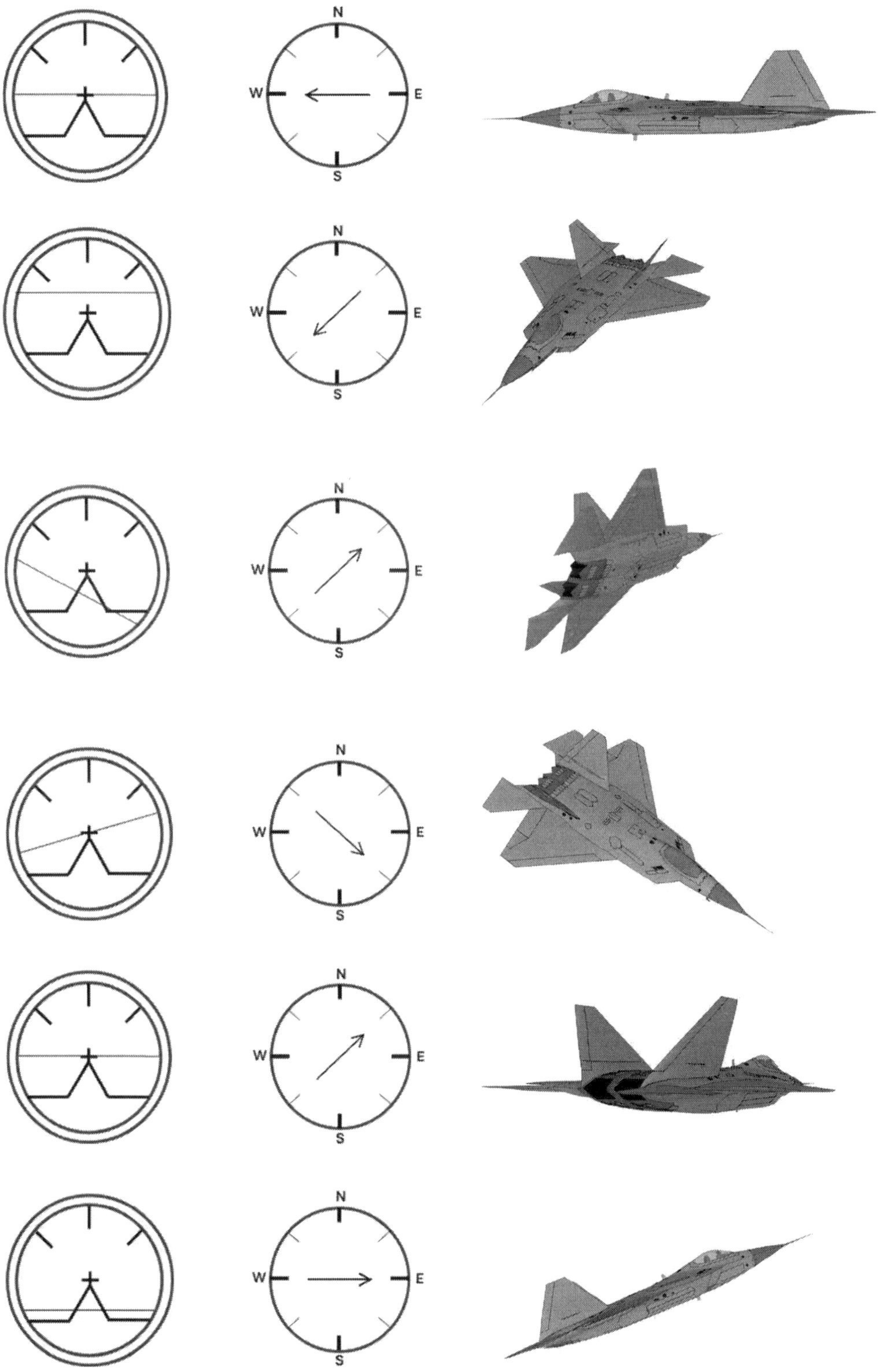

CHAPTER 2. VERBAL ANALOGIES

INTRODUCTION

As you know, an analogy is simply a comparison of two similar things, used primarily for the purpose of clarification or description. The Verbal Analogies section of the AFOQT is the same concept, but you will need to use logic and context to determine the right answer. You will encounter two question formats:

1. COLD is to HOT as UP is to ________.

   A. Down
   B. Left
   C. Colder
   D. Hotter
   E. Inside out

2. COLD is to HOT as ________.

   A. Up is to down
   B. Parent is to child
   C. Red is to green
   D. Hat is to sock
   E. Fingers are to toes

Both examples are elementary level simply to illustrate the format. The answer to both is of course choice A. However, for the second format, you must remember to keep the order of words the same! For example, look at the following question and answer choices:

3. PARENT is to CHILD as_______.

   A. Puppy is to dog
   B. Cat is to kitten
   C. Cake is to bread
   D. Cup is to glass
   E. Down is to up

Answer B is the correct choice. Choice A is incorrect because the order is backwards—it is analogous to child/parent rather than parent/child.

The questions on the AFOQT will be significantly more difficult and require a solid vocabulary as well as the ability to use logic, reason, and context clues to determine the correct answer.

The following list is not exhaustive, but here are some examples of analogies you will encounter:

- **Part to whole:** WHEEL is to CAR as BLADE is to FAN.
- **User to Tool:** COOK is to CHEF as OPERATE is to SURGEON
- **Numerical:** 2 is to 6 as 3 is to 9
- **Geographic:** AUSTIN is to TEXAS as DENVER is to COLORADO
- **Cause to effect:** PUSH is to MOVE as FIRE is to BURN
- **General to specific:** CAR is to CHEVROLET as FAST FOOD is to MCDONALDS
- **Object to function:** SOCCER BALL is to KICK as BASEBALL BAT is to HIT

VERBAL ANALOGIES PRACTICE SET

Select the correct answer from the choices given.

1. TEACHER is to STUDENT as _______.

A) Athlete is to Coach
B) Doctor is to Patient
C) Artist is to Painting
D) Pilot is to Plane
E) Worker is to Boss

2. ACCELERATE is to DECELERATE as EXPAND is to _______.

A) Shrink
B) Contract
C) Reduce
D) Compress
E) Narrow

3. REPRIMAND is to PRAISE as OBSTRUCT is to _______.

A) Block
B) Clear
C) Divert
D) Assist
E) Hinder

4. CLOCK is to TIME as THERMOMETER is to _______.

A) Temperature
B) Weather
C) Mercury
D) Heat
E) Cold

5. TELESCOPE is to STAR as BINOCULARS are to _______.

A) Tree
B) Bird
C) Forest
D) Horizon
E) Mountain

6. EYE is to SEE as EAR is to _______.

A) Noise
B) Listen
C) Hear
D) Sound
E) Voice

7. FISH is to WATER as BIRD is to _______.

A) Air
B) Sky
C) Tree
D) Nest
E) Wing

8. SUGAR is to SWEET as SALT is to _______.

A) Bitter
B) Sour
C) Salty
D) Spicy
E) Bland

9. HAND is to ARM as FOOT is to _______.

A) Knee
B) Toe
C) Shoe
D) Leg
E) Sock

10. WINTER is to COLD as SUMMER is to _______.

A) Heat
B) Warm
C) Rain
D) Sun
E) Vacation

11. TELESCOPE is to STARS as MICROSCOPE is to _______.

A) Cells
B) Atoms
C) Bacteria
D) Molecules
E) Insects

12. TEETH are to CHEW as FEET are to _______.

A) Walk
B) Shoes
C) Kick
D) Run
E) Dance

13. CAT is to FUR as BIRD is to _______.

A) Wings
B) Feathers
C) Fly
D) Nest
E) Eggs

14. PERSIST is to ABANDON as RECALL is to _______.

A) Forget
B) Remember
C) Hesitate
D) Retrieve
E) Retain

15. TRANSLUCENT is to OPAQUE as AMICABLE is to _______.

A) Hostile
B) Cooperative
C) Neutral
D) Friendly
E) Obnoxious

16. AMBIGUOUS is to UNCLEAR as SALIENT is to _______.

A) Vague
B) Inconspicuous
C) Prominent
D) Hidden
E) Obscure

17. PRAGMATIC is to IDEALISTIC as CAUTIOUS is to _______.

A) Reckless
B) Hesitant
C) Bold
D) Fearful
E) Indecisive

18. TEACHER is to SCHOOL as DOCTOR is to _______.

A) Clinic
B) Medicine
C) Hospital
D) Patient
E) Surgery

19. CAR is to DRIVER as PLANE is to _______.

A) Pilot
B) Passenger
C) Wing
D) Engine
E) Mechanic

20. DAY is to LIGHT as NIGHT is to _______.

A) Moon
B) Stars
C) Sleep
D) Darkness
E) Dreams

21. DOG is to BARK as CAT is to _______.

A) Tail
B) Meow
C) Scratch
D) Fur
E) Purr

22. ENGINE is to CAR as BATTERY is to _______.

A) Light
B) Television
C) Phone
D) Fan
E) Motor

23. RAIN is to CLOUD as LIGHT is to _______.

A) Star
B) Fire
C) Sun
D) Leaf
E) Heat

24. FABRICATE is to DESTROY as CONSTRUCT is to _______.

A) Build
B) Dismantle
C) Strengthen
D) Renovate
E) Create

25. VERBOSE is to CONCISE as SUPERFICIAL is to _______.

A) Deep
B) Detailed
C) instinctive
D) Critical
E) Profound

Answers and Explanations

1. B) Doctor is to Patient

Explanation: A teacher instructs a student, just as a doctor cares for a patient. Both relationships involve guidance or service.

2. B) Contract

Explanation: Accelerate is the opposite of decelerate, just as expand is the opposite of contract.

3. D) Assist

Explanation: Reprimand is the opposite of praise, just as obstruct is the opposite of assist.

4. A) Temperature

Explanation: A clock measures time, just as a thermometer measures temperature. Both are tools for measurement.

5. B) Bird

Explanation: A telescope is used to view stars, just as binoculars are used to view birds.

6. C) Hear

Explanation: An eye is used for seeing, just as an ear is used for hearing.

7. A) Air

Explanation: A fish lives in water, just as a bird moves and lives in the air.

8. C) Salty

Explanation: Sugar is associated with sweetness, just as salt is associated with saltiness.

9. D) Leg

Explanation: A hand is a part of an arm, just as a foot is a part of a leg.

10. A) Heat

Explanation: Winter is associated with cold weather, just as summer is associated with hot weather.

11. C) Bacteria

Explanation: A telescope is used to observe stars, just as a microscope is used to observe bacteria.

12. A) Walk

Explanation: Teeth are used for chewing, just as feet are used for walking.

13. B) Feathers

Explanation: A cat is covered in fur, just as a bird is covered in feathers.

14. A) Forget

Explanation: Persist is the opposite of abandon, just as recall is the opposite of forget.

15. A) Hostile

Explanation: Translucent is the opposite of opaque, just as amicable (friendly) is the opposite of hostile.

16. C) Prominent

Explanation: Ambiguous means unclear, just as salient means prominent or noticeable.

17. A) Reckless

Explanation: Pragmatic is the opposite of idealistic, just as cautious is the opposite of reckless.

18. C) Hospital

Explanation: A teacher works in a school, just as a doctor works in a hospital.

19. A) Pilot

Explanation: A car is operated by a driver, just as a plane is operated by a pilot.

20. D) Darkness

Explanation: Day is associated with light, just as night is associated with darkness.

21. B) Meow

Explanation: A dog makes a barking sound, just as a cat makes a meowing sound.

22. C) Phone

Explanation: An engine powers a car, just as a battery powers a phone.

23. C) Sun

Explanation: Rain comes from clouds, just as light comes from the sun.

24. B) Dismantle
Explanation: Fabricate is the opposite of destroy, just as construct is the opposite of dismantle.

25. E) Profound
Explanation: Verbose (wordy) is the opposite of concise, just as superficial (shallow) is the opposite of profound.

CHAPTER 3. TABLE READING

INTRODUCTION

The Table Reading section of the AFOQT consists of 40 multiple-choice questions, and your task is to locate specific numbers on a large grid using X and Y coordinates. On the surface, this task may seem straightforward, but three key factors make it more challenging than it first appears:

- The grid is exceptionally large, filled with numerous rows and columns of tightly packed numbers.
- You are NOT allowed to use a straight edge or any other tools to guide your search for the coordinates.
- The entire section has a strict 7-minute time limit, giving you an average of just over 10 seconds per question.

These constraints can quickly turn this seemingly simple task into a high-pressure exercise. The combination of the grid's size, the time limit, and the prohibition against aids like a straight edge can lead to visual fatigue and mental overload. You must juggle multiple pieces of information in your head: the question number, the X-coordinate, the Y-coordinate, the correct value from the grid, and the matching answer choice. The task is a test of focus, precision, and time management under pressure.

SUCCESS STRATEGIES

While the time pressure is significant, the most critical advice for success in this section is that accuracy is more important than speed. Answering only 35 questions with near-perfect accuracy is far better than rushing to complete all 40 questions but making many errors along the way. Rushing will lead to mistakes, frustration, and forgetfulness. To maintain accuracy and focus, follow these tips:

- Move Methodically: Work through the questions in order without skipping around. Maintain a steady and efficient pace, but don't let the time limit pressure you into rushing.
- Stay Focused: Each question introduces new coordinates and potential distractions. Take a moment to refocus between questions so you don't mix up numbers.
- Practice Efficiently: Practice using grids of similar size and density to familiarize yourself with the task. Set a timer for 7 minutes, go through the table carefully at a brisk pace, and see how many questions you can answer accurately.

Example Question

X-Value

Y-Value	-15	-14	-13	-12	-11	-10	-9	-8	-7	-6	-5	-4	-3	-2	-1	0	1	2	3	4	5	6	7	8	9	10	11	12	13	14	15
15	99	74	92	85	82	71	28	75	14	6	68	1	4	2	30	2	57	56	32	93	67	2	73	49	95	85	29	85	71	37	3
14	47	90	3	69	86	95	18	34	73	31	84	19	30	76	45	3	71	14	19	54	95	10	91	87	92	41	1	61	27	37	48
13	97	79	58	88	0	31	10	21	58	63	57	73	41	85	3	25	35	63	17	43	79	73	61	22	63	9	11	33	49	64	88
12	21	27	19	45	76	29	27	1	73	64	12	61	97	95	37	45	82	43	81	65	22	100	11	23	100	25	38	6	82	70	91
11	10	88	72	97	9	13	83	74	99	5	25	1	44	74	64	28	95	2	39	41	51	61	73	84	55	43	62	85	14	48	18

Question 1: What is the number at X = 7 and Y = 12?

Solution: Locate column 7 under the X-axis, then move down to row 11 on the Y-axis. The intersection of these coordinates contains your target number, which is 11 in the above example. Once you find the number, scan the answer choices to select the correct one.

Question 2: What is the number at X = -4 and Y = 13?

Answer: 73.

TABLE READING PRACTICE SET

Select the correct answer from the choices given. For practice only. This practice set does not reflect the actual number of questions in the test.

X-Value (columns), Y-Value (rows)

Y \ X	-3	-2	-1	0	1	2	3
3	47	40	98	30	47	67	25
2	51	97	56	19	77	7	62
1	23	11	38	100	50	39	95
0	78	16	2	69	4	23	16
-1	10	21	55	100	10	6	59
-2	87	80	83	33	74	83	33
-3	66	71	93	56	51	91	61

Question 1-5 is based on the above table.

1. (-2, -3)

A) 91
B) 80
C) 33
D) 56
E) 71

2. (1, 1)

A) 39
B) 100
C) 47
D) 7
E) 50

3. (0, -1)

A) 33
B) 55
C) 16
D) 100
E) 2

4. (2, 3)

A) 62
B) 77
C) 25
D) 7
E) 67

5. (-1, 0)

A) 69
B) 78
C) 98
D) 2
E) 56

X-Value (columns), Y-Value (rows)

Y \ X	-3	-2	-1	0	1	2	3
3	93	55	89	8	89	7	86
2	97	91	80	100	64	23	79
1	82	55	83	58	1	41	98
0	85	50	53	44	96	94	96
-1	65	73	78	81	40	28	31
-2	55	19	15	47	6	17	5
-3	49	53	18	7	97	43	86

Question 6-10 is based on the above table.

6. (3, 3)

A) 8
B) 7
C) 86
D) 64
E) 79

7. (0, -2)

A) 19
B) 47
C) 44
D) 55
E) 15

8. (-2, 0)

A) 98
B) 50
C) 53
D) 55
E) 82

9. (2, -3)

A) 7
B) 46
C) 17
D) 43
E) 86

10. (-1, 1)

A) 65
B) 83
C) 78
D) 21
E) 81

Answers and Explanations

Question 1-5

1. E. The x-value is -2, and the y-value is -3, so the answer is 71.

2. E. The x-value is 1, and the y-value is 1, so the answer is 50.

3. D. The x-value is 0, and the y-value is -1, so the answer is 100.

4. E. The x-value is 2, and the y-value is 3, so the answer is 67.

5. D. The x-value is -1, and the y-value is 0, so the answer is 2.

Question 6-10

6. C. The x-value is 3, and the y-value is 3, so the answer is 86.

7. B. The x-value is 0, and the y-value is -2, so the answer is 47.

8. B. The x-value is -2, and the y-value is 0, so the answer is 50.

9. D. The x-value is 2, and the y-value is -3, so the answer is 43.

10. B. The x-value is -1, and the y-value is 1, so the answer is 83.

Chapter 4. Aviation Information

Introduction

The Aviation Information section tests your knowledge of general flight and airplane mechanics, procedures, and functions. Since you have a full 8 minutes to complete these 20 questions, there's no need to rush. However, there is a mix of bad and good news here for studying.

The bad news is that this study section will be the most time-intensive unless you're already a certified pilot. You'll need to become familiar with a vast amount of information. The good news is that we are here to help you with distilled and concise aviation information so you can be well-prepared for the test in the shortest time possible. To that goal, we organize the Aviation Information chapter into two sections:

Section 1: General Aviation Knowledge. This section covers a broad spectrum of aviation knowledge, including:

1. Airplane Components
2. Airplane Performance
3. Airplane Aerodynamics
4. Engines
5. Electrical System
6. Flight Instruments
7. Weight and Balance
8. Flight Planning
9. Airspace
10. Aviation Maps
11. Radio Navigation
12. Radio Operations
13. Understanding Weather
14. Airport Operations

Section 2: Essential Aviation Concepts. Information for this section is compiled from authoritative sources such as the Federal Aviation Administration's Pilot's Handbook of Aeronautical Knowledge. We spent countless hours poring through such materials and distilling books thick as a tome into under 50 pages of essential aviation concepts. That's still a considerable amount of information to learn, master, and memorize, but we believe this valuable shortcut is as concise as we can get without becoming too concise. There are reasons it takes hundreds of hours of study and practice to become a pilot!

Another good news regarding the Aviation Information section of the test is that you are not expected to be a pilot already; you do not need to memorize every term verbatim, complete complicated calculations, or demonstrate your proficiency in avionics instruments.

The two-pronged approach of studying general aviation knowledge and memorizing essential aviation concepts will create the most effective solution for the broadest spectrum of readers. If the amount of information looks overwhelming initially, we advise taking your time and working methodically. You can do this!

SECTION 1. GENERAL AVIATION KNOWLEDGE

1. Airplane Components

Becoming a pilot is an exciting journey, but one must invest the time to understand the intricate components of an airplane. Aspiring aviators often find it beneficial to delve into the fundamental aspects of aircraft systems to build a strong foundation for their flying endeavors. Below are the major components of an airplane.

Fuselage

The fuselage is the central structure of the aircraft, housing the cockpit, passenger cabin, cargo holds, and essential systems. It is typically cylindrical and provides the necessary framework to which the wings, tail, and other components are attached. In addition to serving as the aircraft's main body, the fuselage also accommodates various systems, such as avionics and hydraulics, contributing to the overall functionality of the airplane.

Wings

Wings are integral to an aircraft's lift and stability. Wings are aerodynamic and are attached to the fuselage. Wings come in various shapes and sizes, each tailored to specific aircraft designs and purposes.

The primary function of wings is to generate lift, the force that counteracts the aircraft's weight and allows it to ascend into the sky. The carefully crafted curvature on the wing's upper surface, coupled with the flatter lower surface, creates a pressure difference, resulting in upward lift. For pilots, understanding the nuances of wing design is pivotal for achieving optimal performance during takeoff, cruising, and landing.

Wings are equipped with ailerons and control surfaces near the wingtips and are responsible for roll control. By moving in opposite directions, ailerons enable the aircraft to bank left or right, facilitating turns. Flaps, located along the trailing edge of the wings, can be extended during takeoff and landing to increase lift and reduce landing speeds.

The wing's structural integrity is maintained by spars and ribs, forming a framework that supports the aerodynamic surfaces. Materials like aluminum, composites, or a combination are used to balance strength, weight, and durability.

Wing configurations vary, influencing an aircraft's performance and capabilities. High-wing designs position the wings above the fuselage, offering enhanced stability and easier ground visibility. In contrast, low-wing configurations provide improved aerodynamics and increased maneuverability. The swept-wing design, often seen in high-speed or supersonic aircraft, reduces drag and enhances performance at higher velocities.

Empennage

The empennage, or tail section, is located at the aircraft's rear section and consists of the horizontal stabilizer, vertical stabilizer, elevators, and rudder. The horizontal stabilizer helps maintain the aircraft's pitch stability while the elevators control pitch during flight. The vertical stabilizer prevents lateral motion, and the rudder allows the pilot to control yaw. A comprehensive understanding of these components is essential for maintaining control and stability during various flight phases.

Landing Gear

The landing gear facilitates takeoff and landing. It typically includes the main landing gear, which supports most of the aircraft's weight, and the nose gear, located at the front. Landing gear designs can vary, with options like tricycle or tailwheel configurations. Knowledge of the landing gear system is vital for safe ground operations and executing smooth takeoffs and landings.

Powerplant

The powerplant, commonly consisting of one or more engines, generates thrust to propel the aircraft forward. Aircraft engines can be piston, turboprop, or jet-powered. The airplane's engine(s) convert fuel into the thrust that defies gravity. In piston engines, controlled explosions within cylinders set the propeller in motion, offering a blend of reliability and simplicity. On the other hand, turbine engines compress and ignite air, generating a high-speed exhaust that propels the aircraft forward with remarkable power and efficiency. The choice between these engines shapes an airplane's performance and endurance.

Avionics

Avionics refers to the electronic systems on an aircraft, including navigation, communication, and monitoring systems. GPS units, radio transceivers, and altitude sensors intricately weave together, empowering pilots to navigate complex airspace precisely. Communication radios establish a vital link with air traffic control, fostering safe and efficient flights. Cockpit displays, often featuring multifunctional screens, relay critical data such as airspeed, altitude, and engine performance.

Mastering avionics involves understanding the symbiosis of these electronic components. Automation aids in flight management, but foundational knowledge ensures pilots can troubleshoot and navigate manually if needed.

Flight Controls

Flight controls are the mechanisms that enable the pilot to maneuver the aircraft. These include the control yoke or stick, pedals, and various associated linkages. The primary control surfaces include ailerons, elevators, and rudders. Ailerons, located on the wings, control roll by moving in opposite directions – one up, one down – to tilt the aircraft. Elevators on the tail manage pitch, controlling the plane's nose-up or nose-down movement. The rudder, also on the tail, handles yaw, allowing the aircraft to turn left or right.

These control surfaces are manipulated by the control yoke or stick in the cockpit. By tilting the control yoke, pilots engage the ailerons for banking left or right. Pushing or pulling the yoke operates the elevators for pitch control. Foot pedals control the rudder for yaw adjustments.

Advanced aircraft often incorporate fly-by-wire systems, replacing traditional mechanical linkages with electronic interfaces. These systems enhance precision and provide the flexibility to adjust control responses based on various flight conditions.

Autopilot systems, integrated into modern avionics, can assist pilots in maintaining altitude, heading, and airspeed. However, these aids complement, not replace, pilots' hands-on skills.

Fuel Systems

Fuel systems store and deliver the fuel necessary for the engines to operate efficiently and are the lifeblood of an aircraft, ensuring a seamless journey through the skies.

In a typical aircraft, fuel is stored in tanks strategically located within the wings or fuselage. The design considers weight distribution and safety, preventing fuel imbalance during flight. Fuel is fed from the tanks to the engines through a complex network of pipes, pumps, and valves. Gravity assists in low-altitude scenarios, while fuel pumps ensure a consistent supply during climbs or maneuvers. The fuel-air mixture is finely tuned to optimize engine performance, balancing power and efficiency.

Fuel systems also play a pivotal role in the management of fuel types. Aviation fuels, such as Avgas for piston engines or Jet-A for turbines, are carefully selected based on the aircraft's requirements.

Hydraulic and Pneumatic Systems

Hydraulic systems use incompressible fluids, usually oil, to transmit force and control various aircraft components. Brake systems, landing gear deployment, and even the movement of flight control surfaces often rely on hydraulic power. The fluid's inability to compress ensures immediate and precise responses, which is essential for the safety and maneuverability of an aircraft and plays a crucial role in operating various aircraft components, such as landing gear, flaps, and brakes.

In contrast, pneumatic systems utilize compressible gases, commonly air, to perform work. While less common in modern aviation, some aircraft systems, such as de-icing mechanisms or emergency braking, may employ pneumatic power. Compressed air's flexibility allows for quick and adaptable responses, contributing to the overall efficiency of certain aircraft functions. Hydraulic and pneumatic systems are often interconnected, necessitating a comprehensive awareness of their roles.

Maintenance and troubleshooting play a pivotal role in ensuring the reliability of these systems. A leak in a hydraulic line or a drop in pressure in a pneumatic system can have profound consequences, underscoring the importance of regular inspections.

2. Airplane Performance

For a given airplane, its performance, e.g., takeoff, climb, cruise, and landing metrics, is documented in the Pilot's Operating Handbook (POH). A pilot should consider the POH a crucial part of an aircraft's equipment, just like its wings. Mastering the use of performance charts is essential for a pilot's role. Air density is the foundation of all airplane performance, as the thickness of the air interacts with the wings or enters the engine significantly influences how well the aircraft operates. Understanding and leveraging air density is vital to successful aviation, as you'll discover when delving into the performance charts.

Air Density and its Impact

The impact of air density is most evident when considering altitude. As an aircraft ascends to higher altitudes, the thickness of the air decreases. This reduction in air density affects the lift generated by the wings, influencing the overall aerodynamic performance of the aircraft. Pilots must be mindful of this relationship, adjusting their approach during different phases of flight.

Heat and humidity further contribute to variations in air density. High temperatures can diminish engine efficiency, affecting an aircraft's takeoff performance, climb rate, and responsiveness. Additionally, humidity affects the oxygen content in the air, further impacting engine performance. Pilots must consider these factors in their flight planning to ensure safe and efficient operations.

Density altitude is a composite metric that combines temperature and pressure altitude, providing a comprehensive measure of air density at a specific location. High density altitude, often encountered at high elevations or during hot weather, can lead to reduced engine power and diminished lift, requiring adjustments in operational strategies.

Service ceiling is another crucial consideration related to air density. It represents the maximum altitude at which an aircraft can maintain a steady and level flight under standard atmospheric conditions. The engine's efficiency diminishes as air density decreases with altitude, and the wings generate less lift. Understanding the service ceiling is essential for pilots to gauge the limits of their aircraft's performance at higher altitudes.

Best Rate and Best Angle of Climb Speeds

Vx (Best Angle of Climb Speed) and Vy (Best Rate of Climb Speed) are two critical airspeeds that significantly influence an aircraft's climb performance. Understanding how these speeds change with altitude is crucial for pilots seeking to optimize their climb profiles and efficiently navigate various atmospheric conditions.

Vx - Best angle of climb speed: Vx represents the airspeed at which an airplane gains the greatest altitude in a given distance. It is used during a short-field takeoff to clear obstacles such as trees or buildings.

As altitude increases, air density decreases. An aircraft needs a higher indicated airspeed at higher altitudes to maintain the same true airspeed. Consequently, the Vx will increase with altitude. Pilots should refer to performance charts in the aircraft's Pilot's Operating Handbook (POH) to determine the specific Vx for different altitudes.

Pilots use Vx when they need to clear obstacles shortly after takeoff, prioritizing altitude gain over forward speed.

Vy - Best Rate of Climb speed: Vy is the airspeed that provides the highest rate of climb in feet per minute. This speed maximizes the altitude gain over a given period and is essential for efficiency during climb segments.

Similar to Vx, Vy is affected by changes in air density with altitude. As altitude increases, the true airspeed required for Vy will also increase. Pilots must consult performance charts to determine the appropriate Vy for different altitudes. Vy is typically used during the climb phase to optimize the rate of ascent when clearing obstacles is not a primary concern.

Altitude impacts air density, affecting the aircraft's aerodynamic performance and, consequently, the optimal airspeeds for climbing. Pilots must reference the aircraft's performance charts to ensure they are operating at the correct speeds at various altitudes, striking a balance between clearing obstacles and achieving an efficient rate of climb. This knowledge enhances the safety and effectiveness of the climb phase, contributing to overall flight proficiency.

Cruise Climb Speed

Cruise climb speed is a crucial airspeed pilots utilize during transitioning from the initial climb phase to the cruising altitude. It represents the optimal speed that balances achieving altitude efficiently and conserving fuel for sustained flight.

As an aircraft ascends through the airspace, transitioning from the initial climb to cruise, maintaining an efficient climb speed becomes essential. Cruise climb speed allows pilots to capitalize on the climb performance while ensuring fuel economy. The selection of cruise climb speed is influenced by various factors, including the aircraft's weight, engine performance, and aerodynamic characteristics. Pilots refer to performance charts in the aircraft's Pilot's Operating Handbook (POH) to determine the specific cruise climb speed for a given situation.

Takeoff Distance Chart

Takeoff distance charts outline the required runway length based on aircraft weight, altitude, temperature, wind, etc. These charts assist in planning safe takeoffs, considering the airplane's performance in varying conditions.

Landing Distance Performance Charts

Like takeoff charts, landing distance performance charts aid pilots in determining the required runway length for a safe landing. Variables such as approach speed, landing weight, and wind conditions influence these charts.

Time, Fuel, and Distance to Climb Chart

Understanding the time, fuel, and distance required to climb to a specific altitude is vital for flight planning. This chart helps pilots estimate the resources needed for ascending to cruising altitude.

Cruise Performance Chart

Cruise performance charts offer insights into the optimal airspeed and power settings for efficient cruising. Pilots can use these charts to maximize fuel efficiency and endurance during the cruise phase of flight.

Endurance and Range Profile Charts

Endurance and range profile charts provide valuable information for pilots aiming to optimize fuel consumption during flight. Endurance charts help determine the maximum time an aircraft can remain airborne, while range charts assist in planning for the maximum distance achievable with available fuel.

Crosswind Component Chart

Crosswind components are crucial considerations during takeoff and landing. The crosswind component chart aids pilots in assessing thc maximum allowable crosswind for safe operations.

3. Airplane Aerodynamics

Aerodynamics is the science of studying the behavior of air as it interacts with solid objects.

Four forces

Lift, weight, thrust, and drag are present whenever a plane is airborne.

Lift: Lift is the upward force that enables an airplane to defy gravity and stay airborne. The wings generate lift as air flows over and under them. The shape of the wing, known as the airfoil, is crucial in creating lift. The wing's upper surface is curved, creating lower pressure and thus lift.

Weight: Weight is the force pulling the airplane toward the Earth. It is counteracted by lift, providing equilibrium during flight. Proper weight management is critical for balanced flight, ensuring the aircraft remains stable and controllable.

Thrust: Thrust is the forward force the aircraft's engines produce, overcoming drag. It propels the airplane through the air. Pilots control thrust to maintain desired speeds and manage changes in altitude during flight.

Drag: Drag is the aerodynamic resistance that opposes the aircraft's forward motion. It results from the friction of air molecules against the aircraft's surfaces. Minimizing drag is essential for fuel efficiency. Streamlining the aircraft's design, including its fuselage and wing shape, helps reduce drag.

The Wing

The wing is the airplane's workhorse, responsible for generating lift and shaping its aerodynamic performance.

Airfoil: The airfoil, or wing profile, is carefully designed to optimize lift. It has a curved upper surface and a flatter lower surface, creating pressure differences that result in lift.

Angle of Attack: The angle of attack is the angle between the chord line (an imaginary line from the leading edge to the trailing edge of the wing) and the oncoming air. It directly influences lift production. Pilots adjust the angle of attack to control the amount of lift generated by the wing.

Stalls

Stalls occur when the critical angle of attack is exceeded, causing a disruption in smooth airflow over the wing.

Critical Angle of Attack: The critical angle of attack is the point at which airflow over the wing becomes turbulent, leading to a stall. Pilots must be vigilant to avoid exceeding this critical angle during maneuvers.

Recovery from a Stall: To recover from a stall, pilots must reduce the angle of attack, typically by lowering the aircraft's nose. This restores smooth airflow over the wings, allowing the aircraft to regain lift.

Drag

Understanding and managing drag is crucial for optimizing an aircraft's performance.

Parasite Drag: Parasite drag is caused by the aircraft's structure and surfaces interacting with the air. Streamlining the aircraft minimizes this form of drag. Design features such as retractable landing gear and smooth fuselage contours contribute to reducing parasite drag.

Induced Drag: Induced drag is a byproduct of lift production. It increases as the angle of attack increases, such as during slow flight or high angles of climb. Like gliders, wings with a high aspect ratio experience less induced drag. Aspect ratio is the ratio of the wingspan to the mean chord (average width) of the wing. The wingspan is the distance from wingtip to wingtip, and the mean chord is the average width of the wing from the leading edge to the trailing edge.

Primary Flight Controls

The primary flight controls include the ailerons, elevators, and rudder. They govern the three axes of an aircraft's movement: roll, pitch, and yaw respectively.

Ailerons: Ailerons are hinged control surfaces mounted on either side of the wings. They dictate the aircraft's roll, enabling it to tilt or bank left or right. When a pilot deflects the ailerons, the lift on one wing increases while decreasing on the other, initiating a roll about the longitudinal axis.

Elevators: Elevators, typically found on the horizontal stabilizer at the tail, govern the aircraft's pitch. Controlled by the yoke or control stick, the movement of the elevators results in a change in the aircraft's nose position. Pulling back raises the nose, initiating a climb, while pushing forward lowers the nose, prompting a descent. Elevators are crucial for controlling the aircraft's altitude.

Rudder: The rudder is mounted on the vertical stabilizer at the tail and is the primary control for yaw. Yaw refers to the rotation of the aircraft about its vertical axis. The pilot can control the aircraft's heading by deflecting the rudder, compensating for adverse yaw induced during turns.

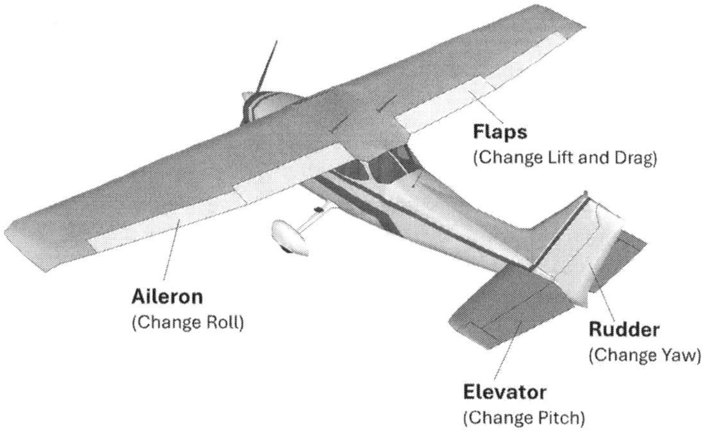

Axis of Rotation	Control Surface	Aircraft Movement
Longitudinal	Ailerons	Roll
Lateral	Elevators	Pitch
Vertical	Rudder	Yaw

Secondary Flight Controls

While primary flight controls govern the fundamental aspects of an aircraft's movement, secondary flight controls fine-tune and enhance its performance. These controls include flaps, slats, spoilers, and trim systems.

Flaps: Flaps are located on the trailing edge of the wings and are deployed during specific phases of flight to augment lift and control. Extended flaps increase wing area during takeoff and landing, providing additional lift at slower speeds, which allows for shorter takeoff distances and lower landing speeds.

Slats: Slats are movable surfaces positioned at the leading edge of the wings. Similar to flaps, slats enhance lift, especially at lower speeds. Extending the slats alters the wing's curvature, improving aerodynamic efficiency during takeoff, landing, and other critical phases of flight.

Spoilers: Spoilers are located on the upper surface of the wings. They disrupt the smooth airflow, reducing lift. Deployed symmetrically, spoilers assist in controlled descent without altering the aircraft's pitch. This capability is particularly valuable during rapid descents or in preparation for landing.

Trim Systems: The airplane trim system is typically located on the trailing edges of control surfaces, such as elevators, ailerons, or rudders. It functions by adjusting the position of these surfaces, allowing pilots to relieve control pressures and maintain a desired attitude.

P-factor and Left Turning Tendencies

Various aerodynamic factors can create imbalances during powered flight. These factors contribute to an aircraft's tendency to turn to the left. One prominent influence is the P-factor, or propeller factor. The p-factor is most noticeable during low speeds or high angles of attack. P-factor occurs due to the descending propeller blade having a higher angle of attack than the ascending blade during a climb. This creates uneven thrust, causing the aircraft to yaw to the left. Pilots counteract these tendencies using the rudder and ailerons.

4. Engines

The engine is often regarded as the heart of an aircraft, encompassing several interconnected systems that harmonize to achieve controlled flight.

The Power Plant: Engine Basics

The engine is at the core of every airplane's mechanical symphony—an intricate assembly of components designed to convert fuel into the thrust that propels the aircraft forward. Aircraft engines come in various types, with piston engines and gas turbine engines being the most common.

Piston Engines: Piston engines, reminiscent of those found in automobiles, are prevalent in smaller aircraft. They operate on the principles of internal combustion, where fuel-air mixtures ignite within cylinders, driving pistons in a repetitive cycle. This reciprocating motion is then converted into rotary motion, ultimately turning the aircraft's propeller.

Gas Turbine Engines: Larger and more powerful aircraft typically employ gas turbine engines, also known as jet engines. They harness the energy generated by the continuous combustion of fuel with compressed air. The resulting high-speed exhaust jet propels the aircraft forward.

Aviation Fuel and the Airplane Fuel System

The lifeblood of any airplane engine is aviation fuel—a specialized blend designed to meet the demanding flight requirements. The airplane fuel system delivers the right amount of fuel to the engine. Fuel is stored in tanks within the aircraft's wings or fuselage. Gravity assists in low-altitude scenarios, while fuel pumps come into play during climbs or high-altitude operations. A precise fuel-air mixture is crucial for engine efficiency, with the carburetor or fuel injector regulating this delicate balance.

Oil System

While engines are in motion, a robust plane oil system ensures critical components remain well-lubricated and protected from friction and heat. Oil circulates through the engine, coating moving parts and reducing wear. The oil system also aids in cooling, removing excess heat generated during combustion.

Engine Cooling System

Efficient engine operation hinges on managing heat effectively, and the engine cooling system plays a pivotal role. Engines generate intense heat during combustion. Without proper cooling mechanisms, components could succumb to overheating.

Air-cooled engines rely on airflow over the engine components to dissipate heat. In contrast, liquid-cooled engines circulate a coolant, typically a mixture of water and antifreeze, to absorb and carry away excess heat. Radiators, analogous to those in automobiles, release this absorbed heat into the surrounding air.

Propellers

At the forefront of the aircraft, the propeller translates rotational motion into forward thrust. Propellers come in various designs, the most common being fixed-pitch and variable-pitch (or controllable-pitch).

Fixed-Pitch Propellers: Fixed-pitch propellers have blades with a fixed angle that cannot be adjusted during flight. They are common in smaller aircraft and are optimized for specific operating conditions.

Variable-Pitch Propellers: Variable-pitch propellers, more prevalent in larger and more complex aircraft, allow pilots to adjust the pitch of the blades during flight.

Synthesis of Systems

Each airplane engine component plays a unique role, yet their synergy is the essence of controlled flight. The engine, fed by aviation fuel and lubricated by a vigilant oil system, propels the aircraft forward. The cooling system prevents components from succumbing to the relentless heat of combustion, and the propeller transforms rotational vigor into directional momentum.

5. Electrical System

At the heart of the airplane's electrical system resides the battery—a robust and reliable source of electrical energy. Batteries in airplanes store electrical energy in chemical form and serve as a crucial power reserve for various aircraft components. Aircraft batteries are typically lead-acid, chosen for their reliability and ability to deliver consistent power.

Battery Potential

The battery potential, measured in volts, represents the force that propels electrical current through the airplane's circuits. As the engine gets turned on, the battery potential ensures a surge of electricity, initiating the intricate dance of avionic instruments, lights, and other electrical components.

Charge-Discharge Ammeter

The charge-discharge ammeter is a vigilant guardian of the battery's health and functionality. This instrument provides real-time feedback on whether the battery is receiving a charge or is discharging its stored energy. Pilots and maintenance crews monitor the ammeter to ensure that the battery remains in a healthy state.

Voltage Regulator

The voltage regulator maintains the balance of electrical potential within the airplane's systems. Its primary role is maintaining a steady output voltage, ensuring that electrical components receive the appropriate power supply. This regulatory function prevents overcharging and safeguards against voltage spikes that could damage sensitive avionic equipment.

Load Meter

The load meter tracks the electrical demands placed on the system. Pilots refer to the load meter to gauge the current drawn by various electrical components, from navigation lights to communication radios. This information aids in optimizing power distribution, ensuring that the load remains within the electrical system's capacity.

Electrical Ground

Grounding in the context of airplane electrical systems is not just about the physical connection to the Earth but involves establishing a reference point for electrical potential. In-flight, aircraft maintain electrical grounding through dedicated systems that connect to the structure, allowing electrical currents to safely dissipate. Ground straps and lightning protection ensure a continuous path for static discharge. Design features prevent water intrusion. Strategic insulation preserves grounding integrity, safeguarding against electrical malfunctions.

Electrical Drain

An electrical drain is an undesired loss of electrical power in the system. Electrical drain can occur due to various factors, including faulty wiring, damaged components, or malfunctioning avionic equipment. Aircraft detect electrical drain through monitoring systems that measure the rate at which electrical power is consumed. Load meters provide real-time data on the electrical load, helping pilots and maintenance crews manage consumption to ensure it remains within the aircraft's electrical system capacity.

The Water Pump

While not typically associated with electrical systems, the water pump is critical in aircraft engines. The water pump circulates coolant through the engine, preventing it from overheating during operation. Although not directly electrical, the water pump indirectly contributes to the overall health and efficiency of the electrical system by maintaining optimal engine temperatures.

6. Flight Instruments

Airplane flight instruments, strategically placed within the cockpit, serve as the pilot's navigational aids, providing crucial information about the aircraft's orientation, altitude, airspeed, and heading.

Airspeed Indicator

The airspeed indicator stands as the aircraft's velocity navigator, offering real-time information about its speed through the air. This essential instrument ensures that the aircraft operates within its designed performance envelope.

The mechanics behind the airspeed indicator involve a diaphragm connected to the pitot-static system. As the aircraft moves through the air, the pitot tube captures the dynamic pressure, while the static port measures ambient air pressure. The diaphragm responds to the pressure differential, translating it into a visual indication of airspeed on the instrument's face.

Attitude Indicator

The attitude indicator, often called the artificial horizon, is the pilot's visual reference to the aircraft's pitch and roll. This instrument is crucial, especially during periods of low visibility or when the natural horizon is obscured.

Mechanically, the attitude indicator consists of a gyroscope mounted in the instrument case. As the aircraft changes pitch or roll, the gyroscope maintains its orientation in space, providing a stable reference point for the pilot. The instrument's display represents the aircraft's position in relation to the artificial horizon, aiding the pilot in maintaining straight and level flight.

Altimeter

The altimeter offers information about its vertical position above sea level. This instrument is instrumental in maintaining proper altitude during various phases of flight. Internally, the altimeter operates based on the principle of barometric pressure. As the aircraft ascends or descends, the atmospheric pressure changes. The altimeter capsule expands or contracts in response to these pressure variations, visually indicating the aircraft's altitude. Pilots often adjust the altimeter setting to account for changes in atmospheric pressure.

Vertical Speed Indicator

The vertical speed indicator complements the altimeter by providing information about the rate of ascent or descent. This instrument is particularly useful during climb and descent phases, helping pilots maintain a controlled vertical profile.

The vertical speed indicator mechanically incorporates a diaphragm connected to the static system. The diaphragm's rate of expansion or contract corresponds to the aircraft's vertical movement. The instrument displays this information in feet per minute, aiding pilots in achieving and maintaining desired rates of climb or descent.

Heading Indicator

The heading indicator, also known as the directional gyro, serves as the compass's mechanical sibling, providing a stable reference for the aircraft's heading. Unlike the magnetic compass, the heading indicator is not susceptible to magnetic variations.

The mechanics involve a gyroscope spinning in a horizontal plane. As the aircraft turns, the gyroscope maintains its orientation, offering a reliable indication of the aircraft's heading. Periodic realignments with the magnetic compass are necessary to correct for gyroscopic drift.

Turn Coordinator

The turn coordinator is the banking advisor, offering information about the rate and quality of turns. This instrument assists pilots in maintaining coordinated turns, ensuring the aircraft's balance during maneuvers.

The turn coordinator integrates a miniature aircraft symbol that represents the aircraft's attitude. A ball within the instrument indicates the aircraft's coordination during turns. This visual cue assists pilots in making precise and balanced turns, enhancing overall flight control.

Magnetic Compass

The magnetic compass is the venerable navigator, relying on the Earth's magnetic field for directional reference. While it may be one of the oldest navigation tools, its simplicity and reliability continue to make it an essential part of the cockpit.

Mechanically, the magnetic compass aligns itself with the Earth's magnetic field. The compass card within the instrument displays the cardinal points, aiding pilots in determining their aircraft's magnetic heading. Pilots must account for magnetic variations, acceleration, and deceleration errors despite the magnetic compass's reliability.

Acceleration And Deceleration Error

Acceleration and deceleration errors are phenomena affecting the accuracy of the magnetic compass, necessitating corrections during changes in the aircraft's speed. When accelerating, the compass tends to indicate a turn to the north, while deceleration causes it to indicate a turn to the south.

These errors result from the inertia of the compass card fluid. As the aircraft accelerates, the fluid lags behind, causing the compass to read erroneously. Pilots apply corrections based on their understanding of these errors to maintain accurate navigation.

Turning Errors

A magnetic compass will provide generally accurate heading information during a straight and level unaccelerated flight. Turning errors occur exclusively on northerly or southerly headings when an airplane banks, which causes the tilting of the compass card.

The Earth's magnetic field is not perfectly horizontal everywhere on the Earth's surface but has an inclination angle with respect to the horizontal plane. To put it differently, imagine the Earth's magnetic field lines not parallel with the Earth's surface but tilt toward and penetrate the Earth's surface. This inclination angle varies depending on the geographic location.

The "dipped magnetic field" concept in aviation is crucial to understanding turning errors. When an aircraft is in a bank or is turning, the magnetic compass needle is influenced by the inclination of the magnetic field lines. This inclination causes the north-seeking end of the compass needle to dip downward. The pendulous properties of the card become ineffective in preventing the north-seeking end of the compass needle from pointing downward toward the dipped magnetic field. Consequently, a transient heading error becomes apparent during these maneuvers. Depending on whether a plane is heading north or south, the compass's turning error manifests as either a lag or lead over the plane's actual heading.

7. Weight and Balance

An aircraft's weight and balance are pivotal in its ability to soar through the skies gracefully and precisely. Each aircraft is meticulously designed with specific weight limitations to ensure that it operates within its engineered parameters. When these limits are exceeded, the consequences can be dire, leading to potential structural damage and compromising the safety of the flight.

Excessive weight can increase loads during takeoff, landing, and turbulence, placing undue strain on the airframe. If prolonged or severe, this stress can lead to structural fatigue, deformation, or, in extreme cases, catastrophic failure.

Center of Gravity

At the heart of weight and balance considerations lies the concept of the center of gravity (CG). The center of gravity is the point at which the aircraft would balance if it were suspended in the air. This point is a critical parameter in ensuring that the aircraft remains stable and controllable during flight.

The center of gravity is not a fixed point; rather, it varies based on the weight distribution within the aircraft. To determine the center of gravity, pilots and operators use a combination of mathematical calculations and reference charts provided by the aircraft manufacturer. These calculations consider each component's weight and arm (distance from the reference point), including passengers, cargo, fuel, and other items on board.

Aft and Forward Center of Gravity Loading

Aircraft are designed with specified limits for the forward and aft center of gravity loading. Loading beyond these limits can have profound effects on the aircraft's stability and control. An aft center of gravity loading, where the center of gravity is toward the rear of the aircraft, can result in difficulty controlling pitch, potentially leading to a stall. Conversely, a forward center of gravity loading can affect the aircraft's ability to rotate during takeoff.

Center of Lift

The wings of an aircraft generate lift as air flows over and under them. The center of lift is the point along the wing's chord line where the lift force is considered to act. This force opposes the aircraft's weight, allowing it to defy gravity and take to the skies.

The wings' aerodynamic characteristics, including the airfoil's shape and the angle of attack, influence the location of the center of lift. Pilots use control surfaces, such as ailerons and elevators, to manipulate the center of lift during different phases of flight.

Stability in Flight

Stability in flight is a delicate equilibrium achieved when the aircraft's center of gravity is appropriately aligned with its center of lift. When the center of gravity and the center of lift are in harmony, the aircraft exhibits a stable and controlled flight.

The position of the center of gravity relative to the center of lift is a crucial determinant of an aircraft's stability. If the center of gravity is too far forward, the aircraft tends to be stable but less maneuverable. Conversely, if the center of gravity is too far aft, the aircraft becomes less stable, with the potential for over-controllability and difficulty recovering from stalls.

Extra Weight

Every additional pound carried by an aircraft contributes to the complexity of weight and balance considerations. Extra weight, whether it be excess cargo, additional fuel, or unexpected baggage, can tip the scales beyond the permissible limits.

<u>Impact on Performance</u>

Extra weight affects an aircraft's performance in multiple ways. It can extend the takeoff distance, reduce the rate of climb, and alter the aircraft's handling characteristics. Pilots must account for these performance changes when planning a flight with additional weight.

<u>Fuel Management</u>

Fuel significantly contributes to an aircraft's weight, and its consumption during flight alters the weight distribution. Pilots carefully manage fuel loads, taking into account the distance to be traveled and the expected fuel burn. This dynamic aspect of weight and balance requires continuous monitoring and adjustments throughout the flight.

<u>Payload Considerations</u>

Passenger and cargo loads are critical components of the overall weight. Pilots work closely with ground crews to ensure that passengers are seated in designated areas and cargo is distributed evenly. The loading of baggage, both in the main cabin and cargo hold, is meticulously planned to maintain the desired center of gravity.

8. Flight Planning

Airplane flight planning is the systematic process of charting an aircraft's course from its origin to its intended destination. This process considers various factors, including weather conditions, airspace restrictions, fuel requirements, and navigational waypoints. The overarching goal is to ensure the flight's safety, efficiency, and compliance while optimizing resource use.

Measuring Direction

Understanding direction is a fundamental aspect of flight planning, and pilots employ specific tools and terminology to navigate accurately. Bearings and headings are critical components in measuring and conveying direction.

<u>Bearings:</u> Bearings represent the direction of one point from another, typically measured in degrees from the north. A bearing of 090 degrees, for example, indicates an eastward direction. Pilots use instruments such as the magnetic compass or navigational aids to ascertain bearings.

<u>Headings</u>: Headings, on the other hand, refer to the direction in which the aircraft is pointed or intends to fly. Pilots set headings on the aircraft's compass or directional gyro to align with their planned route. Adjustments may be made during the flight to account for wind or navigational updates.

Time Management

Effective time management is integral to the success of any flight plan. Pilots meticulously calculate estimated departure and arrival times, accounting for factors such as wind speed, altitude, and airspeed.

Flight planning involves estimating travel time between waypoints and reaching the destination. Pilots continuously compare planned times with actual times during the flight, making adjustments as needed.

Fuel and Time Efficiency: Managing fuel efficiency is intertwined with time considerations. Pilots aim to optimize fuel consumption by adjusting airspeed or altitude based on prevailing conditions, ultimately impacting the overall duration of the flight.

Time Zones: As aircraft traverse vast distances, they often cross multiple time zones, introducing additional complexity to flight planning. Understanding time zones is crucial for coordinating schedules, navigation, and communication.

Standard Time vs. Coordinated Universal Time (UTC): Aviation operates on Coordinated Universal Time (UTC), a global time standard. Pilots convert their planned times, which may be in local time zones, to UTC for consistency and coordination with air traffic control and other aviation entities. A side note: because members of the International Telecommunication Union (ITU) could not come to a unanimous agreement on either the English acronym (CUT) or the French acronym (TUC), the acronym UTC became the agreed-upon compromise.

Longitude and Latitude on Sectional Charts

Sectional charts, essential tools for pilots, depict geographical features, airspace boundaries, and navigation aids. Understanding how to interpret longitude and latitude on these charts is vital to effective flight planning.

Longitude and Latitude Grid: Sectional charts are divided into a grid of longitude and latitude lines. Longitude lines run north-south, while latitude lines run east-west. Each intersection of these lines represents a specific geographical point.

Navigational Waypoints: Pilots use longitude and latitude coordinates to identify waypoints, crucial markers along their route. These waypoints facilitate precise navigation and communication with air traffic control.

Cross-Country Navigation

Cross-country navigation involves planning and executing flights that span significant distances, requiring careful consideration of waypoints, navigation aids, and fuel stops. Pilots follow a structured approach to ensure the success of cross-country journeys.

Pre-flight Planning: Before departure, pilots thoroughly examine the route, considering factors such as weather, airspace restrictions, and available navigational aids. They calculate fuel requirements, identify suitable alternate airports, and review the overall feasibility of the journey.

In-Flight Navigation: During the flight, pilots use a combination of navigational tools, including GPS systems, VOR (VHF Omni-directional Range) stations, and navigational charts, to stay on course. They continuously update their position and make course adjustments as needed.

Fuel Stops and Alternates: Cross-country flights often involve strategic fuel stops to ensure the aircraft has an adequate fuel supply for the entire journey. Pilots also identify alternate airports in case of unexpected diversions or emergencies.

9. Airspace

From controlled and uncontrolled airspace to the classifications of Class A, E, G, and B, C, D airspace, as well as the unique realm of special-use airspace, airspace rules shape how aircraft traverse the skies.

Controlled and Uncontrolled Airspace

Imagine the airspace above us as a complex three-dimensional puzzle, carefully divided to ensure the aircraft's safe and efficient movement. The first fundamental classification lies in whether an airspace is controlled or uncontrolled. Controlled airspace is managed and regulated by air traffic control (ATC) authorities, ensuring the safe separation of aircraft. This type of airspace is typically found around busy airports, where the volume of air traffic requires meticulous coordination.

In contrast, uncontrolled airspace lacks continuous ATC services. This type of airspace is common in remote or less busy areas, allowing pilots greater freedom and responsibility for their navigation. Uncontrolled airspace doesn't mean chaos; pilots adhere to specific rules and use visual references to maintain safe separation from other aircraft.

Class A Airspace: The High Altitude Haven

Class A airspace represents the pinnacle of controlled airspace, reserved for high-altitude operations. Starting at 18,000 feet above sea level and extending upwards, this airspace class is designed to accommodate long-distance and high-altitude flights. Within Class A airspace, pilots rely heavily on instruments for navigation, and they are in constant communication with air traffic control.

Class A airspace is a realm where pressurized cabins become essential. Aircraft flying at these altitudes encounter reduced air pressure, necessitating sealed cabins to maintain a comfortable and safe environment for passengers and crew. Mechanical systems onboard aircraft operating in Class A airspace must be capable of enduring the challenges posed by extreme altitudes and long-duration flights.

Class E Airspace: The Transitional Territory

Class E airspace acts as a transition between controlled and uncontrolled regions, often situated above and below Class A airspace. This airspace connects different areas and provides a buffer zone for arriving and departing flights. While less rigorously managed than Class A, E airspace still demands vigilant communication and adherence to specific rules.

From a mechanical perspective, aircraft operating in Class E airspace should possess reliable communication systems and navigation equipment. The transition between controlled and uncontrolled airspace requires aircraft to seamlessly switch between relying on air traffic control services and maintaining self-navigation, underscoring the importance of robust mechanical systems.

Class G Airspace: The Open Skies

Class G airspace, also known as uncontrolled airspace, is where the skies feel vast and unencumbered. Typically found in less populated areas, this airspace class allows pilots greater autonomy in navigation. Mechanical considerations in Class G airspace include ensuring the aircraft is equipped for visual flight rules (VFR), where pilots navigate using landmarks and visual references rather than relying solely on instruments.

The mechanical systems in planes operating in Class G airspace must support the pilot's ability to maintain situational awareness without the constant guidance of air traffic control. This includes dependable engines, responsive control surfaces, and instrumentation that aids in navigation based on visual cues.

Class B, C, and D Airspace: The Urban Airspaces

Moving closer to the ground, we encounter the intricacies of Class B, C, and D airspace. These classes are commonly found around busy airports and metropolitan areas, requiring meticulous coordination between pilots and air traffic control. The mechanical aspect of planes operating in these airspaces is crucial, as the proximity of other aircraft and ground structures demands precision and responsiveness.

In Class B airspace, where traffic volume is exceptionally high, aircraft must possess advanced avionics and communication systems. These mechanical components enable pilots to follow specific clearance instructions and maintain safe separation from other planes. Similarly, Class C and D airspace emphasizes efficient communication, reliable navigation, and mechanical systems that can handle the complexities of close-quarter flying.

Special-Use Airspace: Navigating Restricted Zones

Beyond the standard classifications, special-use airspace adds more complexity to the skies. These areas are designated for specific purposes, such as military training, aerial demonstrations, or testing of new aircraft. Mechanical considerations in special-use airspace involve adapting aircraft systems to the unique demands of these activities, including rapid altitude changes, high-speed maneuvers, and communication with military controllers.

10. Aviation Maps

Aviation maps, more formally known as aeronautical sectional charts and VFR terminal area charts, serve as the lifeline for pilots, guiding them through the vast expanse of the sky with precision.

Aeronautical Sectional Charts

Aeronautical sectional charts are the cornerstone of aviation navigation. Designed for Visual Flight Rules (VFR) pilots, who rely on visual cues rather than solely on instruments, these maps provide a comprehensive and detailed view of the airspace, resembling a blueprint of the sky.

At first glance, these charts might appear overwhelming, with an intricate network of lines, symbols, and colors. However, they contain a treasure trove of information. The sectional charts are divided into squares, each representing a specific geographic area. These squares, often called "sectors," are the canvas upon which pilots paint their flight paths.

One of the key features of sectional charts is the depiction of topographical information. Elevation contours, rivers, lakes, and even major roadways are intricately laid out, providing pilots with a visual representation of the terrain beneath them. This topographical detail is crucial for pilots who must constantly assess the landscape for potential landing sites in case of emergencies.

Furthermore, aeronautical sectional charts are color-coded to signify different types of airspace. Distinct colors and symbols distinguish restricted areas, controlled airspaces, and special-use airspaces.

VFR Terminal Area Charts

While aeronautical sectional charts provide a broader perspective of airspace, VFR Terminal Area Charts zoom in on specific urban areas, offering a more detailed view. These charts are particularly useful for pilots navigating busy airspace around major airports.

The intricate details provided by VFR terminal area charts include specific landmarks, radio frequencies, and communication procedures, all critical for safe navigation in congested airspace. As planes approach airports, these charts guide pilots through the complex network of runways, taxiways, and terminals.

Understanding the symbology used in VFR terminal area charts is vital for deciphering the information they convey. The depiction of Class B, C, and D airspace and visual reporting points and waypoints allows pilots to maintain situational awareness in the urban aviation landscape.

Topographical Information

A rich layer of topographical information lies beneath the intricate web of lines and symbols on aviation maps. Elevation contours are the most fundamental topographical feature on these maps. By observing the contour lines, pilots can gauge the terrain's steepness and anticipate potential challenges during takeoff and landing. This information is invaluable, especially when flying in mountainous regions where rapid elevation changes can significantly impact aircraft performance.

Water bodies, from lakes to rivers, are also prominently featured on aviation maps.

Roads and highways crisscrossing the landscape provide additional points of reference for pilots. These corridors aid navigation and serve as emergency landing options when needed.

11. Radio Navigation

Electronic elucidation is a sophisticated process that harnesses radio signals to provide pilots with crucial positional information. Think of it as a digital dance between the aircraft and ground-based navigation aids, allowing for precise navigation even when the skies are shrouded in clouds.

VOR Navigation

One key player in this intricate dance is VOR, or Very High-Frequency Omni-Directional Range, navigation. VOR stations, strategically located on the ground, transmit signals that aircraft receive and interpret. These signals provide direction and distance information, allowing pilots to determine their position relative to the VOR station. The airplane constantly triangulates its position based on signals received from multiple VOR stations.

Horizontal Situation Indicator (HSI)

The horizontal situation indicator (HSI) consolidates various navigation information into a single, easy-to-read display. The HSI showcases the aircraft's heading and integrates VOR and other navigational data.

Distance Measuring Equipment (DME)

This technology allows pilots to determine their exact distance from a ground-based DME station. By receiving and interpreting signals, the aircraft calculates its distance in nautical miles from the station, providing a crucial metric for flight planning and situational awareness.

Carry-On Navigation

But what happens when an aircraft ventures beyond the reach of ground-based navigation aids? This is where carry-on navigation systems step in. GPS, or Global Positioning System, has become synonymous with modern air travel. Carry-on navigation offers pilots real-time positioning information, making it an indispensable tool for both en-route navigation and precision approaches during landing.

12. Radio Operation

Airplane radio operations are one key piloting skill. Pilots convey their intentions and ensure a shared understanding of the airspace dynamics. Whether it's announcing their position, requesting permission to take off or land, or communicating with air traffic control (ATC), mastering radio techniques is akin to learning the language of the skies.

Radio Communication

Modern aircraft are equipped with advanced communication systems, integrating VHF radios, transponders, and navigation equipment. Talking the aviation language involves mastering the language and protocols of air traffic control communications. From initial contact with the control tower to receiving clearances and updates during flight, pilots must articulate their messages with clarity and brevity, ensuring a shared understanding among all stakeholders in the airspace.

Controlled Airports

Controlled airports, where air traffic control towers manage the flow of arriving and departing flights, require unique communication skills. Pilots must adhere to established procedures, including obtaining clearance for takeoff and landing, following assigned headings, and coordinating with ground control during taxiing.

VHF Transmissions

Aviation transmissions are usually on the very high frequency (VHF)

portion of the radio spectrum. VHF radios operate within a specific frequency range and facilitate clear and reliable communication between aircraft and ground stations.

ASOS/AWOS

Automated Surface Observation Systems (ASOS) and Automated Weather Observing Systems (AWOS) provide real-time weather information at airports.

Emergency Frequency

The emergency frequency, 121.5 MHz, serves as a universal channel for distress calls and urgent communications. It is a lifeline in critical situations, allowing pilots to swiftly broadcast distress signals and receive assistance.

Radar System

Radar, an acronym for Radio Detection and Ranging, forms the backbone of air traffic control systems. By utilizing microwave radar energy, controllers can detect solid objects, such as aircraft, over significant distances. This technology enables the identification of these objects and provides information regarding their direction, speed, and, if equipped with a Mode-C transponder, altitude. In essence, radar allows controllers to extend their visual range far beyond the limits of their eyesight.

Operating on the echo principle, radar transmits microwave energy pulses through an antenna arrangement. Subsequently, a portion of this energy reflects back to the originating antenna, manifesting as a visual blip on a screen. Through this process, controllers can effectively identify aircraft from afar. Visual Flight Rules (VFR) pilots, who navigate primarily by visual references, can benefit from radar assistance, especially in adverse weather conditions.

Transponders

Transponders, short for transmitter-responder, transmit an aircraft's identification code to radar systems, enhancing its visibility to air traffic controllers. Pilots must understand how to operate transponders and comply with specific codes assigned for different phases of flight.

Clearance Delivery

Clearance delivery is a phase of radio communication before departure. Pilots obtain their clearances, including route information, altitudes, and squawk codes for their transponders.

13. Understanding Weather

The aviation weather charts, laden with symbols and lines, serve as a visual guide to the atmospheric phenomena that pilots navigate, offering insights into the interconnected world of weather and flight.

Atmospheric Circulation

Atmospheric circulation forms the rhythmic heartbeat of Earth's atmosphere. Driven by the unequal heating of the Earth's surface, air masses are set into motion, creating a dynamic flow of winds. The Coriolis Force, a consequence of the Earth's rotation, manifests when observed from an aerial perspective above the North Pole. Due to the Earth's counterclockwise rotation, the air undergoes a curvature or twist to the right in relation to the terrain in the Northern Hemisphere (and conversely, to the left in the Southern Hemisphere). Regardless of the air's initial direction, it experiences an additional rightward curvature or twist in its course of motion.

Pressure and Vertical Air Movement

As air moves across the Earth's surface, differences in temperature and pressure arise, giving birth to vertical air movement. Warm air, being less dense, rises, creating low-pressure areas. Conversely, cold air descends, leading to areas of high pressure. This interplay of pressure and vertical air movement forms the basis for weather patterns and influences the flight dynamics of aircraft.

Getting Water in the Air

The Earth's atmosphere is able to hold and transport water vapor, which leads to various weather patterns. Water enters the air through processes such as evaporation and transpiration. As air rises and cools, it reaches a point where it can no longer hold the moisture, leading to condensation. The result is the formation of clouds and, subsequently, precipitation.

Warm and Cold Fronts

Warm air, less dense, ascends over cold air, leading to the creation of warm fronts. Conversely, cold air wedges under warm air, forming cold fronts. The clash between these air masses becomes a battleground in the sky, influencing weather patterns.

Two Ways to Cool Air: Adiabatic Cooling and Frontal Lifting

Two primary mechanisms, adiabatic cooling and frontal lifting, contribute to cooling the air. Adiabatic cooling occurs as air ascends and expands, leading to a temperature decrease. Frontal lifting, on the other hand, results from the collision of warm and cold air masses, compelling the warmer air to rise and cool.

Relative Humidity and Dew Point

Relative humidity measures the air's moisture content relative to its capacity to hold moisture at a given temperature. While relative humidity enlightens us about the air's capability to retain water vapor, it falls short in predicting the timing of cloud formation, as it merely conveys the current state of the air, offering no insights into its future conditions. Another meteorological parameter, the dew point, allows us to predict the occurrence of clouds. The dew point is the temperature at which air becomes saturated with moisture, leading to condensation. Understanding the dew point is critical for predicting fog, dew formation, and cloud ceilings.

Condensation and Cloud Formation

Condensation marks the transformation of water vapor into liquid droplets, a process central to cloud formation. Clouds, diverse in their shapes and types, become visible indicators of atmospheric conditions.

Lapse Rates and Temperature Inversions

The lapse rate refers to the rate at which the temperature of the air changes with an increase in altitude. The temperature typically decreases with altitude under standard atmospheric conditions, resulting in a negative lapse rate. However, under certain conditions, particularly in the presence of an inversion layer, the temperature may increase with altitude, leading to a positive lapse rate. An inversion occurs when a layer of warm air traps cooler air beneath it, causing an increase in temperature with altitude. This reversal in the usual temperature pattern results in a positive lapse rate for that specific layer. Lapse rates can vary throughout the atmosphere and can be negative, positive, or zero.

Atmospheric Stability

Atmospheric stability, determined by the relationship between air parcels and their surrounding environment, dictates weather patterns. Stable air suppresses vertical motion, leading to clear skies and calm conditions. Unstable air, in contrast, fosters vertical motion, creating clouds, storms, and turbulence.

High and Low-Pressure Areas

High and low-pressure systems are atmospheric features that significantly influence weather patterns. High-pressure areas typically bring clear skies and stable conditions, while low-pressure areas are associated with clouds, precipitation, and potentially stormy weather.

Frontal Systems: Collisions in the Atmosphere

A frontal system is a boundary between two air masses with distinct temperature and humidity characteristics. This meteorological phenomenon prompts significant weather changes, often leading to precipitation and alterations in wind patterns. The collision and interaction of contrasting air masses along these boundaries are fundamental drivers of atmospheric dynamics and weather variability.

Thunderstorm and Fog

A thunderstorm is a weather phenomenon characterized by lightning and thunder. It is typically accompanied by heavy rain, strong winds, and sometimes hail. Thunderstorms are caused by the rapid upward movement of warm, moist air, which creates instability in the atmosphere. When the air reaches a certain height, the water vapor condenses, forming clouds and releasing energy through thunder and lightning.

Fog is a visible aerosol consisting of tiny water droplets suspended in the air. It forms when the air near the ground cools, and the water vapor condenses. Fog can reduce visibility significantly, making it challenging for pilots to navigate. Several factors can contribute to fog formation, including high humidity, calm winds, and clear skies. Fog is common in areas near bodies of water or in valleys surrounded by mountains.

14. Airport Operations

Airport operations are critical in aviation, demanding meticulous adherence to standardized procedures for safe and efficient air traffic flow.

Runway Lighting

Runway lighting provides crucial visual cues for pilots during takeoff and landing. White lights delineate the runway's edges, with red lights indicating the end. Approach lighting systems guide pilots during descent, facilitating safe landings.

Taxiway and Additional Runway Markings

Taxiing across the airport tarmac involves deciphering a visual language of markings. Yellow lines delineate taxiways, guiding aircraft between runways and terminals. Holding position markings indicate where aircraft should pause during ground operations. The array of additional markings, from threshold markings to centerline markings, contributes to a systematic and organized aircraft movement on the ground.

Airport Beacons

Airport beacons, projecting a rotating or flashing green and white light, serve as aerial lighthouses, signaling an airport's location to pilots from a distance. These beacons are instrumental in identifying the airport's position and orientation, assisting pilots in orienting themselves within the airspace.

Traffic Pattern

The traffic pattern is a choreography of aerial movements designed to facilitate safe takeoffs and landings. Pilots adhere to specified altitudes and directions, typically following a rectangular pattern around the runway.

Common Traffic Advisory Frequency (CTAF)

CATF serves as the auditory conduit for airport communication. Pilots use this frequency to broadcast their intentions, from taxiing to takeoff and landing, creating a shared awareness of other aircraft nearby.

Automatic Terminal Information Service (ATIS)

ATIS offers a pre-flight briefing through a continuous loop of recorded information. This includes weather conditions, active runways, and other relevant airport data. Pilots access this information before departure, enabling them to anticipate conditions, plan their routes, and make informed decisions throughout the flight.

Visual Approach Slope Indicator (VASI)

VASI is a visual aid aiding pilots in maintaining the correct approach slope during landing. A combination of red and white lights signals whether the aircraft is above, below, or on the desired glide path. Pilots rely on VASI for precise vertical guidance during the critical landing phase.

Wake Turbulence

Airport wake turbulence, a byproduct of aircraft movement, arises from the wingtip vortices generated by an aircraft's lift creation. As an aircraft slices through the air, these rotating air masses, or vortices, trail behind, creating a zone of turbulence. Understanding and mitigating this wake turbulence is crucial for maintaining safe distances between successive aircraft in airport operations.

SECTION 2. ESSENTIAL AVIATION CONCEPTS

A

Absolute accuracy. The ability to determine present position in space independently, and is most often used by pilots.

Absolute altitude. The actual distance between an aircraft and the terrain over which it is flying.

Absolute pressure. Pressure measured from the reference of zero pressure, or a vacuum.

Acceleration error. A magnetic compass error apparent when the aircraft accelerates while flying on an easterly or westerly heading, causing the compass card to rotate toward North.

Accelerate-go distance. The distance required to accelerate to V1 (see *takeoff decision speed*) with all engines at takeoff power, experience an engine failure at V1, and continue the takeoff on the remaining engine(s). The runway required includes the distance required to climb to 35 feet by which time V2 (see *takeoff safety speed*) speed must be attained.

Accelerate-stop distance. The distance required to accelerate to V1 with all engines at takeoff power, experience an engine failure at V1, and abort the takeoff and bring the airplane to a stop using braking action only (use of thrust reversing is not considered).

Accelerometer. A part of an inertial navigation system (INS) that accurately measures the force of acceleration in one direction.

ADC. See *air data computer*.

ADF. See *automatic direction finder*.

ADI. See *attitude director indicator*.

Adjustable-pitch propeller. A propeller with blades whose pitch can be adjusted on the ground with the engine not running, but which cannot be adjusted in flight. Also referred to as a *ground adjustable propeller*. Sometimes also used to refer to constant-speed propellers that are adjustable in flight.

Adjustable stabilizer. A stabilizer that can be adjusted in flight to trim the airplane, thereby allowing the airplane to fly hands-off at any given airspeed.

ADM. See *aeronautical decision-making*.

ADS-B. See *automatic dependent surveillance-broadcast*.

Advection fog. Fog resulting from the movement of warm, humid air over a cold surface.

Adverse yaw. A condition of flight in which the nose of an airplane tends to yaw toward the outside of the turn. This is caused by the higher induced drag on the outside wing, which is also producing more lift. Induced drag is a by-product of the lift associated with the outside wing.

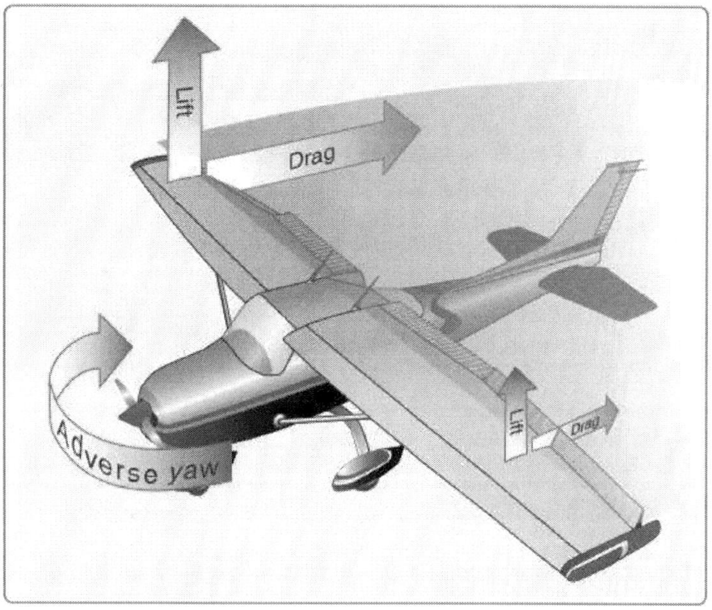

Aerodynamics. The science of the action of air on an object, and with the motion of air on other gases. Aerodynamics deals with the production of lift by the aircraft, the relative wind, and the atmosphere.

Aeronautical chart. A map used in air navigation containing all or part of the following: topographic features, hazards and obstructions, navigation aids, navigation routes, designated airspace, and airports.

Aeronautical decision-making (ADM). A systematic approach to the mental process used by pilots to consistently determine the best course of action in response to a given set of circumstances.

Agonic line. An irregular imaginary line across the surface of the Earth along which the magnetic and geographic poles are in alignment, and along which there is no magnetic variation.

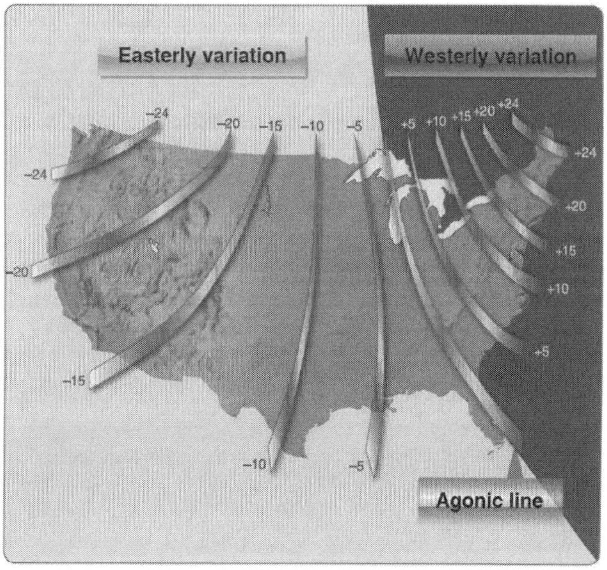

Ailerons. Primary flight control surfaces mounted on the trailing edge of an airplane wing, near the tip. Ailerons control roll about the longitudinal axis.

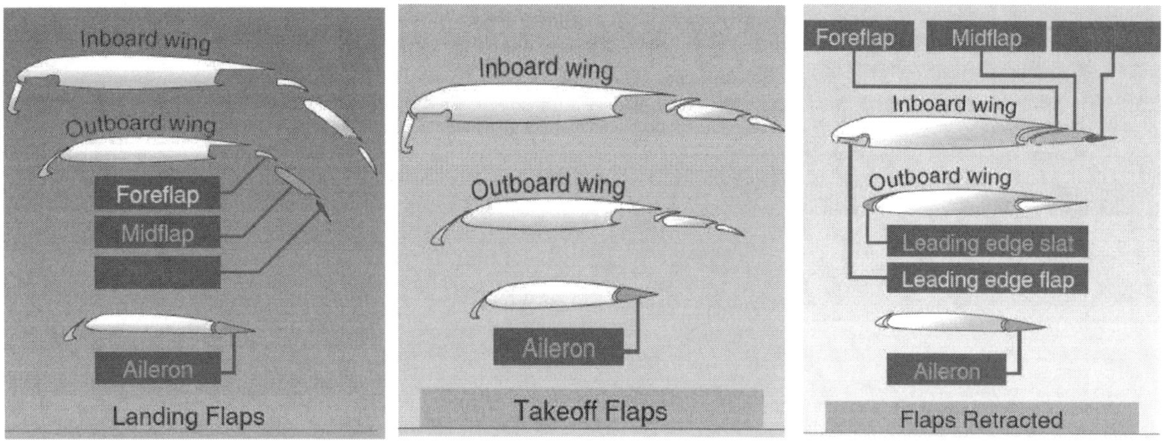

Aircraft altitude. The actual height above sea level at which the aircraft is flying.

Aircraft approach category. A performance grouping of aircraft based on a speed of 1.3 times the stall speed in the landing configuration at maximum gross landing weight.

Air data computer (ADC). An aircraft computer that receives and processes pitot pressure, static pressure, and temperature to calculate very precise altitude, indicated airspeed, true airspeed, and air temperature.

Airfoil. Any surface, such as a wing, propeller, rudder, or even a trim tab, which provides aerodynamic force when it interacts with a moving stream of air.

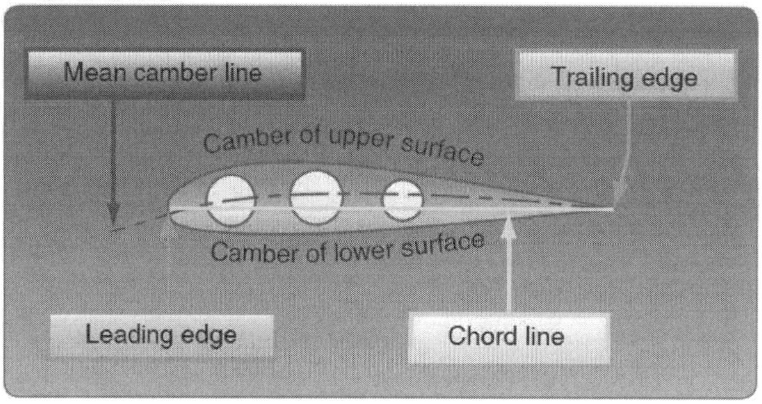

AIRMET. Inflight weather advisory issued as an amendment to the area forecast, concerning weather phenomena of operational interest to all aircraft and that is potentially hazardous to aircraft with limited capability due to lack of equipment, instrumentation, or pilot qualifications.

Airport diagram. The section of an instrument approach procedure chart that shows a detailed diagram of the airport. This diagram includes surface features and airport configuration information.

Airport markings. (Examples and explanations follow in relevant figure.)

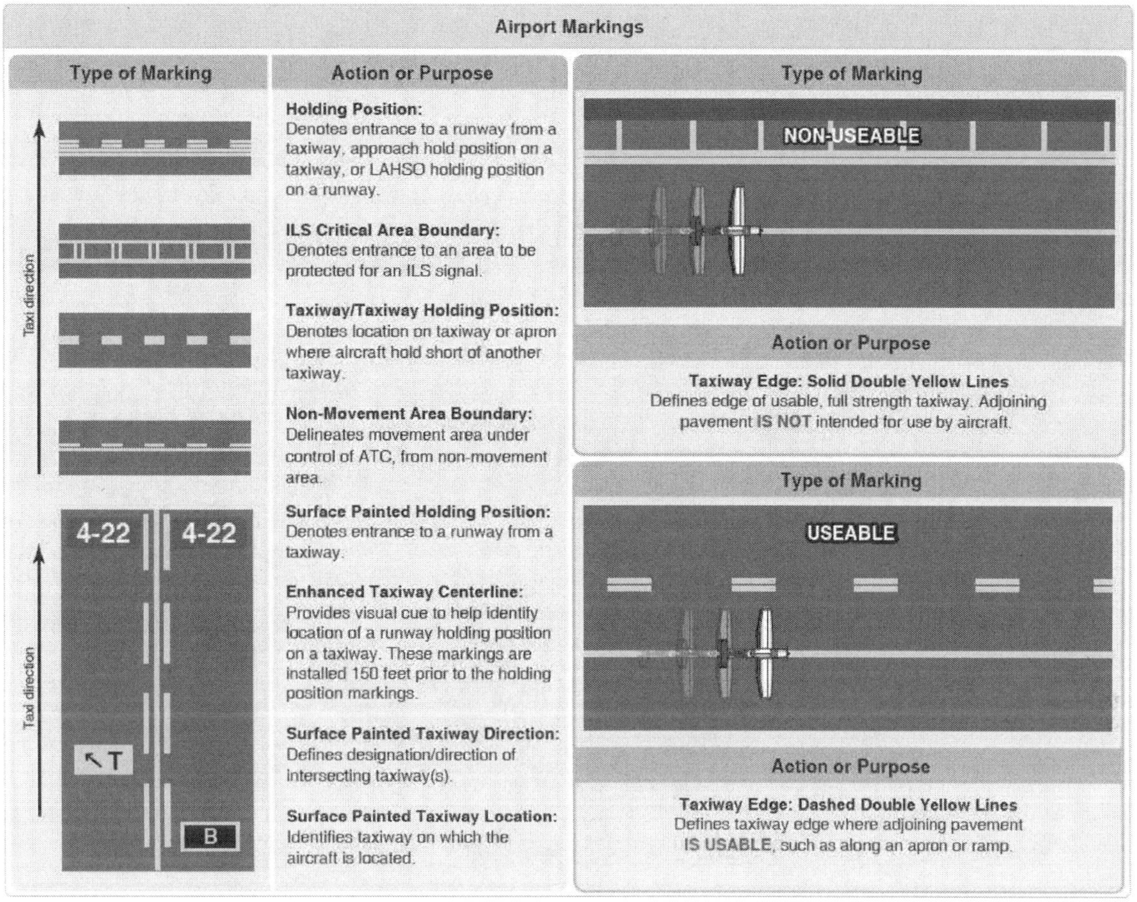

Airport signs. (Examples and explanations follow in relevant figure.)

Airport Signs

Type of Sign	Action or Purpose	Type of Sign	Action or Purpose
A 4-22	**Taxiway/Runway Hold Position:** Holding position for RWY 4-22 on TWY A.		**Runway Safety Area Boundary:** Identifies exit boundary of runway safety area.
26-8	**Runway/Runway Intersection:** Identifies intersecting runways or holding position for LAHSO operations.	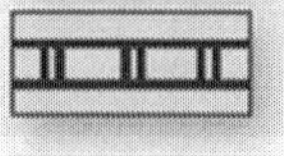	**ILS Critical Area Boundary:** Identifies exit boundary of ILS critical area.
B 8-APCH	**Runway Approach Hold Position:** Runway approach holding position for RWY 8 on TWY B.	↙J↗	**Taxiway Direction:** Defines direction and designation of intersecting taxiway(s).
C ILS	**ILS Critical Area Hold Position:** Holding position for the ILS critical area on TWY C.	←K	**Runway Exit:** Defines direction and designation of exit taxiway from runway.
⊖	**No Entry:** Identifies paved areas where aircraft entry is prohibited.	22↑	**Outbound Destination:** Defines directions to takeoff runway(s).
B	**Taxiway Location:** Identifies taxiway on which aircraft is located.	↖MIL	**Inbound Destination:** Defines directions to destination for arriving aircraft.
22	**Runway Location:** Identifies runway on which aircraft is located.		**Taxiway Ending Marker:** Indicates taxiway does not continue.
4	**Runway Distance Remaining:** Provides remaining runway length in 1,000-foot increments.	↙A G L→	**Direction Sign Array:** Identifies location in conjunction with multiple intersecting taxiways.

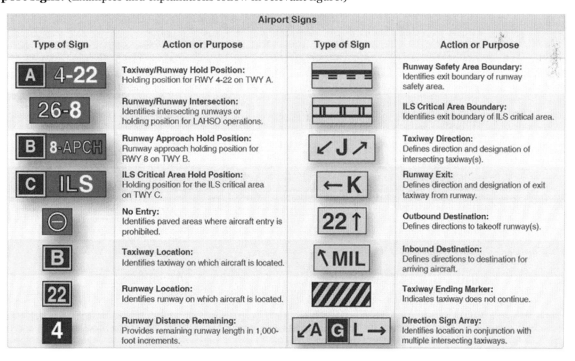

Airport surface detection equipment (ASDE). Radar equipment specifically designed to detect all principal features and traffic on the surface of an airport, presenting the entire image on the control tower console; used to augment visual observation by tower personnel of aircraft and/or vehicular movements on runways and taxiways.

Airport surveillance radar (ASR). Approach control radar used to detect and display an aircraft's position in the terminal area.

Airport surveillance radar approach. An instrument approach in which ATC issues instructions for pilot compliance based on aircraft position in relation to the final approach course and the distance from the end of the runway as displayed on the controller's radar scope.

Air route surveillance radar (ARSR). Air route traffic control center (ARTCC) radar used primarily to detect and display an aircraft's position while en route between terminal areas.

Air route traffic control center (ARTCC). Provides ATC service to aircraft operating on IFR flight plans within controlled airspace and principally during the en route phase of flight.

Airspeed. Rate of the aircraft's progress through the air.

Airspeed indicator. A differential pressure gauge that measures the dynamic pressure of the air through which the aircraft is flying. Displays the craft's airspeed, typically in knots, to the pilot.

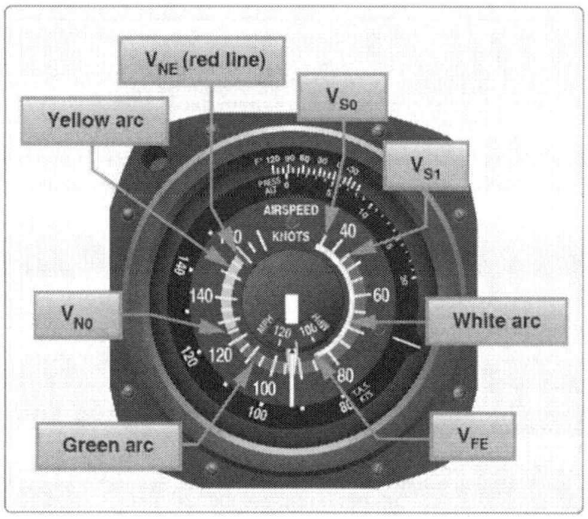

Air Traffic Control (ATC). The system and procedures to ensure safe, efficient movement of aircraft within all airspace classes, including to manage and optimize airspace utilization, provide instructions to and guide pilots, and maintain separation between aircraft to prevent collisions by facilitating the expeditious and orderly flow of miliary aircraft.

Air traffic control radar beacon system (ATCRBS). Sometimes called *secondary surveillance radar* (SSR), which utilizes a transponder in the aircraft. The ground equipment is an interrogating unit, in which the beacon antenna is mounted so it rotates with the surveillance antenna. The interrogating unit transmits a coded pulse sequence that actuates the aircraft transponder. The transponder answers the coded sequence by transmitting a preselected coded sequence back to the ground equipment, providing a strong return signal and positive aircraft identification, as well as other special data.

Airway. An airway is based on a centerline that extends from one navigation aid or intersection to another navigation aid (or through several navigation aids or intersections); used to establish a known route for en route procedures between terminal areas.

AHRS. See *attitude and heading reference system.*

Alert area. An area in which there is a high volume of pilot training or an unusual type of aeronautical activity.

ALS. See *approach lighting system.*

Alternate airport. An airport designated in an IFR flight plan, providing a suitable destination if a landing at the intended airport becomes inadvisable.

Alternate static source valve. A valve in the instrument static air system that supplies reference air pressure to the altimeter, airspeed indicator, and vertical speed indicator if the normal static pickup should become clogged or iced over.

Altimeter. A flight instrument that indicates altitude by sensing pressure changes.

Altimeter setting. Station pressure (the barometric pressure at the location the reading is taken) which has been corrected for the height of the station above sea level.

Altitude engine. A reciprocating aircraft engine having a rated takeoff power that is producible from sea level to an established higher altitude.

Aneroid. The sensitive component in an altimeter or barometer that measures the absolute pressure of the air. It is a sealed, flat capsule made of thin disks of corrugated metal soldered together and evacuated by pumping all the air out of it.

Aneroid barometer. An instrument that measures the absolute pressure of the atmosphere by balancing the weight of the air above it against the spring action of the aneroid.

Angle of attack. The angle of attack is the angle at which relative wind meets an airfoil. It is the angle that is formed by the chord of the airfoil and the direction of the relative wind or between the chord line and the flight path. The angle of attack changes during a flight as the pilot changes the direction of the aircraft and is related to the amount of lift being produced.

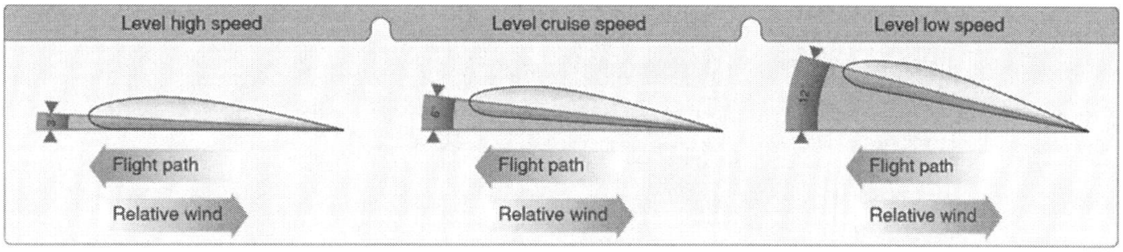

Angle of incidence. The acute angle formed between the chord line of an airfoil and the longitudinal axis of the aircraft on which it is mounted.

Anhedral. A downward slant from root to tip of an aircraft's wing or horizontal tail surface.

Antiservo tab. An adjustable tab attached to the trailing edge of a stabilator that moves in the same direction as the primary control. It is used to make the stabilator less sensitive.

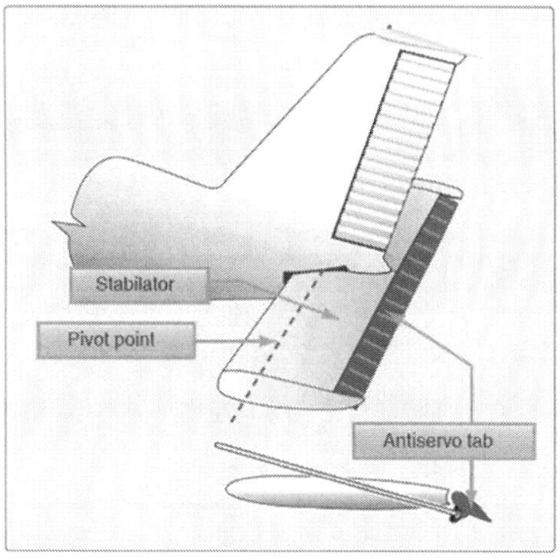

Approach lighting system (ALS). Provides lights that will penetrate the atmosphere far enough from touchdown to give directional, distance, and glidepath information for safe transition from instrument to visual flight.

Area chart. Part of the low-altitude en route chart series, this chart furnishes terminal data at a larger scale for congested areas.

Area forecast (FA). A report that gives a picture of clouds, general weather conditions, and visual meteorological conditions (VMC) expected over a large area encompassing several states.

Area navigation (RNAV). Allows a pilot to fly a selected course to a predetermined point without the need to overfly ground-based navigation facilities, by using waypoints.

ARSR. See air route surveillance radar.

ARTCC. See *air route traffic control center*.

ASDE. See *airport surface detection equipment*.

ASOS. See *Automated Surface Observing System*.

Aspect ratio. Span of a wing divided by its average chord.

ASR. See *airport surveillance radar.*

Asymmetric thrust. Also known as *P-factor*. A tendency for an aircraft to yaw to the left due to the descending propeller blade on the right producing more thrust than the ascending blade on the left. This occurs when the aircraft's longitudinal axis is in a climbing attitude in relation to the relative wind. The P-factor would be to the right if the aircraft had a counterclockwise rotating propeller.

ATC. See *Air Traffic Control.*

ATCRBS. See *air traffic control radar beacon system.*

ATIS. See *automatic terminal information service.*

Attitude and heading reference system (AHRS). A system composed of three-axis sensors that provide heading, attitude, and yaw information for aircraft. Such systems are designed to replace traditional mechanical gyroscopic flight instruments and provide superior reliability and accuracy.

Attitude director indicator (ADI). An aircraft attitude indicator that incorporates flight command bars to provide pitch and roll commands.

Attitude indicator. The foundation for all instrument flight, this instrument reflects the airplane's attitude in relation to the horizon.

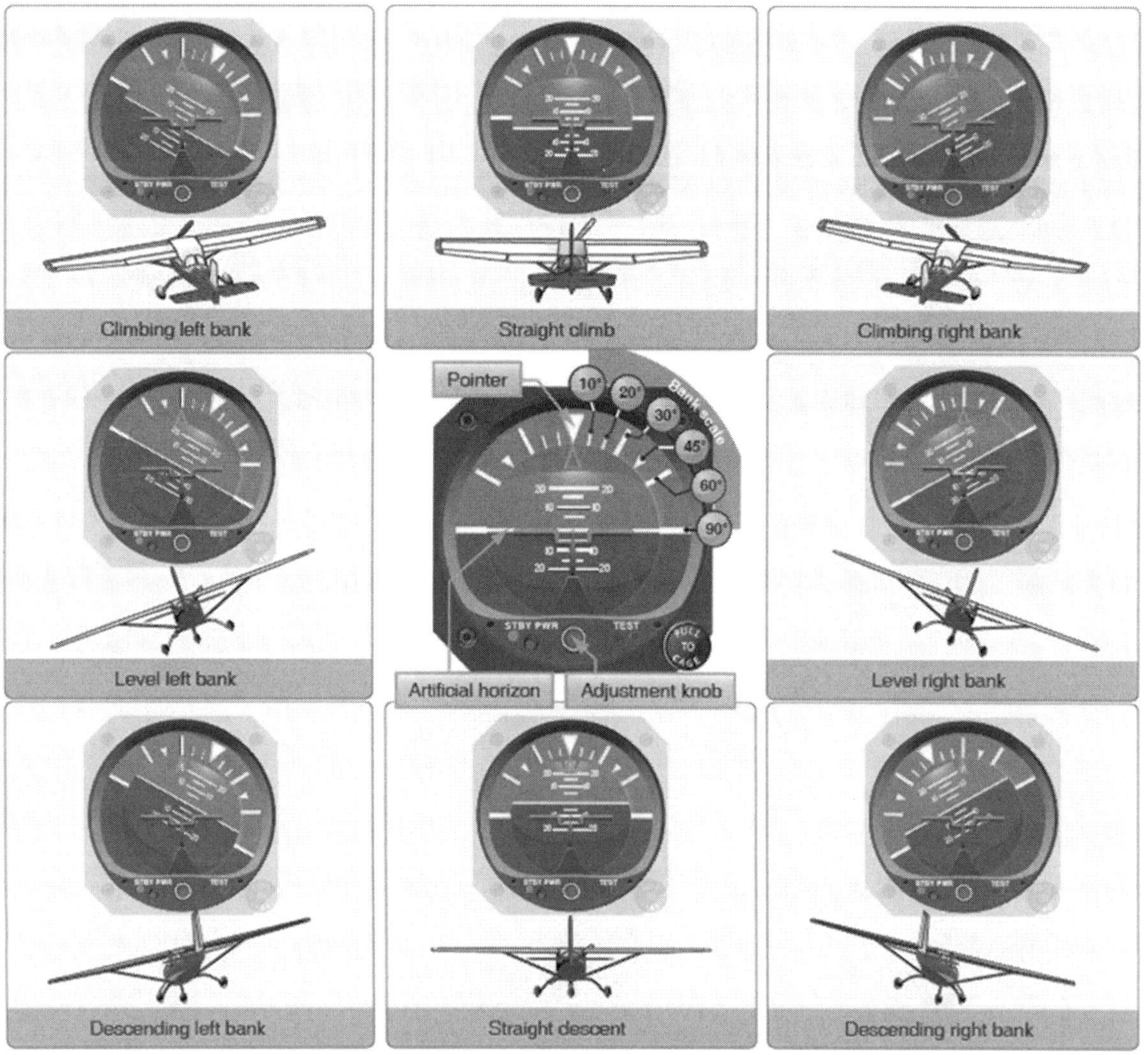

Attitude instrument flying. Controlling the aircraft by reference to the instruments rather than by outside visual cues.

Autokinesis. Nighttime visual illusion that a stationary light is moving, which becomes apparent after several seconds of staring at the light.

Automated Surface Observing System (ASOS). Weather reporting system which provides surface observations every minute via digitized voice broadcasts and printed reports.

Automated Weather Observing System (AWOS). Automated weather reporting system consisting of various sensors, a processor, a computer-generated voice subsystem, and a transmitter to broadcast weather data.

Automatic dependent surveillance-broadcast (ADS-B). A function on an aircraft or vehicle that periodically broadcasts its state vector (i.e., horizontal and vertical position, horizontal and vertical velocity) and other information.

Automatic direction finder (ADF). Electronic navigation equipment that operates in the low- and medium-frequency bands. Used in conjunction with the ground-based nondirectional beacon (NDB), the instrument displays the number of degrees clockwise from the nose of the aircraft to the station being received.

Automatic terminal information service (ATIS). The continuous broadcast of recorded non- control information in selected terminal areas. Its purpose is to improve controller effectiveness and relieve frequency congestion by automating repetitive transmission of essential but routine information.

Aviation Routine Weather Report (METAR). Observation of current surface weather reported in a standard international format.

AWOS. See *Automated Weather Observing System*.

Axes of an aircraft. Three imaginary lines that pass through an aircraft's center of gravity. The axes can be considered as imaginary axles around which the aircraft rotates. The three axes pass through the center of gravity at 90° angles to each other. The axis that runs from nose to tail is called the *longitudinal axis* (controlling roll), The axis from wingtip to wingtip is the lateral axis (pitch), and the axis that passes vertically through the center of gravity is the vertical axis (yaw).

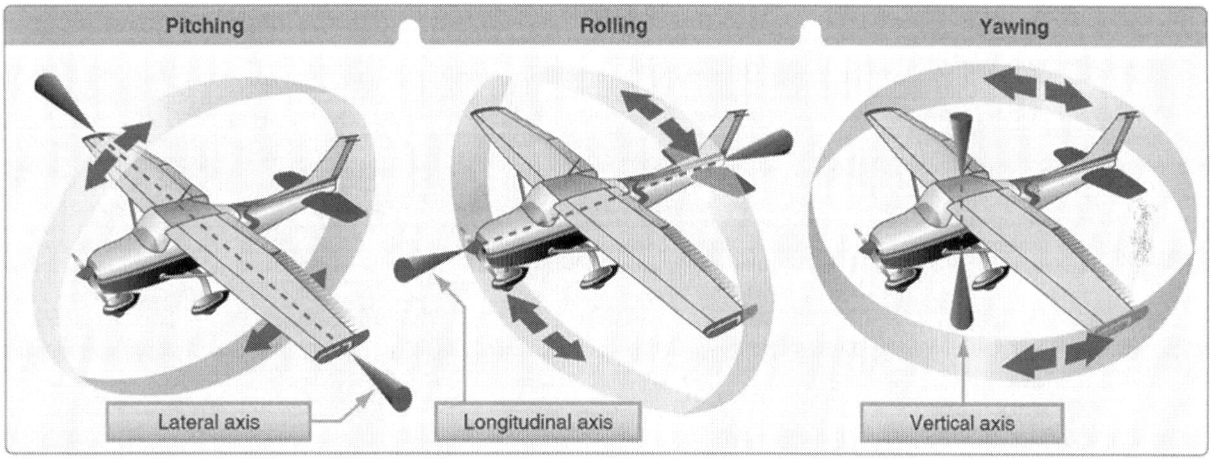

Axial flow compressor. A type of compressor used in a turbine engine in which the airflow through the compressor is essentially linear. An axial-flow compressor is made up of several stages of alternate rotors and stators. The compressor ratio is determined by the decrease in area of the succeeding stages.

B

Back course (BC). The reciprocal of the localizer course for an ILS. When flying a back-course approach, an aircraft approaches the instrument runway from the end at which the localizer antennas are installed.

Balance tab. An auxiliary control mounted on a primary control surface, which automatically moves in the direction opposite the primary control to provide an aerodynamic assist in the movement of the control.

Barometric scale. A scale on the dial of an altimeter to which the pilot sets the barometric pressure level from which the altitude shown by the pointers is measured.

Basic empty weight (GAMA standard). Basic empty weight includes the standard empty weight plus optional and special equipment that has been installed.

BC. See *back course.*

Bernoulli's principle. A principle that explains how the pressure of a moving fluid varies with its speed of motion. An increase in the speed of movement causes a decrease in the fluid's pressure.

Block altitude. A block of altitudes assigned by ATC to allow altitude deviations; for example, "Maintain block altitude 9 to 11 thousand."

C

Cabin altitude. Cabin pressure in terms of equivalent altitude above sea level.

Cage. The black markings on the ball instrument indicating its neutral position.

Calibrated orifice. A hole of specific diameter used to delay the pressure change in the case of a vertical speed indicator.

Calibrated airspeed. The speed at which the aircraft is moving through the air, found by correcting IAS for instrument and position errors.

Camber. The camber of an airfoil is the characteristic curve of its upper and lower surfaces. The upper camber is more pronounced, while the lower camber is comparatively flat. This causes the velocity of the airflow immediately above the wing to be much higher than that below the wing.

Canard. A horizontal surface mounted ahead of the main wing to provide longitudinal stability and control. It may be a fixed, movable, or variable geometry surface, with or without control surfaces.

Canard configuration. A configuration in which the span of the forward wings is substantially less than that of the main wing.

Cantilever. A wing designed to carry loads without external struts.

CAS. See *calibrated airspeed.*

CDI. See *course deviation indicator.*

Ceiling. The height above the earth's surface of the lowest layer of clouds, which is reported as broken or overcast, or the vertical visibility into an obscuration.

Center of gravity (CG). The point at which an airplane would balance if it were possible to suspend it at that point. It is the mass center of the airplane, or the theoretical point at which the entire weight of the airplane is assumed to be concentrated. It may be expressed in inches from the reference datum, or in percentage of mean aerodynamic chord (MAC). The location depends on the distribution of weight in the airplane.

Center of gravity limits. The specified forward and aft points within which the CG must be located during flight. These limits are indicated on pertinent airplane specifications.

Center of gravity range. The distance between the forward and aft CG limits indicated on pertinent airplane specifications.

Center of pressure. A point along the wing chord line where lift is considered to be concentrated. For this reason, the center of pressure is commonly referred to as the *center of lift.*

Centrifugal flow compressor. An impeller-shaped device that receives air at its center and slings the air outward at high velocity into a diffuser for increased pressure. Also referred to as a *radial outflow compressor.*

Centrifugal force. An outward force that opposes centripetal force, resulting from the effect of inertia during a turn.

Centripetal force. A center-seeking force directed inward toward the center of rotation created by the horizontal component of lift in turning flight.

CG. See *center of gravity.*

Changeover point (COP). A point along the route or airway segment between two adjacent navigation facilities or waypoints where changeover in navigation guidance should occur.

Chord line. An imaginary straight line drawn through an airfoil from the leading edge to the trailing edge.

Circling approach. A maneuver initiated by the pilot to align the aircraft with a runway for landing when a straight-in landing from an instrument approach is not possible or is not desirable.

CL. See *coefficient of lift*.

Class A airspace. Airspace from 18,000 feet MSL up to and including FL 600, including the airspace overlying the waters within 12 NM of the coast of the 48 contiguous states and Alaska; and designated international airspace beyond 12 NM of the coast of the 48 contiguous states and Alaska within areas of domestic radio navigational signal or ATC radar coverage, and within which domestic procedures are applied.

Class B airspace. Airspace from the surface to 10,000 feet MSL surrounding the nation's busiest airports in terms of IFR operations or passenger numbers. The configuration of each Class B airspace is individually tailored and consists of a surface area and two or more layers, and is designed to contain all published instrument procedures once an aircraft enters the airspace. For all aircraft, an ATC clearance is required to operate in the area, and aircraft so cleared receive separation services within the airspace.

Class C airspace. Airspace from the surface to 4,000 feet above the airport elevation (charted in MSL) surrounding those airports having an operational control tower, serviced by radar approach control, and having a certain number of IFR operations or passenger numbers.

Although the configuration of each Class C airspace area is individually tailored, the airspace usually consists of a 5 NM radius core surface area that extends from the surface up to 4,000 feet above the airport elevation, and a 10 NM radius shelf area that extends from 1,200 feet to 4,000 feet above the airport elevation.

Class D airspace. Airspace from the surface to 2,500 feet above the airport elevation (charted in MSL) surrounding those airports that have an operational control tower. The configuration of each Class D airspace area is individually tailored, and when instrument procedures are published, the airspace is normally designed to contain the procedures.

Class E airspace. Airspace that is not Class A, Class B, Class C, or Class D, and is controlled airspace.

Class G airspace. Airspace that is uncontrolled, except when associated with a temporary control tower, and has not been designated as Class A, Class B, Class C, Class D, or Class E airspace.

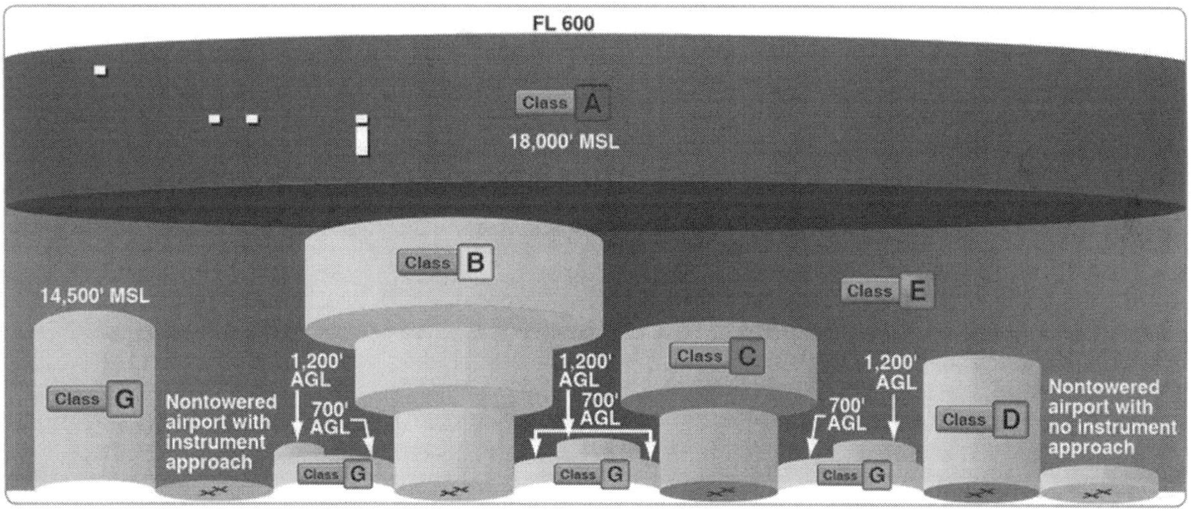

Clean configuration. A configuration in which all flight control surfaces have been placed to create minimum drag. In most aircraft this means flaps and gear retracted.

Clearance. ATC permission for an aircraft to proceed under specified traffic conditions within controlled airspace, for the purpose of providing separation between known aircraft.

Clearance delivery. Control tower position responsible for transmitting departure clearances to IFR flights.

Clearance limit. The fix, point, or location to which an aircraft is cleared when issued an air traffic clearance.

Clearance on request. An IFR clearance not yet received after filing a flight plan.

Clearance void time. Used by ATC, the time at which the departure clearance is automatically canceled if takeoff has not been made. The pilot must obtain a new clearance or cancel the IFR flight plan if not off by the specified time.

Coefficient of lift (CL). The ratio between lift pressure and dynamic pressure.

Cold front. The boundary between two air masses where cold air is replacing warm air.

Compass course. A true course corrected for variation and deviation errors.

Compass locator. A low-power, low- or medium-frequency (L/MF) radio beacon installed at the site of the outer or middle marker of an ILS.

Compass rose. A small circle graduated in 360° increments, to show direction expressed in degrees.

Complex aircraft. An aircraft with retractable landing gear, flaps, and a controllable-pitch propeller.

Compressor stall. In gas turbine engines, a condition in an axial-flow compressor in which one or more stages of rotor blades fail to pass air smoothly to the succeeding stages. A stall condition is caused by a pressure ratio that is incompatible with the engine rpm. Compressor stall will be indicated by a rise in exhaust temperature or rpm fluctuation, and if allowed to continue, may result in flameout and physical damage to the engine.

Computer navigation fix. A point used to define a navigation track for an airborne computer system such as GPS or FMS.

Concentric rings. Dashed-line circles depicted in the plan view of IAP charts, outside of the reference circle, that show en route and feeder facilities.

Cone of confusion. A cone-shaped volume of airspace directly above a VOR station where no signal is received, causing the CDI to fluctuate.

Configuration. This is a general term, which normally refers to the position of the landing gear and flaps.

Constant-speed propeller. A controllable-pitch propeller whose pitch is automatically varied in flight by a governor to maintain a constant rpm despite varying air loads.

Continuous flow oxygen system. System that supplies a constant supply of pure oxygen to a rebreather bag that dilutes the pure oxygen with exhaled gases and thus supplies a healthy mix of oxygen and ambient air to the mask. Primarily used in passenger cabins of commercial airliners.

Control and performance. A method of attitude instrument flying in which one instrument is used for making attitude changes, and the other instruments are used to monitor the progress of the change.

Control display unit. A display interfaced with the master computer, providing the pilot with a single control point for all navigations systems, thereby reducing the number of required flight deck panels.

Controllability. A measure of the response of an aircraft relative to the pilot's flight control inputs.

Controllable-pitch propeller (CPP). A type of propeller with blades that can be rotated around their long axis to change their pitch. If the pitch can be set to negative values, the reversible propeller can also create reverse thrust for braking or reversing without the need of changing the direction of shaft revolutions.

Controlled airspace. An airspace of defined dimensions within which ATC service is provided to IFR and VFR flights in accordance with the airspace classification. It includes Class A, Class B, Class C, Class D, and Class E airspace.

Control pressures. The amount of physical exertion on the control column necessary to achieve the desired attitude.

Control Surfaces. Moveable surfaces on an airplane's wings and tail allow a pilot to maneuver an airplane and control its attitude or orientation. (The examples in the following figure show two different aircraft control-surface configurations.)

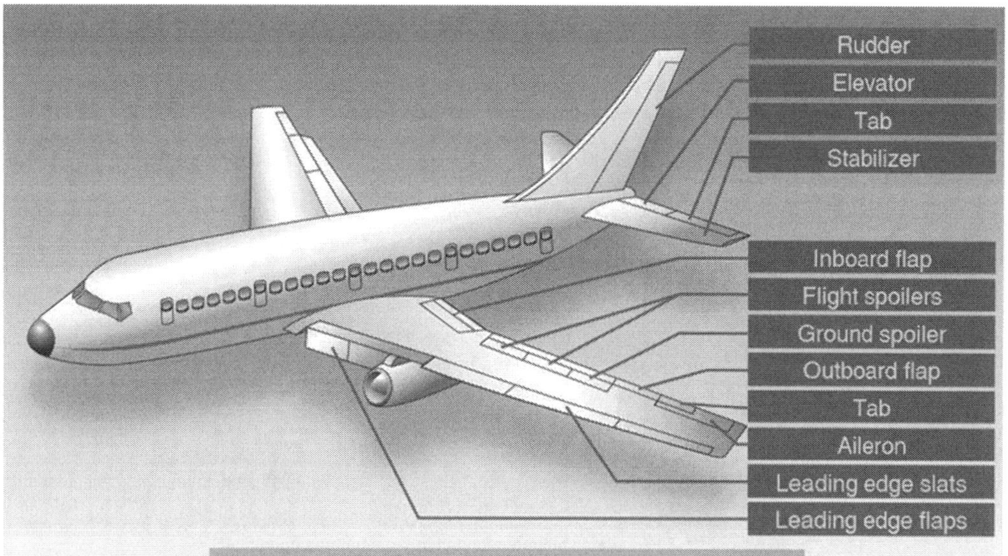

737 Control Surfaces

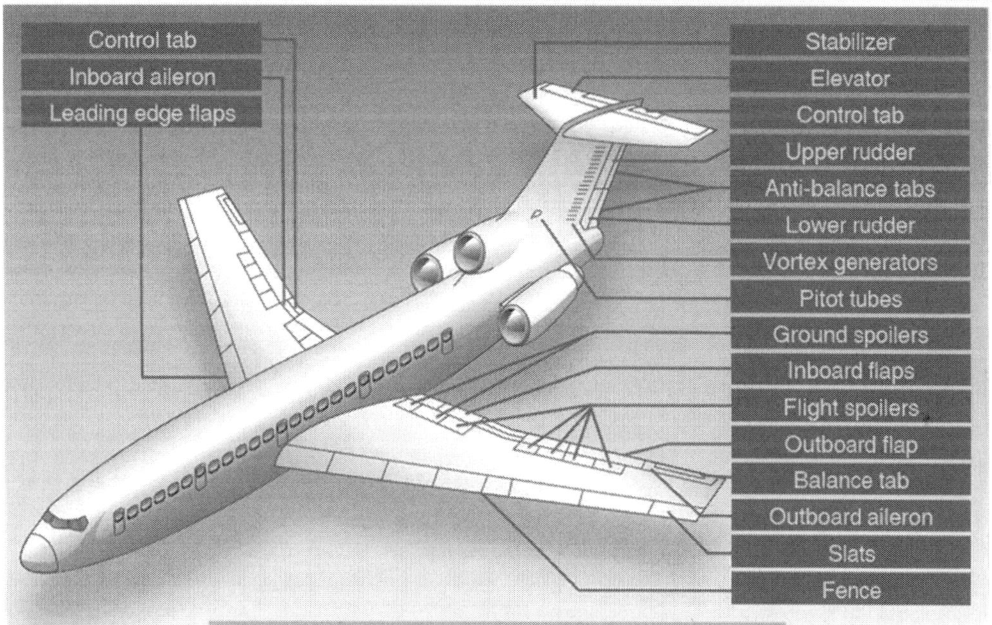

727 Control Surfaces

Convective weather. Unstable, rising air found in cumuliform clouds.

Convective SIGMET. Weather advisory concerning convective weather significant to the safety of all aircraft, including thunderstorms, hail, and tornadoes.

Conventional landing gear. Landing gear employing a third rear-mounted wheel. These airplanes are also sometimes referred to as *tailwheel airplanes*.

Coordinated flight. Flight with a minimum disturbance of the forces maintaining equilibrium, established via effective control use.

COP. See *changeover point*.

Coriolis illusion. The illusion of rotation or movement in an entirely different axis, caused by an abrupt head movement, while in a prolonged constant-rate turn that has ceased to stimulate the brain's motion sensing system.

Coupled ailerons and rudder. Rudder and ailerons are connected with interconnected springs to counteract adverse yaw. Can be overridden if it becomes necessary to slip the aircraft.

Course. The intended direction of flight in the horizontal plane measured in degrees from north.

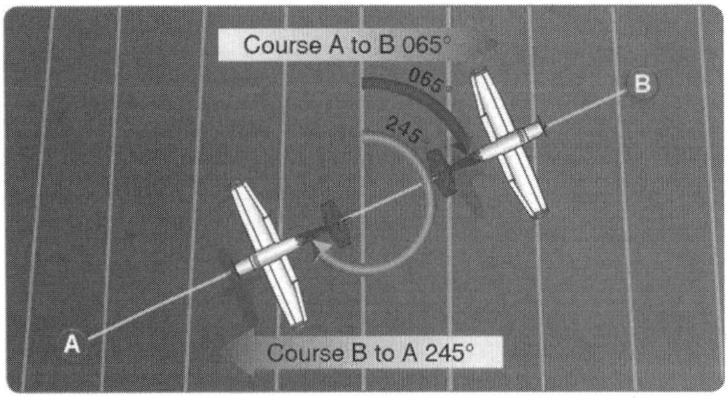

Course deviation indicator (CDI), This navigation instrument is used to display aircraft positions relative to the desired course. CDI provides visual cues, typically in the form of a needle or bar, to indicate whether an aircraft is staying on course or deviating to the left or right. Helps pilots maintain accurate navigation and course tracking during flight.

Cowl flaps. Shutter-like devices arranged around certain air-cooled engine cowlings, which may be opened or closed to regulate the flow of air around the engine.

CPP. See *controllable-pitch propeller*.

Crew resource management (CRM). The application of team management concepts in the flight deck environment, including practices and training applied to optimize teamwork, communication, and decision-making within aircraft crews to improve cooperation, situational awareness, and problem-solving among pilots, navigators, engineers, and other crew to enhance flight safety, mission success, and operational effectiveness.

Critical altitude. The maximum altitude under standard atmospheric conditions at which a turbocharged engine can produce its rated horsepower.

Critical angle of attack. The angle of attack at which a wing stalls regardless of airspeed, flight attitude, or weight.

Critical areas. Areas where disturbances to the ILS localizer and glideslope courses may occur when surface vehicles or aircraft operate near the localizer or glideslope antennas.

CRM. See *crew resource management*.

Cross-check. The first fundamental skill of instrument flight, also known as *scan*, the continuous and logical observation of instruments for attitude and performance information.

Cruise clearance. An ATC clearance issued to allow a pilot to conduct flight at any altitude from the minimum IFR altitude up to and including the altitude specified in the clearance. Also authorizes a pilot to proceed to and make an approach at the destination airport.

Current induction. An electrical current being induced into, or generated in, any conductor that is crossed by lines of flux from any magnet.

D

DA. See *decision altitude*.

Dead reckoning. Navigation of an airplane solely by means of computations based on airspeed, course, heading, wind direction and speed, groundspeed, and elapsed time.

Deceleration error. A magnetic compass error that occurs when the aircraft decelerates while flying on an easterly or westerly heading, causing the compass card to rotate toward South.

Decision altitude (DA). A specified altitude in the precision approach, charted in feet MSL, at which a missed approach must be initiated if the required visual reference to continue the approach has not been established.

Decision height (DH). A specified altitude in the precision approach, charted in height above threshold elevation, at which a decision must be made either to continue the approach or to execute a missed approach.

Density altitude. Pressure altitude corrected for nonstandard temperature. Density altitude is used in computing the performance of an aircraft and its engines.

Departure procedure (DP). Preplanned IFR ATC departure, published for pilot use, in textual and graphic format.

Deviation. A magnetic compass error caused by local magnetic fields within the aircraft. Deviation error is different on each heading.

DG. See *directional gyro.*

DGPS. See *Differential Global Positioning System.*

DH. See *decision height.*

Differential ailerons. Control surface rigged such that the aileron moving up moves a greater distance than the aileron moving down. The up aileron produces extra parasite drag to compensate for the additional induced drag caused by the down aileron. This balancing of the drag forces helps minimize adverse yaw.

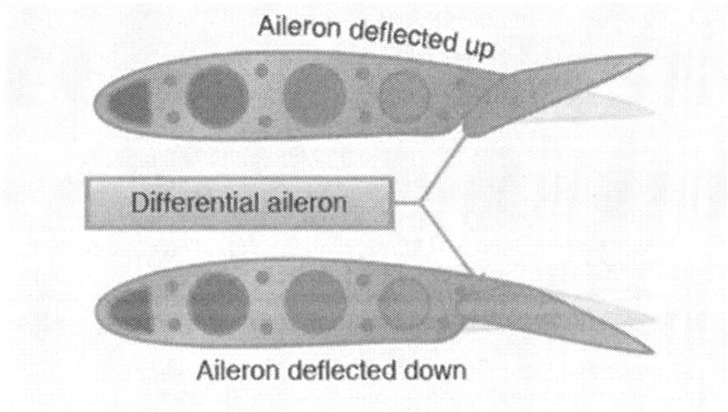

Differential Global Positioning System (DGPS). A system that improves the accuracy of Global Navigation Satellite Systems (GNSS) by measuring changes in variables to provide satellite positioning corrections.

Differential pressure. A difference between two pressures. The measurement of airspeed is an example of the use of differential pressure.

Dihedral. The positive acute angle between the lateral axis of an airplane and a line through the center of a wing or horizontal stabilizer. Dihedral contributes to the lateral stability of an airplane.

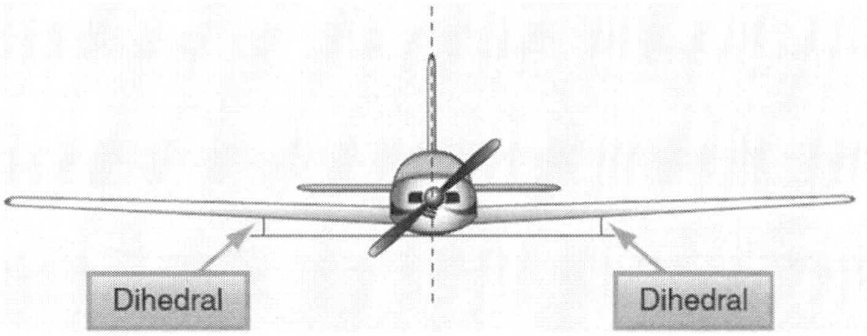

Diluter-demand oxygen system. An oxygen system that delivers oxygen mixed or diluted with air to maintain a constant oxygen partial pressure as the altitude changes.

Direct indication. The true and instantaneous reflection of aircraft pitch-and-bank attitude by the miniature aircraft, relative to the horizon bar of the attitude indicator.

Direct User Access Terminal System (DUATS). A system that provides current FAA weather and flight plan filing services to certified civil pilots, via personal computer, modem, or telephone access to the system. Pilots can request specific types of weather briefings and other pertinent data for planned flights.

Directional gyro (DG). An instrument which senses airplane movement and displays heading based on a 360° azimuth, with the final zero omitted. The directional gyro, also called *heading indicator*, is fundamentally a mechanical instrument designed to facilitate the use of the magnetic compass. The directional gyro is not affected by the forces that make the magnetic compass difficult to interpret.

Directional stability. Stability about the vertical axis of an aircraft, whereby an aircraft tends to return, on its own, to flight aligned with the relative wind when disturbed from that equilibrium state. The vertical tail is the primary contributor to directional stability, causing an airplane in flight to align with the relative wind.

Distance circle. See *reference circle.*

Distance measuring equipment (DME). A pulse-type electronic navigation system that shows the pilot, by an instrument-panel indication, the number of nautical miles between the aircraft and a ground station or waypoint.

DME. See *distance measuring equipment.*

DME arc. A flight track that is a constant distance from the station or waypoint.

Doghouse. A turn-and-slip indicator dial mark in the shape of a doghouse. See *slipping turn.*

Domestic reduced vertical separation minimum (DRVSM). Additional flight levels between FL 290 and FL 410 to provide operational, traffic, and airspace efficiency.

DP. See *departure procedure.*

Drag. The net aerodynamic force parallel to the relative wind, usually the sum of two components: induced drag and parasite drag.

Drag curve. The curve created when plotting induced drag and parasite drag.

Drift angle. Angle between heading and track.

DRVSM. See *Domestic Reduced Vertical Separation Minimum.*

DUATS. See *Direct User Access Terminal System.*

Duplex. Transmitting on one frequency and receiving on a separate frequency.

Dutch roll. A combination of rolling and yawing oscillations that normally occurs when the dihedral effects of an aircraft are more powerful than the directional stability. Usually dynamically stable but objectionable in an airplane because of the oscillatory nature.

Dynamic hydroplaning. A condition that exists when landing on a surface with standing water deeper than the tread depth of the tires. When the brakes are applied, there is a possibility that the brake will lock up and the tire will ride on the surface of the water, much like a water ski.

When the tires are hydroplaning, directional control and braking action are virtually impossible. An effective anti-skid system can minimize the effects of hydroplaning.

Dynamic stability. The property of an aircraft that causes it, when disturbed from straight-and- level flight, to develop forces or moments that restore the original condition of straight and level.

E

Eddy currents. Current induced in a metal cup or disc when it is crossed by lines of flux from a moving magnet.

Eddy current damping. The decreased amplitude of oscillations by the interaction of magnetic fields. In the case of a vertical card magnetic compass, flux from the oscillating permanent magnet produces eddy currents in a damping disk or cup. The magnetic flux produced by the eddy currents opposes the flux from the permanent magnet and decreases the oscillations.

EFC. See *expect-further-clearance.*

EFD. See *electronic flight display.*

Electronic flight display (EFD). For standardization purposes, any flight instrument display that uses LCD or other image-producing system (cathode ray tube [CRT], etc.)

Elevator. The horizontal, movable primary control surface in the tail section, or empennage, of an airplane. The elevator is hinged to the trailing edge of the fixed horizontal stabilizer.

Elevator illusion. The sensation of being in a climb or descent, caused by the kind of abrupt vertical accelerations that result from up- or downdrafts.

Empennage. The section of the airplane that consists of the vertical stabilizer, the horizontal stabilizer, and the associated control surfaces.

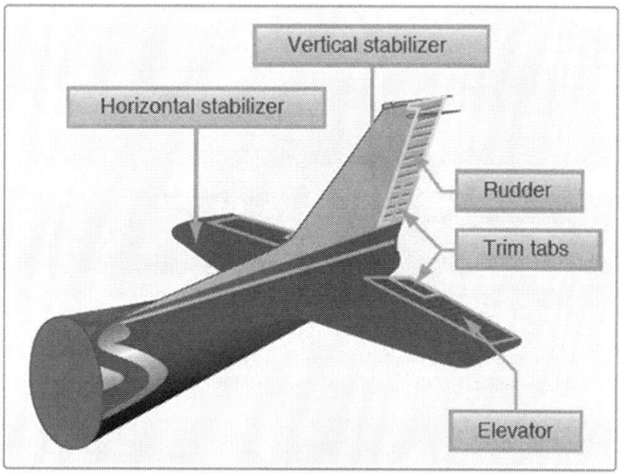

Emphasis error. The result of giving too much attention to a particular instrument during the cross-check, instead of relying on a combination of instruments necessary for attitude and performance information.

Empty-field myopia. Induced nearsightedness that is associated with flying at night, in instrument meteorological conditions and/or reduced visibility. With nothing to focus on, the eyes automatically focus on a point just slightly ahead of the airplane.

Encoding altimeter. A special type of pressure altimeter used to send a signal to the air traffic controller on the ground, showing the pressure altitude the aircraft is flying.

Engine pressure ratio (EPR). The ratio of turbine discharge pressure divided by compressor inlet pressure, which is used as an indication of the amount of thrust being developed by a turbine engine.

En route facilities ring. Depicted in the plan view of IAP charts, a circle which designates NAVAIDs, fixes, and intersections that are part of the en route low altitude airway structure.

En route high-altitude charts. Aeronautical charts for en route instrument navigation at or above 18,000 feet MSL.

En route low-altitude charts. Aeronautical charts for en route IFR navigation below 18,000 feet MSL.

EPR. See *engine pressure ratio.*

Equilibrium. A condition that exists within a body when the sum of the moments of all the forces acting on the body is equal to zero. In aerodynamics, equilibrium is when all opposing forces acting on an aircraft are balanced (steady, unaccelerated flight conditions).

Equivalent airspeed. Airspeed equivalent to CAS in standard atmosphere at sea level. As the airspeed and pressure altitude increase, the CAS becomes higher than it should be, and a correction for compression must be subtracted from the CAS.

Expect-further-clearance (EFC). The time a pilot can expect to receive clearance beyond a clearance limit.

Explosive decompression. A change in cabin pressure faster than the lungs can decompress. Lung damage is possible.

F

FA. See *area forecast.*

FAF. See *final approach fix.*

False horizon. Inaccurate visual information for aligning the aircraft, caused by various natural and geometric formations that disorient the pilot from the actual horizon.

FB. See *winds and temperature aloft forecast.*

FDI. See *flight director indicator.*

Federal airways. Class E airspace areas that extend upward from 1,200 feet to, but not including, 18,000 feet MSL, unless otherwise specified.

Feeder facilities. Used by ATC to direct aircraft to use intervening fixes between the en route structure and the initial approach fix.

Final approach. Part of an instrument approach procedure in which alignment and descent for landing are accomplished.

Final approach fix (FAF). The fix from which the IFR final approach to an airport is executed, and which identifies the beginning of the final approach segment. An FAF is designated on government charts by a Maltese cross symbol for nonprecision approaches, and a lightning bolt symbol for precision approaches.

Fixating. Staring at a single instrument, thereby interrupting the cross-check process.

Fixed-pitch propellers. Propellers with fixed blade angles. Fixed-pitch propellers are designed as climb propellers, cruise propellers, or standard propellers.

Fixed slot. A fixed, nozzle shaped opening near the leading edge of a wing that ducts air onto the top surface of the wing. Its purpose is to increase lift at higher angles of attack.

FL. See *flight level.*

Flameout. A condition in the operation of a gas turbine engine in which the fire in the engine goes out due to either too much or too little fuel sprayed into the combustors.

Flaps. Hinged portion of the trailing edge between the ailerons and fuselage. In some aircraft ailerons and flaps are interconnected to produce full-span "flaperons." In either case, flaps change the lift and drag on the wing.

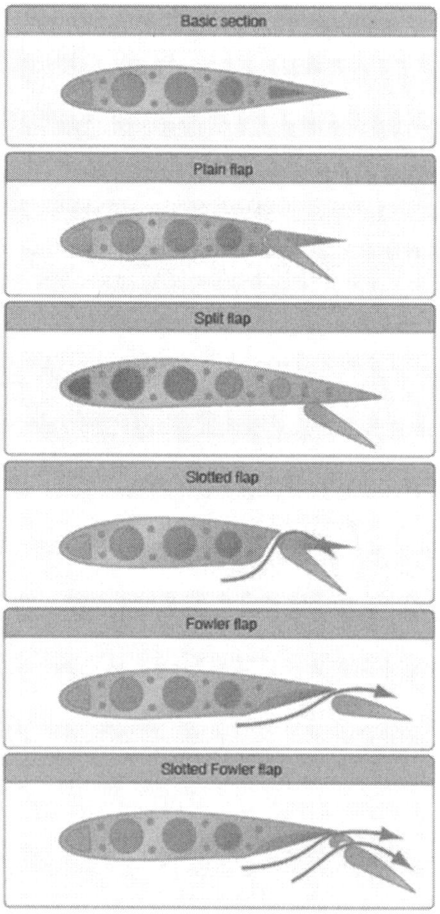

Floor load limit. The maximum weight the floor can sustain per square inch/foot as provided by the manufacturer.

Flight configurations. Adjusting the aircraft control surfaces (including flaps and landing gear) in a manner that will achieve a specified attitude.

Flight director indicator (FDI). One of the major components of a flight director system, it provides steering commands that the pilot (or the autopilot, if coupled) follows.

Flight level (FL). A measure of altitude (in hundreds of feet) used by aircraft flying above 18,000 feet with the altimeter set at 29.92 "Hg.

Flight management system (FMS). Provides pilot and crew with highly accurate and automatic long-range navigation capability, blending available inputs from long- and short-range sensors.

Flight path. The line, course, or track along which an aircraft is flying or is intended to be flown.

Flight patterns. Basic maneuvers, flown by reference to the instruments rather than outside visual cues, for the purpose of practicing basic attitude flying. The patterns simulate maneuvers encountered on instrument flights such as holding patterns, procedure turns, and approaches.

Flight strips. Paper strips containing instrument flight information, used by ATC when processing flight plans.

FMS. See *flight management system.*

FOD. See *foreign object damage.*

Force (F). The energy applied to an object that attempts to cause the object to change its direction, speed, or motion. In aerodynamics, force is expressed as F, T (thrust), L (lift), W (weight), or D (drag), usually in pounds.

Foreign object damage (FOD). Damage to a gas turbine engine caused by some object being sucked into the engine while it is running. Debris from runways or taxiways can cause foreign object damage during ground operations, and the ingestion of ice and birds can cause FOD in flight.

Form drag. The drag created because of the shape of a component or the aircraft.

Frise-type aileron. Aileron having the nose portion projecting ahead of the hinge line. When the trailing edge of the aileron moves up, the nose projects below the wing's lower surface and produces some parasite drag, decreasing the amount of adverse yaw.

Fuel load. The expendable part of the load of the airplane. Fuel load includes only usable fuel, not fuel required to fill the lines or that which remains trapped in the tank sumps.

Fuselage. The section of the airplane that consists of the cabin and/or cockpit, containing seats for the occupants and the controls for the airplane.

G

GAMA. The General Aviation Manufacturers Association, the trade association that represents general aviation aircraft manufacturers and related companies globally. GAMA establishes and promotes measurement standards for general aviation aircraft to ensure industry safety and consistency.

Glideslope (GS). Part of the ILS that projects a radio beam upward at an angle of approximately 3° from the approach end of an instrument runway. The glideslope provides vertical guidance to aircraft on the final approach course for the aircraft to follow when making an ILS approach along the localizer path.

Glideslope intercept altitude. The minimum altitude of an intermediate approach segment prescribed for a precision approach that ensures obstacle clearance.

Global landing system (GLS). An instrument approach with lateral and vertical guidance with integrity limits (similar to barometric vertical navigation, or BARO VNAV).

Global navigation satellite system (GNSS). Satellite navigation system that provides autonomous geospatial positioning with global coverage. It allows small electronic receivers to determine their location (longitude, latitude, and altitude) to within a few meters using time signals transmitted along a line of sight by radio from satellites.

Global positioning system (GPS). Navigation system that uses satellite rather than ground- based transmitters for location information.

GLS. See *global landing system.*

GNSS. See *global navigation satellite system.*

GPS. See *global positioning system.*

GPS Approach Overlay Program. An authorization for pilots to use GPS avionics under IFR for flying designated existing nonprecision instrument approach procedures, except for LOC, LDA, and SDF procedures.

GPWS. See *ground proximity warning system.*

Graveyard spiral. The illusion of the cessation of a turn while still in a prolonged, coordinated, constant rate turn, which can lead a disoriented pilot to a loss of control of the aircraft.

Great circle route. The shortest distance across the surface of a sphere (the Earth) between two points on the surface.

Ground adjustable trim tab. Non-movable metal trim tab on a control surface. Bent in one direction or another while on the ground to apply trim forces to the control surface.

Ground effect. The condition of slightly increased air pressure below an airplane wing or helicopter rotor system that increases the amount of lift produced. It exists within approximately one wingspan or one rotor diameter from the ground. It results from a reduction in upwash, downwash, and wingtip vortices, and provides a corresponding decrease in induced drag.

Ground proximity warning system (GPWS). A system designed to determine an aircraft's clearance above the Earth and provides limited predictability about aircraft position relative to rising terrain.

Groundspeed. Speed over the ground, either closing speed to the station or waypoint, or speed over the ground in whatever direction the aircraft is currently going, depending upon the navigation system used.

GS. See *glideslope.*

GWPS. See *ground proximity warning system.*

Gyroscopic precession. An inherent quality of rotating bodies, which causes an applied force to be manifested 90° in the direction of rotation from the point where the force is applied.

H

HAA. See *height above airport.*

HAL. See *height above landing.*

HAT. See *height above touchdown elevation.*

Hazardous attitudes. Five aeronautical decision-making attitudes that may contribute to poor pilot judgment: anti-authority, impulsivity, invulnerability, machismo, and resignation.

Hazardous Inflight Weather Advisory Service (HIWAS). An en route FSS service providing continuously updated automated alerts of hazardous weather within 150 nautical miles of selected VORs, available only in the conterminous 48 states.

Head-up display (HUD). A special type of flight viewing screen that allows the pilot to watch the flight instruments and other data while looking through the windshield of the aircraft for other traffic, the approach lights, or the runway.

Heading. The direction in which the nose of the aircraft is pointing during flight.

Heading indicator. An instrument which senses airplane movement and displays heading based on a 360° azimuth, with the final zero omitted. The heading indicator, also called a *directional gyro* (DG), is fundamentally a mechanical instrument designed to facilitate the use of the magnetic compass. The heading indicator is not affected by the forces that make the magnetic compass difficult to interpret.

Height above airport (HAA). The height of the MDA above the published airport elevation.

Height above landing (HAL). The height above a designated helicopter landing area used for helicopter instrument approach procedures.

Height above touchdown elevation (HAT). The DA/DH or MDA above the highest runway elevation in the touchdown zone (first 3,000 feet of the runway).

HF. See *high frequency.*

High frequency (HF). Range of radio frequencies typically between 3 and 30 megahertz (MHz) used for long-distance communication. HF radios are employed for military operations, air traffic control, and coordination among forces]

High performance aircraft. An aircraft with an engine of more than 200 horsepower.

Histotoxic hypoxia. The inability of cells to effectively use oxygen. Plenty of oxygen is being transported to the cells that need it, but they are unable to use it.

HIWAS. See *Hazardous Inflight Weather Advisory Service.*

Holding. A predetermined maneuver that keeps aircraft within a specified airspace while awaiting further clearance from ATC.

Holding pattern. A racetrack pattern, involving two turns and two legs, used to keep an aircraft within a prescribed airspace with respect to a geographic fix. A standard pattern uses right turns; nonstandard patterns use left turns.

Homing. Flying the aircraft on any heading required to keep the needle pointing to the 0° relative bearing position.

Horizontal situation indicator (HSI). A flight navigation instrument that combines the heading indicator with a CDI, to provide the pilot with better situational awareness of location with respect to the courseline. (See the following figure for illustration.)

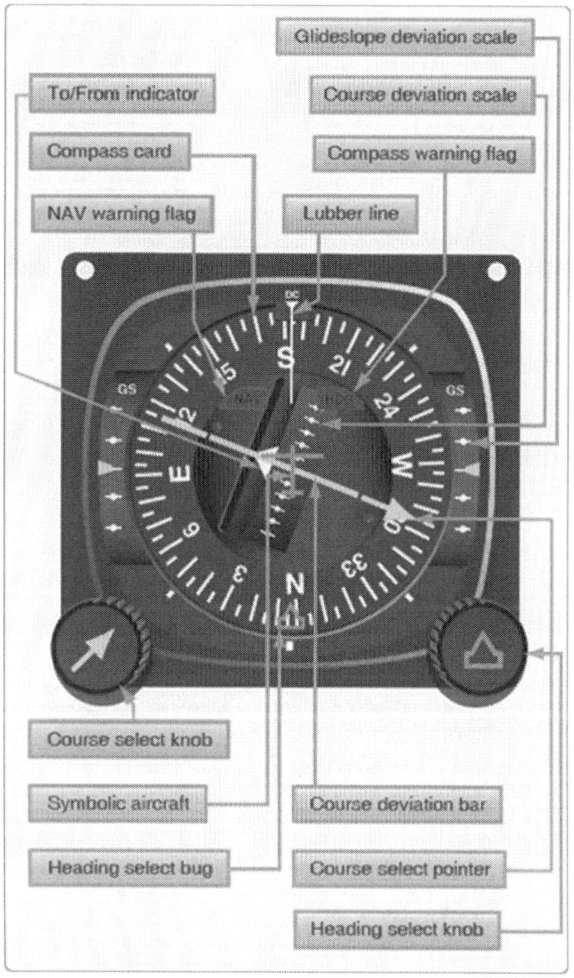

Horsepower. The term, originated by inventor James Watt, means the amount of work a horse could do in one second. One horsepower equals 550 foot-pounds per second, or 33,000 foot- pounds per minute.

HSI. See *horizontal situation indicator*.

HUD. See *head-up display*.

Hydroplaning. A condition that exists when landing on a surface with standing water deeper than the tread depth of the tires. When the brakes are applied, there is a possibility that the brake will lock up and the tire will ride on the surface of the water, much like a water ski. When the tires are hydroplaning, directional control and braking action are virtually impossible. An effective anti-skid system can minimize the effects of hydroplaning.

Hypemic hypoxia. A type of hypoxia that is a result of oxygen deficiency in the blood, rather than a lack of inhaled oxygen. It can be caused by a variety of factors. *Hypemic* means "not enough blood."

Hypoxia. A state of oxygen deficiency in the body sufficient to impair functions of the brain and other organs.

Hypoxic hypoxia. This type of hypoxia is a result of insufficient oxygen available to the lungs. A decrease of oxygen molecules at sufficient pressure can lead to hypoxic hypoxia.

I

IAF. See *initial approach fix*.

IAP. See *instrument approach procedures*.

IAS. See *indicated airspeed*.

ICAO. See *International Civil Aviation Organization*.

Ident. Air Traffic Control request for a pilot to push the button on the transponder to identify return on the controller's scope.

IFR. See *instrument flight rules.*

ILS. See *instrument landing system.*

ILS categories. Categories of instrument approach procedures allowed at airports equipped with the following types of instrument landing systems:

- **ILS Category I:** Provides for approach to a height above touchdown of not less than 200 feet, and with runway visual range of not less than 1,800 feet.
- **ILS Category II:** Provides for approach to a height above touchdown of not less than 100 feet and with runway visual range of not less than 1,200 feet.
- **ILS Category IIIA:** Provides for approach without a decision height minimum and with runway visual range of not less than 700 feet.
- **ILS Category IIIB:** Provides for approach without a decision height minimum and with runway visual range of not less than 150 feet.
- **ILS Category IIIC:** Provides for approach without a decision height minimum and without runway visual range minimum.

IMC. See *instrument meteorological conditions.*

Inclinometer. An instrument consisting of a curved glass tube, housing a glass ball, and damped with a fluid that resembles kerosene. It may be used to indicate inclination, as a level, or, as used in the turn indicators, to show the relationship between gravity and centrifugal force in a turn.

Indicated airspeed (IAS). Shown on the dial of the instrument airspeed indicator on an aircraft. Indicated airspeed (IAS) is the airspeed indicator reading uncorrected for instrument, position, and other errors. Indicated airspeed means the speed of an aircraft as shown on its pitot static airspeed indicator calibrated to reflect standard atmosphere adiabatic compressible flow at sea level uncorrected for airspeed system errors. Calibrated airspeed (CAS) is IAS corrected for instrument errors, position error (due to incorrect pressure at the static port) and installation errors.

Indicated altitude. The altitude read directly from the altimeter (uncorrected) when it is set to the current altimeter setting.

Indirect indication. A reflection of aircraft pitch-and-bank attitude by instruments other than the attitude indicator.

Induced drag. Drag caused by the same factors that produce lift; its amount varies inversely with airspeed. As airspeed decreases, the angle of attack must increase, in turn increasing induced drag.

Inertial navigation system (INS). A computer-based navigation system that tracks the movement of an aircraft via signals produced by onboard accelerometers. The initial location of the aircraft is entered into the computer, and all subsequent movement of the aircraft is sensed and used to keep the position updated. An INS does not require any inputs from outside signals.

Initial approach fix (IAF). The fix depicted on IAP charts where the instrument approach procedure (IAP) begins unless otherwise authorized by ATC.

Inoperative components. Higher minimums are prescribed when the specified visual aids are not functioning; this information is listed in the Inoperative Components Table found in the United States Terminal Procedures Publications.

INS. See *inertial navigation system.*

Instantaneous vertical speed indicator (IVSI). Assists in interpretation by instantaneously indicating the rate of climb or descent at a given moment with little or no lag as displayed in a vertical speed indicator (VSI).

Instrument approach procedures (IAP). A series of predetermined maneuvers for the orderly transfer of an aircraft under IFR from the beginning of the initial approach to a landing or to a point from which a landing may be made visually.

Instrument flight rules (IFR). Rules and regulations established by the Federal Aviation Administration to govern flight under conditions in which flight by outside visual reference is not safe. IFR flight depends upon flying by reference to instruments in the flight deck, and navigation is accomplished by reference to electronic signals.

Instrument landing system (ILS). An electronic system that provides both horizontal and vertical guidance to a specific runway, used to execute a precision instrument approach procedure.

Instrument meteorological conditions (IMC). Meteorological conditions expressed in terms of visibility, distance from clouds, and ceiling less than the minimums specified for visual meteorological conditions, requiring operations to be conducted under IFR.

Instrument takeoff. Using the instruments rather than outside visual cues to maintain runway heading and execute a safe takeoff.

Interference drag. Drag generated by the collision of airstreams creating eddy currents, turbulence, or restrictions to smooth flow.

International Civil Aviation Organization (ICAO). The United Nations agency for developing the principles and techniques of international air navigation, and fostering planning and development of international civil air transport.

International standard atmosphere (IAS). A model of standard variation of pressure and temperature.

Interpolation. The estimation of an intermediate value of a quantity that falls between marked values in a series. Example: In a measurement of length, with a rule that is marked in eighths of an inch, the value falls between 3/8 inch and 1/2 inch. The estimated (interpolated) value might then be said to be 7/16 inch.

Inversion. An increase in temperature with altitude.

Inversion illusion. The feeling that the aircraft is tumbling backwards, caused by an abrupt change from climb to straight and-level flight while in situations lacking visual reference.

Isobars. Lines which connect points of equal barometric pressure.

Isogonic lines. Lines drawn across aeronautical charts to connect points having the same magnetic variation.

IVSI. See *instantaneous vertical speed indicator*.

J

Jet route. A route designated to serve flight operations from 18,000 feet MSL up to and including FL 450.

Jet stream. A high-velocity narrow stream of winds, usually found near the upper limit of the troposphere, which flows generally from west to east.

K

KIAS. Knots indicated airspeed.

Knot. The knot is a unit of speed equal to one nautical mile (1.852 km) per hour, approximately 1.151 mph.

Kollsman window. A barometric scale window of a sensitive altimeter used to adjust the altitude for the altimeter setting.

L

LAAS. See *local area augmentation system*.

Lag. The delay that occurs before an instrument needle attains a stable indication.

Land breeze. A coastal breeze flowing from land to sea caused by temperature differences when the sea surface is warmer than the adjacent land. The land breeze usually occurs at night and alternates with the sea breeze that blows in the opposite direction by day.

Land as soon as possible. Land without delay at the nearest suitable area, such as an open field, at which a safe approach and landing is assured.

Land as soon as practical. The landing site and duration of flight are at the discretion of the pilot. Extended flight beyond the nearest approved landing area is not recommended.

Land immediately. The urgency of the landing is paramount. The primary consideration is to ensure the survival of the occupants. Landing in trees, water, or other unsafe areas should be considered only as a last resort.

Lateral axis. An imaginary line passing through the center of gravity of an airplane and extending across the airplane from wingtip to wingtip.

Lateral stability (rolling). The stability about the longitudinal axis of an aircraft. Rolling stability or the ability of an airplane to return to level flight due to a disturbance that causes one of the wings to drop.

Latitude. Measurement north or south of the equator in degrees, minutes, and seconds. Lines of latitude are also referred to as *parallels.*

LDA. See *localizer-type directional aid.*

Lead radial. Radial at which the turn from the DME arc to the inbound course is started.

Leading edge. The part of an airfoil that meets the airflow first.

Leading edge devices. High lift devices which are found on the leading edge of the airfoil. The most common types are fixed slots, movable slats, and leading edge flaps.

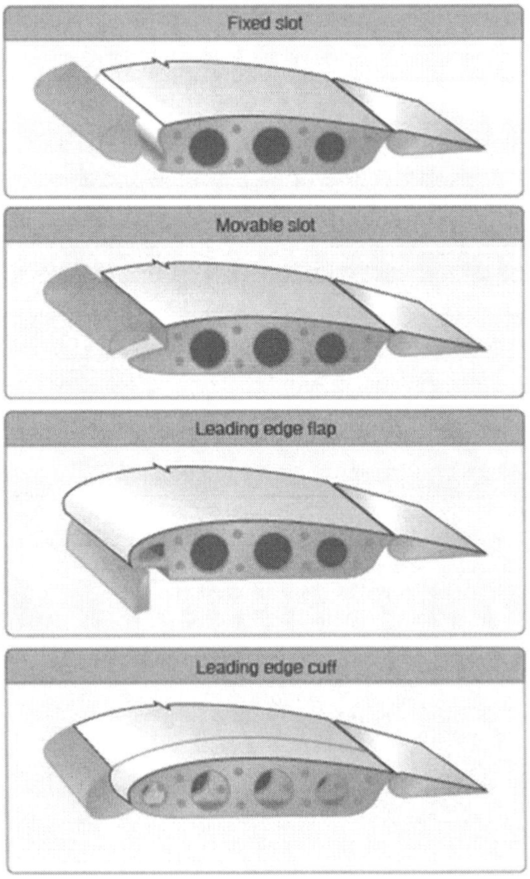

Leading edge flap. A portion of the leading edge of an airplane wing that folds downward to increase the camber, lift, and drag of the wing. The leading-edge flaps are extended for takeoffs and landings to increase the amount of aerodynamic lift that is produced at any given airspeed.

Leans, the. A physical sensation caused by an abrupt correction of a banked attitude entered too slowly to stimulate the motion sensing system in the inner ear. The abrupt correction can create the illusion of banking in the opposite direction.

Licensed empty weight. The empty weight that consists of the airframe, engine(s), unusable fuel, and undrainable oil plus standard and optional equipment as specified in the equipment list. Some manufacturers used this term prior to GAMA standardization.

Lift. A component of the total aerodynamic force on an airfoil and acts perpendicular to the relative wind.

Limit load factor. Amount of stress, or load factor, that an aircraft can withstand before structural damage or failure occurs.

Lines of flux. Invisible lines of magnetic force passing between the poles of a magnet.

L/MF. See *low or medium frequency.*

LMM. See *locator middle marker.*

Load factor. The ratio of a specified load to the total weight of the aircraft. The specified load is expressed in terms of any of the following: aerodynamic forces, inertial forces, or ground or water reactions.

LOC. See *localizer.*

Local area augmentation system (LAAS). A Differential Global Positioning System (DGPS) that improves the accuracy of the system by determining position error from the GPS satellites, then transmitting the error, or corrective factors, to the airborne GPS receiver.

Localizer (LOC). The portion of an ILS that gives left/right guidance information down the centerline of the instrument runway for final approach.

Localizer-type directional aid (LDA). A NAVAID used for nonprecision instrument approaches with utility and accuracy comparable to a localizer, but which is not a part of a complete ILS and is not aligned with the runway. Some LDAs are equipped with a glideslope.

Locator middle marker (LMM). Nondirectional radio beacon (NDB) compass locator, collocated with a middle marker (MM).

Locator outer marker (LOM). NDB compass locator, collocated with an outer marker (OM).

LOM. See *locator outer marker.*

Longitude. Measurement east or west of the Prime Meridian in degrees, minutes, and seconds. The Prime Meridian is 0° longitude and runs through Greenwich, England. Lines of longitude are also referred to as *meridians.*

Longitudinal axis. An imaginary line through an aircraft from nose to tail, passing through its center of gravity. The longitudinal axis is also called the *roll axis* of the aircraft. Movement of the ailerons rotates an airplane about its longitudinal axis.

Longitudinal stability (pitching). Stability about the lateral axis. A desirable characteristic of an airplane whereby it tends to return to its trimmed angle of attack after displacement.

Low or medium frequency (L/MF). A frequency range between 190 and 535 kHz with the medium frequency above 300 kHz. Generally associated with nondirectional beacons transmitting a continuous carrier with either a 400 or 1,020 Hz modulation.

Lubber line. The reference line used in a magnetic compass or heading indicator.

M

MAA. See *maximum authorized altitude.*

MAC. See *mean aerodynamic chord.*

Mach number. The ratio of the true airspeed of the aircraft to the speed of sound in the same atmospheric conditions, named in honor of Ernst Mach, late 19th century physicist.

Mach meter. The instrument that displays the ratio of the speed of sound to the true airspeed an aircraft is flying.

Magnetic bearing (MB). The direction to or from a radio transmitting station measured relative to magnetic north.

Magnetic compass. A device for determining direction measured from magnetic north.

Magnetic dip. A vertical attraction between a compass needle and the magnetic poles. The closer the aircraft is to a pole, the more severe the effect.

Magnetic heading (MH). The direction an aircraft is pointed with respect to magnetic north.

Magnus effect. Lifting force produced when a rotating cylinder produces a pressure differential. This is the same effect that makes a baseball curve or a golf ball slice.

Mandatory altitude. An altitude depicted on an instrument approach chart with the altitude value both underscored and overscored. Aircraft are required to maintain altitude at the depicted value.

Mandatory block altitude. An altitude depicted on an instrument approach chart with two underscored and overscored altitude values between which aircraft are required to maintain altitude.

Maneuverability. Ability of an aircraft to change directions along a flight path and withstand the stresses imposed upon it.

Maneuvering speed (VA). The design maneuvering speed. Operating at or below design maneuvering speed does not provide structural protection against multiple full control inputs in one axis or full control inputs in more than one axis at the same time.

Manifold absolute pressure. The absolute pressure of the fuel/air mixture within the intake manifold, usually indicated in inches of mercury.

MAP. See *missed approach point.*

Margin identification. The top and bottom areas on an instrument approach chart that depict information about the procedure, including airport location and procedure identification.

Marker beacon. A low-powered transmitter that directs its signal upward in a small, fan-shaped pattern. Used along the flight path when approaching an airport for landing, marker beacons indicate both aurally and visually when the aircraft is directly over the facility.

Maximum altitude. An altitude depicted on an instrument approach chart with overscored altitude value at which or below aircraft are required to maintain altitude.

Maximum authorized altitude (MAA). A published altitude representing the maximum usable altitude or flight level for an airspace structure or route segment.

Maximum landing weight. The greatest weight that an airplane normally is allowed to have at landing.

Maximum ramp weight. The total weight of a loaded aircraft, including all fuel. It is greater than the takeoff weight due to the fuel that will be burned during the taxi and runup operations. Ramp weight may also be referred to as *taxi weight.*

Maximum takeoff weight. The maximum allowable weight for takeoff.

Maximum weight. The maximum authorized weight of the aircraft and all its equipment as specified in the Type Certificate Data Sheets (TCDS) for the aircraft.

Maximum zero fuel weight (GAMA standard). The maximum weight, exclusive of usable fuel.

MB. See *magnetic bearing.*

MCA. See *minimum crossing altitude.*

MDA. See *minimum descent altitude.*

MEA. See *minimum en route altitude.*

Mean aerodynamic chord (MAC). The average distance from the leading edge to the trailing edge of the wing.

Mean sea level (MSL). The average height of the surface of the sea at a particular location for all stages of the tide over a 19-year period.

MEL. See *minimum equipment list.*

Meridians. Lines of longitude.

Mesosphere. A layer of the atmosphere directly above the stratosphere.

METAR. See *Aviation Routine Weather Report.*

MFD. See *multi-function display.*

MH. See *magnetic heading.*

MHz. Megahertz.

Microburst. A strong downdraft which normally occurs over horizontal distances of 1 NM or less and vertical distances of less than 1,000 feet. Despite its small horizontal scale, an intense microburst could induce windspeeds greater than 100 knots and downdrafts as strong as 6,000 feet per minute.

Microwave landing system (MLS). A precision instrument approach system operating in the microwave spectrum which normally consists of an azimuth station, elevation station, and precision distance measuring equipment.

Mileage breakdown. A fix indicating a course change that appears on the chart as an "x" at a break between two segments of a federal airway.

Military operations area (MOA). Airspace established for the purpose of separating certain military training activities from IFR traffic.

Military training route (MTR). Airspace of defined vertical and lateral dimensions established for the conduct of military training at airspeeds beyond 250 knots indicated airspeed (KIAS).

Minimum altitude. An altitude depicted on an instrument approach chart with the altitude value underscored. Aircraft are required to maintain altitude at or above the depicted value.

Minimum crossing altitude (MCA). The lowest allowed altitude at certain fixes an aircraft must cross when proceeding in the direction of a higher minimum en route altitude (MEA).

Minimum descent altitude (MDA). The lowest altitude (in feet MSL) to which descent is authorized on final approach, or during circle-to-land maneuvering in execution of a nonprecision approach.

Minimum drag. The point on the total drag curve where the lift-to-drag ratio is the greatest. At this speed, total drag is minimized.

Minimum en route altitude (MEA). The lowest published altitude between radio fixes that ensures acceptable navigational signal coverage and meets obstacle clearance requirements between those fixes.

Minimum equipment list (MEL). A document that outlines the specific (minimum) equipment and systems required for an aircraft to be considered airworthy, operational, and functioning properly for safe flight operations. Also provides guidelines for maintenance actions and procedures in the event of equipment failures or discrepancies.

Minimum obstruction clearance altitude (MOCA). The lowest published altitude in effect between radio fixes on VOR airways, off-airway routes, or route segments, which meets obstacle clearance requirements for the entire route segment and ensures acceptable navigational signal coverage only within 25 statute (22 nautical) miles of a VOR.

Minimum reception altitude (MRA). The lowest altitude at which an airway intersection can be determined.

Minimum safe altitude (MSA). The minimum altitude depicted on approach charts which provides at least 1,000 feet of obstacle clearance for emergency use within a specified distance from the listed navigation facility.

Minimum vectoring altitude (MVA). An IFR altitude lower than the minimum en route altitude (MEA) that provides terrain and obstacle clearance.

Minimums section. The area on an IAP chart that displays the lowest altitude and visibility requirements for the approach.

Missed approach. A maneuver conducted by a pilot when an instrument approach cannot be completed to a landing.

Missed approach point (MAP). A point prescribed in each instrument approach at which a missed approach procedure shall be executed if the required visual reference has not been established.

MLS. See *microwave landing system.*

MM. Middle marker.

MOA. See *military operations area.*

MOCA. See *minimum obstruction clearance altitude.*

Mode C. Altitude reporting transponder mode.

Monocoque. A shell-like fuselage design in which the stressed outer skin is used to support most imposed stresses. Monocoque fuselage design may include bulkheads but not stringers.

Movable slat. A movable auxiliary airfoil on the leading edge of a wing. It is closed in normal flight but extends at high angles of attack. This allows air to continue flowing over the top of the wing and delays airflow separation.

MRA. See *minimum reception altitude.*

MSA. See *minimum safe altitude.*

MSL. See *mean sea level.*

MTR. See *military training route.*

Multi-function display (MFD). Small screen (CRT or LCD) in an aircraft that can be used to display information to the pilot in numerous configurable ways. Often an MFD will be used in concert with a primary flight display.

MVA. See *minimum vectoring altitude.*

N

N1. Rotational speed of the low-pressure compressor in a turbine engine.

N2. Rotational speed of the high-pressure compressor in a turbine engine.

Nacelle. A streamlined enclosure on an aircraft in which an engine is mounted. On multiengine propeller-driven airplanes, the nacelle is normally mounted on the leading edge of the wing.

NACG. See *National Aeronautical Charting Group.*

NAS. See *National Airspace System.*

National Airspace System (NAS). The common network of United States airspace—air navigation facilities, equipment and services, airports or landing areas; aeronautical charts, information and services; rules, regulations and procedures, technical information; and manpower and material.

National Aeronautical Charting Group (NACG). A Federal agency operating under the FAA, responsible for publishing charts such as the terminal procedures and en route charts.

National Oceanic and Atmospheric Administration (NOAA). U.S. government agency that provides weather forecasts, climate information, and environmental data. Supports military operations by providing accurate weather information, storm tracking, and environmental intelligence to aid in mission planning and execution.

National Route Program (NRP). A set of rules and procedures designed to increase the flexibility of user flight planning within published guidelines.

National Security Area (NSA). Areas consisting of airspace of defined vertical and lateral dimensions established at locations where there is a requirement for increased security and safety of ground facilities. Pilots are requested to voluntarily avoid flying through the depicted NSA. When it is necessary to provide a greater level of security and safety, flight in NSAs may be temporarily prohibited. Regulatory prohibitions are disseminated via NOTAMs.

National Transportation Safety Board (NTSB). A United States Government independent organization responsible for investigations of accidents involving aviation, highways, waterways, pipelines, and railroads in the United States. NTSB is charged by congress to investigate every civil aviation accident in the United States.

National Weather Service (NWS). U.S. government agency that provides weather forecasts, warnings, and other meteorological information to the public. That information aids in mission planning, flight operations, and aviation safety.

NAVAID. Navigational aid.

NAV/COM. Navigation and communication radio.

NDB. See *nondirectional radio beacon.*

Negative static stability. The initial tendency of an aircraft to continue away from the original state of equilibrium after being disturbed.

Neutral static stability. The initial tendency of an aircraft to remain in a new condition after its equilibrium has been disturbed.

NM. Nautical mile.

NOAA. See *National Oceanic and Atmospheric Administration.*

No-gyro approach. A radar approach that may be used in case of a malfunctioning gyro- compass or directional gyro. Instead of providing the pilot with headings to be flown, the controller observes the radar track and issues control instructions "turn right/left" or "stop turn," as appropriate.

Nondirectional radio beacon (NDB). A ground-based radio transmitter that transmits radio energy in all directions.

Nonprecision approach. A standard instrument approach procedure in which only horizontal guidance is provided.

No procedure turn (NoPT). Term used with the appropriate course and altitude to denote that the procedure turn is not required.

NoPT. See *no procedure turn.*

NOTAM. See *Notice to Airmen.*

Notice to Airmen (NOTAM). A notice filed with an aviation authority to alert aircraft pilots of any hazards en route or at a specific location. The authority in turn provides means of disseminating relevant NOTAMs to pilots.

NRP. See *National Route Program.*

NSA. See *National Security Area.*

NTSB. See *National Transportation Safety Board.*

NWS. See *National Weather Service.*

O

OAT. See *outside air temperature.*

Obstacle departure procedures (ODP). A preplanned instrument flight rule (IFR) departure procedure printed for pilot use in textual or graphic form to provide obstruction clearance via the least onerous route from the terminal area to the appropriate en route structure. ODPs are recommended for obstruction clearance and may be flown without ATC clearance unless an alternate departure procedure (SID or radar vector) has been specifically assigned by ATC.

Obstruction lights. Lights that can be found both on and off an airport to identify obstructions.

ODP. See *obstacle departure procedures.*

OM. Outer marker.

Orientation. Awareness of the position of the aircraft and of oneself in relation to a specific reference point.

Outer marker. A marker beacon at or near the glideslope intercept altitude of an ILS approach. It is normally located four to seven miles from the runway threshold on the extended centerline of the runway.

Outside air temperature (OAT). The measured or indicated air temperature (IAT) corrected for compression and friction heating. Also referred to as *true air temperature.*

Overcontrolling. Using more movement in the control column than is necessary to achieve the desired pitch-and-bank condition.

Overboost. A condition in which a reciprocating engine has exceeded the maximum manifold pressure allowed by the manufacturer. Can cause damage to engine components.

Overpower. To use more power than required for the purpose of achieving a faster rate of airspeed change.

P

P-static. See *precipitation static.*

PAPI. See *precision approach path indicator.*

PAR. See *precision approach radar.*

Parallels. Lines of latitude.

Parasite drag. Drag caused by the friction of air moving over the aircraft structure; its amount varies directly with the airspeed.

Payload (GAMA standard). The weight of occupants, cargo, and baggage.

P-factor. A tendency for an aircraft to yaw to the left due to the descending propeller blade on the right producing more thrust than the ascending blade on the left. This occurs when the aircraft's longitudinal axis is in a climbing attitude in relation to the relative wind. The P-factor would be to the right if the aircraft had a counterclockwise rotating propeller.

PFD. See *primary flight display.*

Phugoid oscillations. Long-period oscillations of an aircraft around its lateral axis. It is a slow change in pitch accompanied by equally slow changes in airspeed. Angle of attack remains constant, and the pilot often corrects for phugoid oscillations without even being aware of them.

PIC. See *pilot in command.*

Pilotage. Navigation by visual reference to landmarks.

Pilot in command (PIC). The pilot responsible for the operation and safety of an aircraft.

Pilot report (PIREP). Report of meteorological phenomena encountered by aircraft.

Pilot's Operating Handbook/Airplane Flight Manual (POH/AFM). FAA-approved documents published by the airframe manufacturer that list the operating conditions for a particular model of aircraft.

PIREP. See *pilot report.*

Pitot pressure. Ram air pressure used to measure airspeed.

Pitot-static head. A combination pickup used to sample pitot pressure and static air pressure.

Plan view. The overhead view of an approach procedure on an instrument approach chart. The plan view depicts the routes that guide the pilot from the en route segments to the IAF.

Planform. The shape (or form) of a wing as viewed from above. It may be long and tapered, short and rectangular, or various other shapes.

POH/AFM. See *Pilot's Operating Handbook/Airplane Flight Manual.*

Point-in-space approach. A type of helicopter instrument approach procedure to a missed approach point more than 2,600 feet from an associated helicopter landing area.

Position error. Error in the indication of the altimeter, ASI, and VSI caused by the air at the static system entrance not being absolutely still.

Position report. A report over a known location as transmitted by an aircraft to ATC.

Positive static stability. The initial tendency to return to a state of equilibrium when disturbed from that state.

Powerplant. A complete engine and propeller combination with accessories.

Precipitation static (P-static). A form of radio interference caused by rain, snow, or dust particles hitting the antenna and inducing a small radio-frequency voltage into it.

Precision approach. A standard instrument approach procedure in which both vertical and horizontal guidance is provided.

Precision approach path indicator (PAPI). A system of lights similar to the *visual approach slope indicator* (VASI), but consisting of one row of lights in two- or four-light systems. A pilot on the correct glideslope will see two white lights and two red lights. See *VASI.*

Precision approach radar (PAR). A type of radar used at an airport to guide an aircraft through the final stages of landing, providing horizontal and vertical guidance. The radar operator directs the pilot to change heading or adjust the descent rate to keep the aircraft on a path that allows it to touch down at the correct spot on the runway.

Precision runway monitor (PRM). System allows simultaneous, independent instrument flight rules (IFR) approaches at airports with closely spaced parallel runways.

Preferred IFR routes. Routes established in the major terminal and en route environments to increase system efficiency and capacity. IFR clearances are issued based on these routes, listed in the Chart Supplement U.S. except when severe weather avoidance procedures or other factors dictate otherwise.

Pressure altitude. Altitude above the standard 29.92 "Hg plane.

Pressure demand oxygen system. A demand oxygen system that supplies 100 percent oxygen at sufficient pressure above the altitude where normal breathing is adequate. Also referred to as a *pressure breathing system.*

Prevailing visibility. The greatest horizontal visibility equaled or exceeded throughout at least half the horizon circle (which is not necessarily continuous).

Primary and supporting. A method of attitude instrument flying using the instrument that provides the most direct indication of attitude and performance.

Primary flight display (PFD). A display that provides increased situational awareness to the pilot by replacing the traditional six instruments used for instrument flight with an easy-to-scan display that provides the horizon, airspeed, altitude, vertical speed, trend, trim, and rate of turn among other key relevant indications.

PRM. See *precision runway monitor.*

Procedure turn. A maneuver prescribed when it is necessary to reverse direction to establish an aircraft on the intermediate approach segment or final approach course.

Profile view. Side view of an IAP chart illustrating the vertical approach path altitudes, headings, distances, and fixes.

Prohibited area. Designated airspace within which flight of aircraft is prohibited.

Propeller. A device for propelling an aircraft that, when rotated, produces by its action on the air, a thrust approximately perpendicular to its plane of rotation. It includes the control components normally supplied by its manufacturer.

Propeller/rotor modulation error. Certain propeller rpm settings or helicopter rotor speeds can cause the VOR course deviation indicator (CDI) to fluctuate as much as

±6°. Slight changes to the rpm setting will normally smooth out this roughness.

R

Rabbit, the. High-intensity flasher system installed at many large airports. The flashers consist of a series of brilliant blue-white bursts of light flashing in sequence along the approach lights, giving the effect of a ball of light traveling toward the runway.

Radar. A system that uses electromagnetic waves to identify the range, altitude, direction, or speed of both moving and fixed objects such as aircraft, weather formations, and terrain. The term RADAR was coined in 1941 as an acronym for Radio Detection and Ranging. The term has since entered the English language as a standard word, radar, losing the capitalization in the process.

Radar approach. The controller provides vectors while monitoring the progress of the flight with radar, guiding the pilot through the descent to the airport/heliport or to a specific runway.

Radar summary chart. A weather product derived from the national radar network that graphically displays a summary of radar weather reports.

Radar weather report (SD). A report issued by radar stations at 35 minutes after the hour, and special reports as needed. Provides information on the type, intensity, and location of the echo tops of the precipitation.

Radials. The courses oriented from a station.

Radio or radar altimeter. An electronic altimeter that determines the height of an aircraft above the terrain by measuring the time needed for a pulse of radio-frequency energy to travel from the aircraft to the ground and return.

Radio frequency (RF). A term that refers to alternating current (AC) that has characteristics such that, if the current is input to antenna, an electromagnetic (EM) field is generated suitable for wireless broadcasting and/or communications.

Radio magnetic indicator (RMI). An electronic navigation instrument that combines a magnetic compass with an ADF or VOR. The card of the RMI acts as a gyro-stabilized magnetic compass, and shows the magnetic heading the aircraft is flying.

Radiosonde. A weather instrument that observes and reports meteorological conditions from the upper atmosphere. This instrument is typically carried into the atmosphere by some form of weather balloon.

Radio wave. An electromagnetic (EM) wave with frequency characteristics useful for radio transmission.

RAIM. See *receiver autonomous integrity monitoring*.

RAM recovery. The increase in thrust due to ram air pressures and density on the front of the engine caused by air velocity.

Random RNAV routes. Direct routes, based on area navigation capability, between waypoints defined in terms of latitude/longitude coordinates, degree-distance fixes, or offsets from established routes/airways at a specified distance and direction.

Ranging signals. Transmitted from the GPS satellite, signals allowing the aircraft's receiver to determine range (distance) from each satellite.

Rapid decompression. The almost instantaneous loss of cabin pressure in aircraft with a pressurized cockpit or cabin.

RB. See *relative bearing*.

RBI. See *relative bearing indicator*.

RCO. See *remote communications outlet.*

Receiver autonomous integrity monitoring (RAIM). A system used to verify the usability of the received GPS signals and warns the pilot of any malfunction in the navigation system. This system is required for IFR-certified GPS units.

Recommended altitude. An altitude depicted on an instrument approach chart with the altitude value neither underscored nor overscored. The depicted value is an advisory value.

Receiver-transmitter (RT). A system that receives and transmits a signal and an indicator.

Reduced vertical separation minimum (RVSM). Reduces the vertical separation between flight levels (FL) 290 and 410 from 2,000 feet to 1,000 feet, and makes six additional FLs available for operation. Also see *domestic vertical separation minimum* (DRVSM.)

Reference circle (also, distance circle). The circle depicted in the plan view of an IAP chart that typically has a 10 NM radius, within which chart the elements are drawn to scale.

Regions of command. The "regions of normal and reversed command" refers to the relationship between speed and the power required to maintain or change that speed in flight.

Region of reverse command. Flight regime in which flight at a higher airspeed requires a lower power setting and a lower airspeed requires a higher power setting to maintain altitude.

REIL. See *runway end identifier lights.*

Relative bearing (RB). The angular difference between the aircraft heading and the direction to the station, measured clockwise from the nose of the aircraft.

Relative bearing indicator (RBI). Also known as the *fixed-card ADF*, zero is always indicated at the top of the instrument and the needle indicates the relative bearing to the station.

Relative wind. Direction of the airflow produced by an object moving through the air. The relative wind for an airplane in flight flows in a direction parallel with and opposite to the direction of flight; therefore, the actual flight path of the airplane determines the direction of the relative wind.

Remote communications outlet (RCO). An unmanned communications facility that is remotely controlled by air traffic personnel.

Required navigation performance (RNP). A specified level of accuracy defined by a lateral area of confined airspace in which an RNP-certified aircraft operates.

Restricted area. Airspace designated under 14 CFR part 73 within which the flight of aircraft, while not wholly prohibited, is subject to restriction.

Reverse sensing. The VOR needle appearing to indicate the reverse of normal operation.

RF. See *radio frequency.*

Rigging. The final adjustment and alignment of an aircraft and its flight control system that provides the proper aerodynamic characteristics.

Rigidity. The characteristic of a gyroscope that prevents its axis of rotation tilting as the Earth rotates.

Rigidity in space. The principle that a wheel with a heavily weighted rim spinning rapidly will remain in a fixed position in the plane in which it is spinning.

Risk elements. There are four fundamental risk elements in aviation: the pilot, the aircraft, the environment, and the type of operation that comprise any given aviation situation.

RMI. See *radio magnetic indicator.*

RNAV. See *area navigation.*

RNP. See *required navigation performance.*

RT. See *receiver-transmitter.*

Rudder. The movable primary control surface mounted on the trailing edge of the vertical fin of an airplane. Movement of the rudder rotates the airplane about its vertical axis.

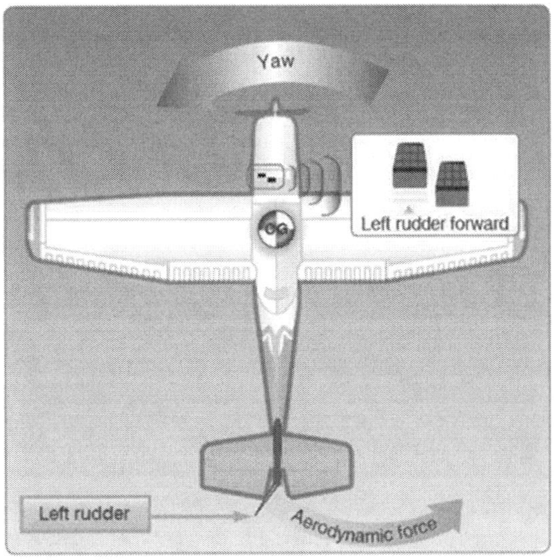

Ruddervator. A pair of control surfaces on the tail of an aircraft arranged in the form of a V. These surfaces, when moved together by the control wheel, serve as elevators, and when moved differentially by the rudder pedals, serve as a rudder.

Runway centerline lights. Runway lighting which consists of flush centerline lights spaced at 50-foot intervals beginning 75 feet from the landing threshold.

Runway edge lights. A component of the runway lighting system that is used to outline the edges of runways at night or during low visibility conditions. These lights are classified according to the intensity they can produce.

Runway end identifier lights (REIL). A pair of synchronized flashing lights, located laterally on each side of the runway threshold, providing rapid and positive identification of the approach end of a runway.

Runway visibility value (RVV). The visibility determined for a particular runway by a transmissometer.

Runway visual range (RVR). The instrumentally derived horizontal distance a pilot should be able to see down the runway from the approach end, based on either the sighting of high- intensity runway lights, or the visual contrast of other objects.

RVR. See *runway visual range.*

RVSM. See *reduced vertical separation minimum.*

RVV. See *runway visibility value.*

S

SA. See *selective availability.*

St. Elmo's Fire. A corona discharge which lights up the aircraft surface areas where maximum static discharge occurs.

Satellite ephemeris data. Data broadcast by the GPS satellite containing very accurate orbital data for that satellite, atmospheric propagation data, and satellite clock error data.

Scan. The first fundamental skill of instrument flight, also known as *cross-check*; the continuous and logical observation of instruments for attitude and performance information.

Sea breeze. A coastal breeze blowing from sea to land caused by the temperature difference when the land surface is warmer than the sea surface. The sea breeze usually occurs during the day and alternates with the land breeze that blows in the opposite direction at night.

Sea level engine. A reciprocating aircraft engine having a rated takeoff power that is producible only at sea level.

Secondary surveillance radar (SSR). Sometimes called *air traffic control radar beacon system* (ATCRBS), which utilizes a transponder in the aircraft. The ground equipment is an interrogating unit, in which the beacon antenna is mounted so it rotates with the surveillance antenna. The interrogating unit transmits a coded pulse sequence that actuates the aircraft transponder. The transponder answers the coded sequence by transmitting a preselected coded sequence back to the ground equipment, providing a strong return signal and positive aircraft identification, as well as other special data.

Sectional aeronautical charts. Designed for visual navigation of slow- or medium-speed aircraft. Topographic information on these charts features the portrayal of relief, and a judicious selection of visual check points for VFR flight. Aeronautical information includes visual and radio aids to navigation, airports, controlled airspace, restricted areas, obstructions and related data.

SDF. See *simplified directional facility*.

Selective availability (SA). A satellite technology permitting the Department of Defense (DOD) to create, in the interest of national security, a significant clock and ephemeris error in the satellites, resulting in a navigation error.

Sensitive altimeter. A form of multipointer pneumatic altimeter with an adjustable barometric scale that allows the reference pressure to be set to any desired level.

Service ceiling. The maximum density altitude where the best rate-of-climb airspeed will produce a 100-feet-per-minute climb at maximum weight while in a clean configuration with maximum continuous power.

Servo. A motor or other form of actuator which receives a small signal from the control device and exerts a large force to accomplish the desired work.

Servo tab. An auxiliary control mounted on a primary control surface, which automatically moves in the direction opposite the primary control to provide an aerodynamic assist in the movement of the control.

SIDS. See *standard instrument departure procedures*.

SIGMET. The acronym for significant meteorological information. A weather advisory in abbreviated plain language concerning the occurrence or expected occurrence of potentially hazardous en route weather phenomena that may affect the safety of aircraft operations.

SIGMET is warning information, hence it is of highest priority among other types of meteorological information provided to the aviation users.

Signal-to-noise ratio. An indication of signal strength received compared to background noise, which is a measure of the adequacy of the received signal.

Significant weather prognostic. Presents four panels showing forecast significant weather.

Simplex. Transmission and reception on the same frequency.

Simplified directional facility (SDF). A NAVAID used for non-precision instrument approaches. The final approach course is similar to that of an ILS localizer; however, the SDF course may be offset from the runway, generally not more than 3°, and the course may be wider than the localizer, resulting in a lower degree of accuracy.

Single-pilot resource management (SRM). The ability for a pilot to manage all resources effectively to ensure the outcome of the flight is successful.

Situational awareness. Pilot knowledge of where the aircraft is location-wise, air traffic control, weather, regulations, aircraft status, and other factors that may affect flight.

Skidding turn. An uncoordinated turn in which the rate of turn is too great for the angle of bank, pulling the aircraft to the outside of the turn. (see turn-and-slip indicator for diagram)

Skin friction drag. Drag generated between air molecules and the solid surface of the aircraft.

Slant range. The horizontal distance from the aircraft antenna to the ground station, due to line- of-sight transmission of the DME signal.

Slaved compass. A system whereby the heading gyro is "slaved to," or continuously corrected to bring its direction readings into agreement with a remotely located magnetic direction sensing device (usually a flux valve or flux gate compass).

Slipping turn. An uncoordinated turn in which the aircraft is banked too much for the rate of turn, so the horizontal lift component is greater than the centrifugal force, pulling the aircraft toward the inside of the turn (see *turn and slip indicator*).

Small airplane. An airplane of 12,500 pounds or less maximum certificated takeoff weight.

Somatogravic illusion. The misperception of being in a nose-up or nose-down attitude, caused by a rapid acceleration or deceleration while in flight situations that lack visual reference.

Spatial disorientation. The state of confusion due to misleading information being sent to the brain from various sensory organs, resulting in a lack of awareness of the aircraft position in relation to a specific reference point.

Special use airspace. Airspace in which flight activities are subject to restrictions that can create limitations on the mixed use of airspace. Consists of prohibited, restricted, warning, military operations, and alert areas.

Special fuel consumption. The amount of fuel in pounds per hour consumed or required by an engine per brake horsepower or per pound of thrust.

Spin. An aggravated stall that results in an airplane descending in a helical, or corkscrew path.

Spiral instability. A condition that exists when the static directional stability of the airplane is very strong as compared to the effect of its dihedral in maintaining lateral equilibrium.

Spiraling slipstream. The slipstream of a propeller-driven airplane rotates around the airplane. This slipstream strikes the left side of the vertical fin, causing the aircraft to yaw slightly. Rudder offset is sometimes used by aircraft designers to counteract this tendency.

Spoilers. High-drag devices that can be raised into the air flowing over an airfoil, reducing lift and increasing drag. Spoilers are used for roll control on some aircraft. Deploying spoilers on both wings at the same time allows the aircraft to descend without gaining speed. Spoilers are also used to shorten the ground roll after landing.

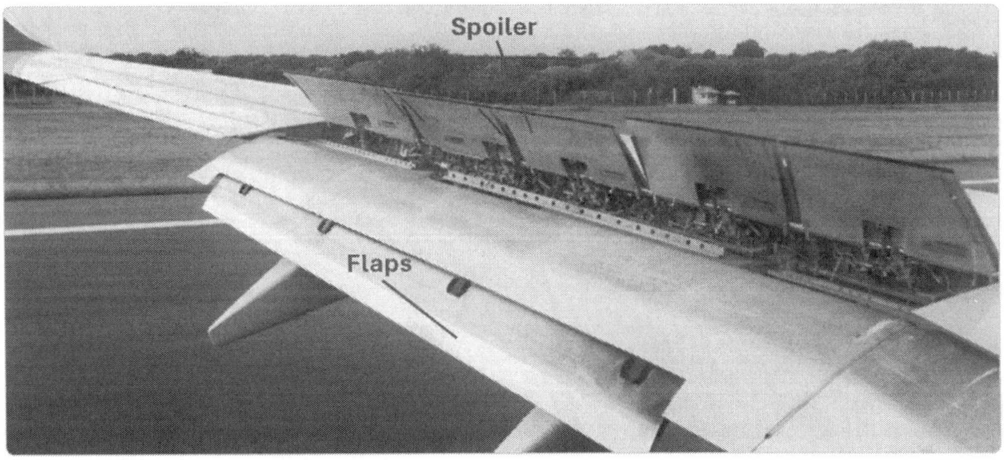

SRM. See *single-pilot resource management.*

SSR. See *secondary surveillance radar.*

SSV. See *standard service volume.*

Stabilator. A single-piece horizontal tail surface on an airplane that pivots around a central hinge point. A stabilator serves the purposes of both the horizontal stabilizer and the elevators.

Stability. The inherent quality of an airplane to correct for conditions that may disturb its equilibrium, and to return or to continue on the original flight path. It is primarily an airplane design characteristic.

Stagnant hypoxia. A type of hypoxia that results when the oxygen-rich blood in the lungs is not moving to the tissues that need it.

Stall. A rapid decrease in lift caused by the separation of airflow from the wing's surface, brought on by exceeding the critical angle of attack. A stall can occur at any pitch attitude or airspeed.

Standard atmosphere. At sea level, the standard atmosphere consists of a barometric pressure of 29.92 inches of mercury ("Hg) or 1013.2 millibars, and a temperature of 15 °C (59 °F). Pressure and temperature normally decrease as altitude increases. The standard lapse rate in the lower atmosphere for each 1,000 feet of altitude is approximately 1 "Hg and 2 °C (3.5 °F). For example, the standard pressure and temperature at 3,000 feet mean sea level (MSL) are 26.92 "Hg (29.92 "Hg– 3 "Hg) and 9 °C (15 °C–6 °C).

Standard empty weight (GAMA standard). This weight consists of the airframe, engines, and all items of operating equipment that have fixed locations and are permanently installed in the airplane including fixed ballast, hydraulic fluid, unusable fuel, and full engine oil.

Standard holding pattern. A holding pattern in which all turns are made to the right.

Standard instrument departure procedures (SIDS). Published procedures to expedite clearance delivery and to facilitate transition between takeoff and en route operations.

Standard rate turn. A turn in which an aircraft changes its direction at a rate of 3° per second (360° in 2 minutes) for low- or medium-speed aircraft. For high-speed aircraft, the standard rate turn is 1½° per second (360° in 4 minutes).

Standard service volume (SSV). Defines the limits of the volume of airspace which the VOR serves.

Standard terminal arrival route (STAR). A preplanned IFR ATC arrival procedure published for pilot use in graphic and/or textual form.

Standard weights. Weights established for numerous items involved in weight and balance computations. These weights should not be used if actual weights are available.

STAR. See *standard terminal arrival route.*

Static longitudinal stability. The aerodynamic pitching moments required to return the aircraft to the equilibrium angle of attack.

Static pressure. Pressure of air that is still or not moving, measured perpendicular to the surface of the aircraft.

Static stability. The initial tendency an aircraft displays when disturbed from a state of equilibrium.

Station. A location in the airplane that is identified by a number designating its distance in inches from the datum. The datum is, therefore, identified as station zero. An item located at station +50 would have an arm of 50 inches.

Stationary front. A front that is moving at a speed of less than 5 knots.

Steep turns. In instrument flight, any turn greater than standard rate; in visual flight, anything greater than a 45° bank.

Stepdown fix. The point after which additional descent is permitted within a segment of an IAP.

Stratosphere. A layer of the atmosphere above the tropopause extending to a height of approximately 160,000 feet.

Supercooled water droplets. Water droplets that have been cooled below the freezing point, but are still in a liquid state.

Surface analysis chart. A report that depicts an analysis of the current surface weather. Shows the areas of high and low pressure, fronts, temperatures, dewpoints, wind directions and speeds, local weather, and visual obstructions.

Synchro. A device used to transmit indications of angular movement or position from one location to another.

Synthetic vision. A realistic display depiction of the aircraft in relation to terrain and flight path.

T

TAA. See *terminal arrival area.*

TACAN. See *tactical air navigation.*

Tactical air navigation (TACAN). An electronic navigation system used by military aircraft, providing both distance and direction information.

TAF. See *terminal aerodrome forecast.*

Takeoff decision speed (V1). Per 14 CFR section 23.51, "The calibrated airspeed on the ground at which, as a result of engine failure or other reasons, the pilot assumed to have made a decision to continue or discontinue the takeoff."

Takeoff distance. The distance required to complete an all-engines operative takeoff to the 35- foot height. It must be at least 15 percent less than the distance required for a one-engine inoperative engine takeoff. This distance is not normally a limiting factor as it is usually less than the one- engine inoperative takeoff distance.

Takeoff safety speed (V2). Per 14 CFR part 1, "A referenced airspeed obtained after lift-off at which the required one engine-inoperative climb performance can be achieved."

TAWS. See *Terrain Awareness and Warning System.*

Taxiway lights. Omnidirectional lights that outline the edges of the taxiway and are blue in color.

Taxiway turnoff lights. Lights that are flush with the runway which emit a steady green color.

TCAS. See *Traffic Alert Collision Avoidance System.*

TCH. See *threshold crossing height.*

TDZE. See *touchdown zone elevation.*

TEC. See *Tower En Route Control.*

Telephone information briefing service (TIBS). An FSS service providing continuously updated automated telephone recordings of area and/or route weather, airspace procedures, and special aviation-oriented announcements.

Temporary flight restriction (TFR). Restriction to flight imposed to do the following:

- Protect persons and property in the air or on the surface from an existing or imminent flight associated hazard;
- Provide a safe environment for the operation of disaster relief aircraft;
- Prevent an unsafe congestion of sightseeing aircraft above an incident;
- Protect the President, Vice President, or other public figures; and
- Provide a safe environment for space agency operations.

Pilots are expected to check appropriate NOTAMs during flight planning when conducting flight in an area where a temporary flight restriction is in effect.

Tension. Maintaining an excessively strong grip on the control column, usually resulting in an overcontrolled situation.

Terminal aerodrome forecast (TAF). A report established for the 5-statute-mile radius around an airport. Utilizes the same descriptors and abbreviations as the METAR report.

Terminal arrival area (TAA). A procedure to provide a new transition method for arriving aircraft equipped with FMS and/or GPS navigational equipment. The TAA contains a "T" structure that normally provides a NoPT for aircraft using the approach.

Terminal instrument approach procedure (TERP). Prescribes standardized methods for use in designing instrument flight procedures.

TERP. See *terminal instrument approach procedure.*

Terminal radar service areas (TRSA). Areas where participating pilots can receive additional radar services. The purpose of the service is to provide separation between all IFR operations and participating VFR aircraft.

Terrain Awareness and Warning System (TAWS). A timed-based system that provides information concerning potential hazards with fixed objects by using GPS positioning and a database of terrain and obstructions to provide true predictability of the upcoming terrain and obstacles.

TFR. See *temporary flight restriction.*

Thermosphere. The last layer of the atmosphere that begins above the mesosphere and gradually fades away into space.

Threshold crossing height (TCH). The theoretical height above the runway threshold at which the aircraft's glideslope antenna would be if the aircraft maintained the trajectory established by the mean ILS glideslope or MLS glidepath.

Thrust. The force which imparts a change in the velocity of a mass. This force is measured in pounds but has no element of time or rate. The term *thrust required* is generally associated with jet engines. A forward force which propels the airplane through the air.

Thrust (aerodynamic force). The forward aerodynamic force produced by a propeller, fan, or turbojet engine as it forces a mass of air to the rear, behind the aircraft.

Thrust line. An imaginary line passing through the center of the propeller hub, perpendicular to the plane of the propeller rotation.

Time and speed table. A table depicted on an instrument approach procedure chart that identifies the distance from the FAF to the MAP, and provides the time required to transit that distance based on various groundspeeds.

Timed turn. A turn in which the clock and the turn coordinator are used to change heading a definite number of degrees in a given time.

TIBS. See *telephone information briefing service.*

TIS. See *traffic information service.*

Title 14 of the Code of Federal Regulations (14 CFR). Includes the federal aviation regulations governing the operation of aircraft, airways, and airmen.

Total drag. The sum of the parasite drag and induced drag.

Touchdown zone elevation (TDZE). The highest elevation in the first 3,000 feet of the landing surface, TDZE is indicated on the instrument approach procedure chart when straight-in landing minimums are authorized.

Touchdown zone lights. Two rows of transverse light bars disposed symmetrically about the runway centerline in the runway touchdown zone.

Tower En Route Control (TEC). The control of IFR en route traffic within delegated airspace between two or more adjacent approach control facilities, designed to expedite traffic, and reduce control and pilot communication requirements.

TPP. See *United States Terminal Procedures Publication.*

Track. The actual path made over the ground in flight.

Tracking. Flying a heading that will maintain the desired track to or from the station regardless of crosswind conditions.

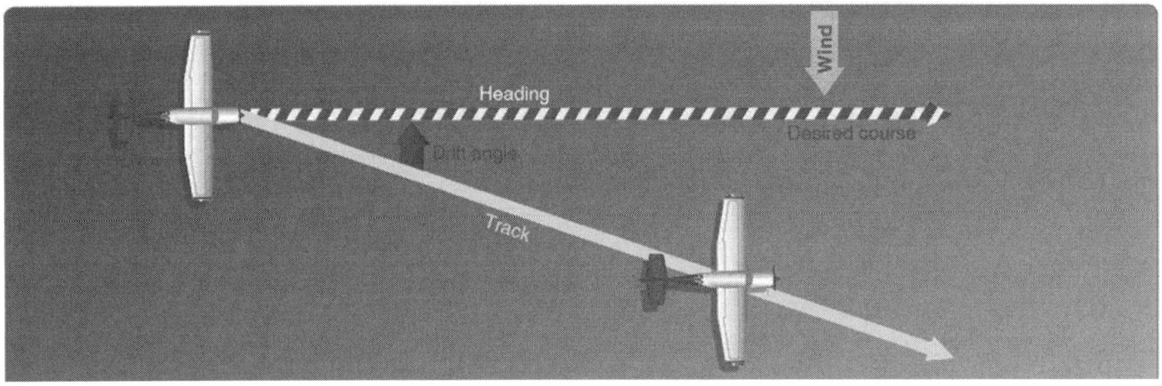

Traffic Alert Collision Avoidance System (TCAS). An airborne system developed by the FAA that operates independently from the ground-based Air Traffic Control system. Designed to increase flight deck awareness of proximate aircraft and to serve as a "last line of defense" for the prevention of midair collisions.

Traffic information service (TIS). A ground-based service providing information to the flight deck via data link using the S-mode transponder and altitude encoder to improve the safety and efficiency of "see and avoid" flight through an automatic display that informs the pilot of nearby traffic.

Trailing edge. The portion of the airfoil where the airflow over the upper surface rejoins the lower surface airflow.

Transcribed Weather Broadcast (TWEB). An FSS service, available in Alaska only, providing continuously updated automated broadcast of meteorological and aeronautical data over selected L/MF and VOR NAVAIDs.

Transponder. The airborne portion of the ATC radar beacon system.

Transponder code. One of 4,096 four-digit discrete codes ATC assigns to distinguish between aircraft.

Trend. Immediate indication of the direction of aircraft movement, as shown on instruments.

Tricycle gear. Landing gear employing a third wheel located on the nose of the aircraft.

Trim. To adjust the aerodynamic forces on the control surfaces so that the aircraft maintains the set attitude without any control input.

Trim tab. A small auxiliary hinged portion of a movable control surface that can be adjusted during flight to a position resulting in a balance of control forces.

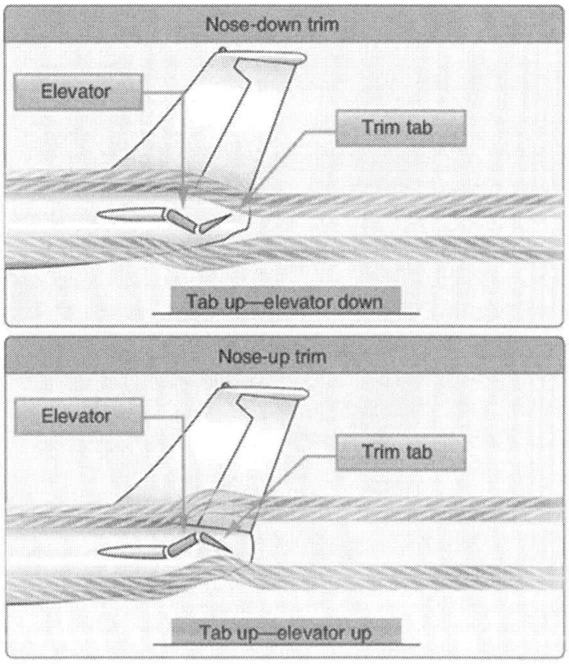

Tropopause. The boundary layer between the troposphere and the stratosphere which acts as a lid to confine most of the water vapor, and the associated weather, to the troposphere.

Troposphere. The layer of the atmosphere extending from the surface to a height of 20,000 to 60,000 feet, depending on latitude.

True airspeed. Actual airspeed, determined by applying a correction for pressure altitude and temperature to the CAS.

True altitude. The vertical distance of the airplane above sea level—the actual altitude. It is often expressed as feet above mean sea level (MSL). Airport, terrain, and obstacle elevations on aeronautical charts are true altitudes.

T-tail. An aircraft with the horizontal stabilizer mounted on the top of the vertical stabilizer, forming a T.

Turn-and-slip indicator. A flight instrument consisting of a rate gyro to indicate the rate of yaw and a curved glass inclinometer to indicate the relationship between gravity and centrifugal force. The turn-and-slip indicator indicates the relationship between angle of bank and rate of yaw. Also called a *turn-and-bank indicator* (see illustration for an example).

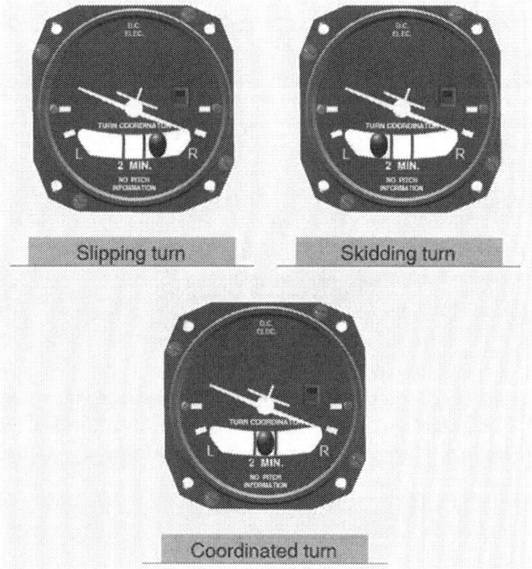

Turn coordinator. A rate gyro that senses both roll and yaw due to the gimbal being canted. Has largely replaced the turn-and-slip indicator in modern aircraft.

TWEB. See *Transcribed Weather Broadcast.*

U

UHF. See *ultra-high frequency.*

Ultra-high frequency (UHF). The range of electromagnetic frequencies between 300 MHz and 3,000 MHz.

Ultimate load factor. In stress analysis, the load that causes physical breakdown in an aircraft or aircraft component during a strength test, or the load that according to computations, should cause such a breakdown.

Uncaging. Unlocking the gimbals of a gyroscopic instrument, making it susceptible to damage by abrupt flight maneuvers or rough handling.

Uncontrolled airspace. Class G airspace that has not been designated as Class A, B, C, D, or

E. It is airspace in which air traffic control has no authority or responsibility to control air traffic; however, pilots should remember there are VFR minimums which apply to this airspace.

Underpower. Using less power than required for the purpose of achieving a faster rate of airspeed change.

United States Terminal Procedures Publication (TPP). Booklets published in regional format by FAA Aeronautical Navigation Products (AeroNav Products) that include DPs, STARs, IAPs, and other information pertinent to IFR flight.

Unusual attitude. An unintentional, unanticipated, or extreme aircraft attitude.

Useful load. The weight of the pilot, copilot, passengers, baggage, usable fuel, and drainable oil. It is the basic empty weight subtracted from the maximum allowable gross weight. This term applies to general aviation aircraft only.

User-defined waypoints. Waypoint location and other data which may be input by the user, this is the only GPS database information that may be altered (edited) by the user.

V

V1. See *takeoff decision speed.*

V2. See *takeoff safety speed.*

VA. See *maneuvering speed.*

Variation. Compass error caused by the difference in the physical locations of the magnetic north pole and the geographic north pole.

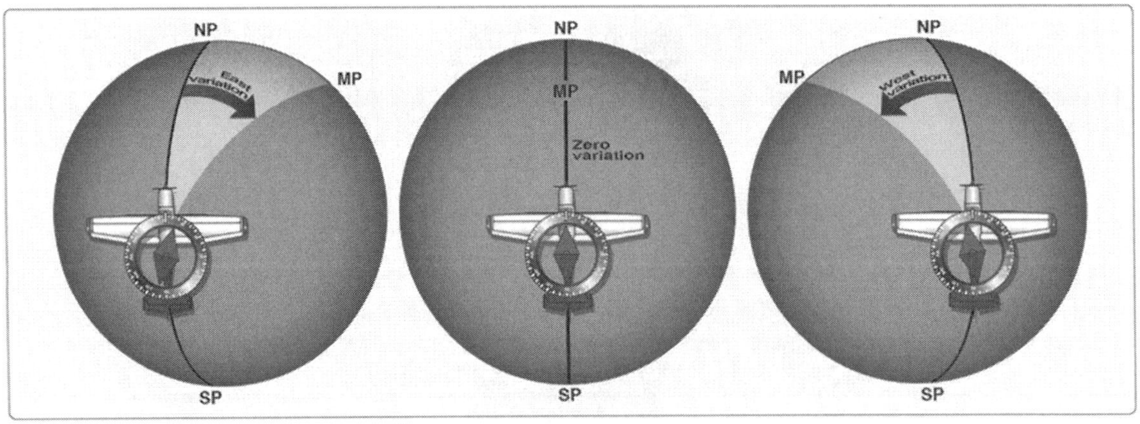

VASI. See *visual approach slope indicator.*

VDP. See *visual descent point.*

Vector. A force vector is a graphic representation of a force and shows both the magnitude and direction of the force.

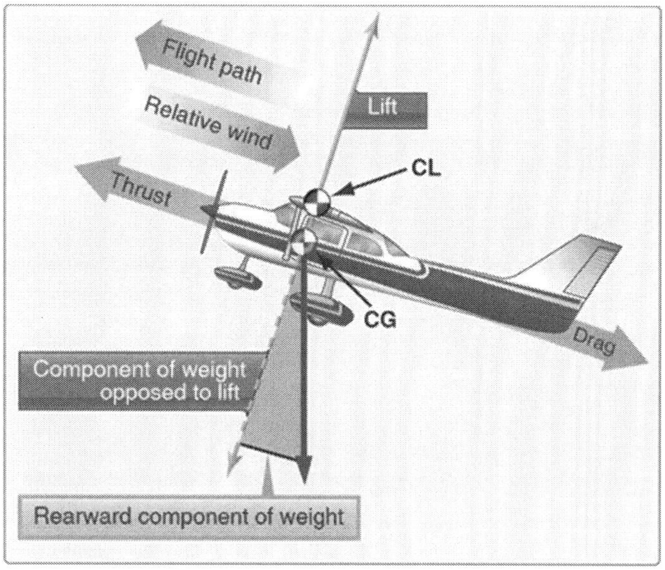

Vectoring. Navigational guidance by assigning headings.

VEF. Calibrated airspeed at which the critical engine of a multi-engine aircraft is assumed to fail.

Velocity. The speed or rate of movement in a certain direction.

Venturi tube. A specially shaped tube attached to the outside of an aircraft to produce suction to allow proper operation of gyro instruments.

Vertical axis. An imaginary line passing vertically through the center of gravity of an aircraft. The vertical axis is called the *z-axis* or the *yaw axis*.

Vertical card compass. A magnetic compass that consists of an azimuth on a vertical card, resembling a heading indicator with a fixed miniature airplane to accurately present the heading of the aircraft. The design uses eddy current damping to minimize lead and lag during turns.

Vertical speed indicator (VSI). A rate-of-pressure change instrument that gives an indication of any deviation from a constant pressure level.

Vertical stability. Stability about an aircraft's vertical axis. Also called *yawing* or *directional stability*.

Very-high frequency (VHF). A band of radio frequencies falling between 30 and 300 MHz.

Very-high frequency omnidirectional range (VOR). Electronic navigation equipment in which the flight deck instrument identifies the radial or line from the VOR station, measured in degrees clockwise from magnetic north, along which the aircraft is located.

Vestibule. The central cavity of the bony labyrinth of the ear, or the parts of the membranous labyrinth that it contains.

VFE. The maximum speed with the flaps extended. The upper limit of the white arc.

VFR. See *visual flight rules.*

VFR on top. ATC authorization for an IFR aircraft to operate in VFR conditions at any appropriate VFR altitude.

VFR over the top. A VFR operation in which an aircraft operates in VFR conditions on top of an undercast.

VFR terminal area chart. At a scale of 1:250,000, a chart that depicts Class B airspace, which provides for the control or segregation of all the aircraft within the Class B airspace. The chart depicts topographic information and aeronautical information including visual and radio aids to navigation, airports, controlled airspace, restricted areas, obstructions, and related data.

V-G diagram. A chart that relates velocity to load factor. It is valid only for a specific weight, configuration and altitude and shows the maximum amount of positive or negative lift the airplane can generate at a given speed. Also shows the safe load factor limits and the load factor that the aircraft can sustain at various speeds.

VHF. See *very-high frequency.*

Victor airways. Airways based on a centerline that extends from one VOR or VORTAC navigation aid or intersection, to another navigation aid (or through several navigation aids or intersections); used to establish a known route for en route procedures between terminal areas.

Visual approach slope indicator (VASI). A visual aid of lights arranged to provide descent guidance information during the approach to the runway. A pilot on the correct glideslope will see red lights over white lights.

Visual descent point (VDP). A defined point on the final approach course of a nonprecision straight-in approach procedure from which normal descent from the MDA to the runway touchdown point may be commenced, provided the runway environment is clearly visible to the pilot.

Visual flight rules (VFR). Flight rules adopted by the FAA governing aircraft flight using visual references. VFR operations specify the amount of ceiling and the visibility the pilot must have to operate according to these rules. When the weather conditions are such that the pilot cannot operate according to VFR, he/she must use instrument flight rules (IFR).

Visual meteorological conditions (VMC). Meteorological conditions expressed in terms of visibility, distance from cloud, and ceiling meeting or exceeding the minimums specified for VFR.

VLE. Landing gear extended speed. The maximum speed at which an airplane can be safely flown with the landing gear extended.

VLO. Landing gear operating speed. The maximum speed for extending or retracting the landing gear if using an airplane equipped with retractable landing gear.

VMC. Minimum control airspeed. This is the minimum flight speed at which a light, twin-engine airplane can be satisfactorily controlled when an engine suddenly becomes inoperative, and the remaining engine is at takeoff power.

VMC. See *visual meteorological conditions.*

VNE. The never-exceed speed. Operating above this speed is prohibited since it may result in damage or structural failure. The red line on the airspeed indicator.

VNO. The maximum structural cruising speed. Do not exceed this speed except in smooth air. The upper limit of the green arc.

VOR. See *very-high frequency omnidirectional range.*

VORTAC. A facility consisting of two components, VOR and TACAN, which provides three individual services: VOR azimuth, TACAN azimuth, and TACAN distance (DME) at one site.

VOR test facility (VOT). A ground facility which emits a test signal to check VOR receiver accuracy. Some VOTs are available to the user while airborne, while others are limited to ground use only.

VOT. See *VOR test facility.*

VSI. See *vertical speed indicator.*

VSo. The stalling speed or the minimum steady flight speed in the landing configuration. In small airplanes, this is the power-off stall speed at the maximum landing weight in the landing configuration (gear and flaps down). The lower limit of the white arc.

VS1. The stalling speed or the minimum steady flight speed obtained in a specified configuration. For most airplanes, this is the power-off stall speed at the maximum takeoff weight in the clean configuration (gear up, if retractable, and flaps up). The lower limit of the green arc.

V-tail. A design which utilizes two slanted tail surfaces to perform the same functions as the surfaces of a conventional elevator and rudder configuration. The fixed surfaces act as both horizontal and vertical stabilizers.

VX. Best angle-of-climb speed. The airspeed at which an airplane gains the greatest amount of altitude in a given distance. It is used during a short-field takeoff to clear an obstacle.

VY. Best rate-of-climb speed. This airspeed provides the most altitude gain in a given period of time.

VYSE. Best rate-of-climb speed with one engine inoperative. This airspeed provides the most altitude gain in a given period in a light, twin-engine airplane following an engine failure.

W

WAAS. See *wide area augmentation system.*

Wake turbulence. Wingtip vortices that are created when an airplane generates lift. When an airplane generates lift, air spills over the wingtips from the high-pressure areas below the wings to the low-pressure areas above them. This flow causes rapidly rotating whirlpools of air called *wingtip vortices* or *wake turbulence*. This effect is strongest in heavy aircraft, especially when they are moving slowly in clean configuration and at high angles of attack, such as shortly after takeoff.

Warm front. The boundary area formed when a warm air mass contacts and flows over a colder air mass. Warm fronts cause low ceilings and rain.

Warning area. An area containing hazards to any aircraft not participating in the activities being conducted in the area. Warning areas may contain intensive military training, gunnery exercises, or special weapons testing.

WARP. See *weather and radar processor.*

Waste gate. A controllable valve in the tailpipe of an aircraft reciprocating engine equipped with a turbocharger. The valve is controlled to vary the amount of exhaust gases forced through the turbocharger turbine.

Waypoint. A designated geographical location used for route definition or progress-reporting purposes and is defined in terms of latitude/longitude coordinates.

WCA. See *wind correction angle.*

Weather and radar processor (WARP). A device that provides real-time, accurate, predictive, and strategic weather information presented in an integrated manner in the National Airspace System (NAS).

Weather depiction chart. Details surface conditions as derived from METAR and other surface observations.

Weight. The force exerted by an aircraft from the pull of gravity.

Wide area augmentation system (WAAS). A Differential Global Positioning System (DGPS) that improves the accuracy of the system by determining position error from the GPS satellites, then transmitting the error, or corrective factors, to the airborne GPS receiver.

Wind correction angle (WCA). The angle between the desired track and the heading of the aircraft necessary to keep the aircraft tracking over the desired track.

Wind direction indicators. Indicators that include a windsock, wind tee, or tetrahedron. Visual reference will determine wind direction and runway in use.

Wind shear. A sudden, drastic shift in windspeed, direction, or both that may occur in the horizontal or vertical plane.

Winds and temperature aloft forecast (FB). A twice daily forecast that provides wind and temperature forecasts for specific locations in the contiguous United States.

Wing area. The total surface of the wing (in square feet), which includes control surfaces and may include wing area covered by the fuselage (main body of the airplane), and engine nacelles.

Wings. Airfoils attached to each side of the fuselage and are the main lifting surfaces that support the airplane in flight.

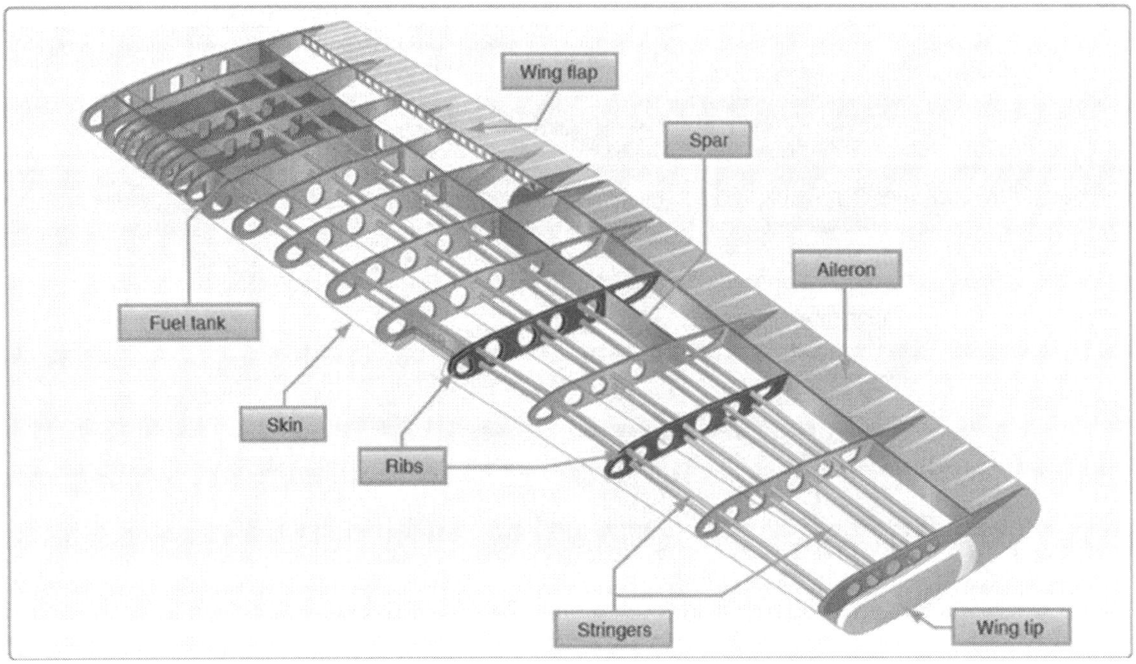

Wing root. The wing root is the part of the wing on a fixed-wing aircraft that is closest to the fuselage. Wing roots usually bear the highest bending forces in flight and during landing, and they often have fairings to reduce interference drag between the wing and the fuselage. The opposite end of a wing from the wing root is the wing tip.

Wingspan. The maximum distance from wingtip to wingtip.

Wingtip vortices. The rapidly rotating air that spills over an airplane's wings during flight. The intensity of the turbulence depends on the airplane's weight, speed, and configuration. Also referred to as *wake turbulence*. Vortices from heavy aircraft may be extremely hazardous to small aircraft.

Wing twist. A design feature incorporated into some wings to improve aileron control effectiveness at high angles of attack during an approach to a stall.

World Aeronautical Charts (WAC). A standard series of aeronautical charts covering land areas of the world at a size and scale convenient for navigation (1:1,000,000) by moderate speed aircraft. Topographic information includes cities and towns, principal roads, railroads, distinctive landmarks, drainage, and relief. Aeronautical information includes visual and radio aids to navigation, airports, airways, restricted areas, obstructions and other pertinent data.

Z

Zone of confusion. Volume of space above the station where a lack of adequate navigation signal directly above the VOR station causes the needle to deviate.

Zulu time. A term used in aviation for coordinated universal time (UTC) which places the entire world on one time standard.

CHAPTER 5. BLOCK COUNTING

The Block Counting section of the AFOQT tests your ability to visualize in three-dimensional space. In simple terms, you must look at a specific block in a stack like the one below and determine how many others are touching it. The task seems simple on the surface, but we suggest pausing to think carefully before answering. There are 30 questions with a 4- to 5-minute time limit, so you only have 9 seconds to answer each question. That is not a lot of time, so practice thinking through the questions systematically. Efficiency is the key, not speed.

Remember that ALL the blocks in each arrangement are the same size. There are no "half blocks," squares, or anything else. All blocks run from end-to-end, just like the parts of the blocks you can see. Likewise, know that blocks that only contact each other at the corners are NOT considered to be touching. For instance, in the following example, only TWO blocks are touching block #1 (the two gray blocks).

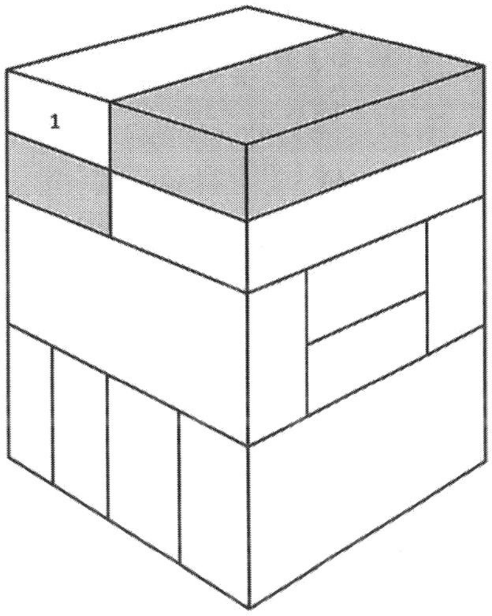

The key to this section is visualizing the parts of the cube that you cannot see. A good method is to start counting the blocks touching the top, then the side, then the bottom, and then the last side. Whenever possible, move in the same direction—either clockwise or counter-clockwise—as you count so that your mind internalizes the process as systematic. A common pitfall is losing track of whether you already counted a block.

To help illustrate this concept, we have highlighted in gray the eight blocks that are touching block #2 in the following figure. Start by counting the two blocks touching the top, then move on to the two blocks touching the back side, and then you will count the four blocks stacked across the bottom. Obviously, since block #2 is on the outside of the cube, no blocks are touching its fourth face.

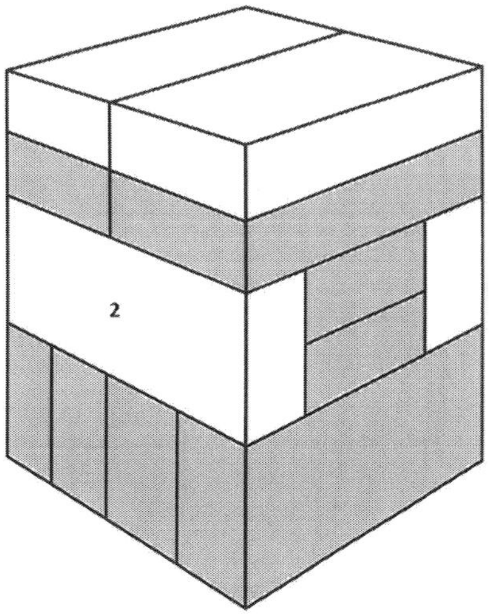

You can find block counting exercises in **Chapter 13: Practice Test** of this book. Unlike some other sections of the AFOQT that require memorization and can be forgotten over time, block counting is a skill that becomes second nature once mastered. Much like riding a bicycle or swimming, your block counting skill, once developed, is difficult to lose. With practice, you'll find that this skill becomes instinctive, allowing you to approach block counting questions with confidence and ease.

CHAPTER 6. MATHEMATICS KNOWLEDGE

INTRODUCTION

The AFOQT Mathematics Knowledge subtest assesses your understanding of high school-level mathematics concepts. This subtest focuses on algebra, geometry, and basic mathematical principles, including number theory, equations, and mathematical operations. You'll need to apply these concepts to solve straightforward problems without the aid of a calculator.

What to Expect

The Mathematics Knowledge subtest assesses both your computational abilities and your capacity to think critically about mathematical relationships. Key topics covered include the following.

Algebraic Concepts: This includes solving linear and quadratic equations, working with inequalities, and simplifying polynomials. You'll encounter problems that require basic algebraic manipulation, such as solving for unknown variables or interpreting algebraic expressions. You may also need to understand functions and their properties.

Geometry: Expect questions on calculating the area, perimeter, volume, and surface area of various geometric shapes. You will also need to apply the Pythagorean theorem, identify properties of angles and triangles, and solve problems involving the relationships between angles in polygons.

Number Theory and Arithmetic Operations: This includes a range of topics such as working with fractions, decimals, percentages, ratios, and proportions. Understanding prime numbers, factors, multiples, and the basic properties of integers is essential. You'll also be expected to perform basic operations like addition, subtraction, multiplication, and division accurately and efficiently.

Mathematical Properties and Rules: Understanding mathematical properties such as the distributive property, associative and commutative properties, and rules for working with exponents and square roots will be important for certain questions.

Data Analysis and Probability: Though not the primary focus, you may encounter basic questions on interpreting data from graphs or charts, calculating averages, and understanding probability.

Types of Questions

Questions on this subtest are generally straightforward, requiring the direct application of mathematical formulas and principles. You won't need to solve complex word problems (these are on the Arithmetic Reasoning subtest), but you'll be tested on your ability to work with numbers and equations directly. The questions often resemble typical problems you'd find in high school math exams, where you solve for unknowns or work through geometric problems step by step.

For example:

Algebraic question: Solve for x in the equation: $2x + 3 = 11$.

Geometry question: Calculate the volume of a cylinder with a given radius and height.

Arithmetic question: Simplify a fraction or convert a decimal to a percentage.

Arithmetic Reasoning vs Mathematics Knowledge Subtests

While both subtests assess your mathematical ability, their focus differs.

Arithmetic Reasoning: This subtest evaluates your ability to solve word problems that require mathematical reasoning and practical application, often focusing on real-world scenarios.

Mathematics Knowledge: In contrast, this subtest focuses on abstract, theoretical math concepts and the ability to solve equations and work through mathematical principles directly. This subtest includes both algebra and geometry. It involves solving equations, understanding inequalities, working with exponents, and applying geometric principles.

Now, let's start to get you well-prepared by reviewing some basic math concepts first!

§1. Mathematics Fundamentals

Whole Number Operations

Whole numbers are non-negative integers that include zero and all positive integers. They form the basic building blocks of arithmetic and are essential for mathematical operations. Whole numbers don't include fractions or decimals, making them suitable for simple arithmetic tasks. Examples of whole numbers include 0, 1, 2, 3, 56, 327, etc. They can be used in various contexts, such as counting objects, keeping track of scores, or measuring discrete quantities. For instance, if you have five apples, three friends, and two books, each of these counts can be expressed as whole numbers, making it easier to perform addition, subtraction, and other arithmetic operations.

Order of Operations (PEMDAS)

Order of Operations, commonly remembered by the acronym PEMDAS, dictates the order in which mathematical operations should be performed to ensure consistency and accuracy. PEMDAS stands for:

- Parentheses: Operations inside parentheses are performed first.
- Exponents: After parentheses, handle powers and roots.
- Multiplication and Division: Next, solve any multiplication or division from left to right.
- Addition and Subtraction: Finally, perform addition and subtraction from left to right.

Examples:

1. $3 + 2 \times 4$: According to PEMDAS, multiplication is done before addition. So, $2 \times 4 = 8$. Then, apply the addition: $3 + 8 = 11$.
2. $(2 + 3)^2$: Operations inside the parentheses come first. So, $2 + 3 = 5$. Then, apply the exponent: $5^2 = 25$.
3. $8 \div 2 \times 3$: Multiplication and division are handled from left to right. So, $8 \div 2 = 4$, and then $4 \times 3 = 12$.

Estimation

Estimation in mathematics is a method of roughly calculating an answer or checking the accuracy of a solution. It provides a way to arrive at a reasonable approximation quickly and efficiently.

Example 1: A tablet costs 244.99, but there is a 20% discount. Approximately how much does it cost now?

Solution: In this problem, the word "approximately" indicates that estimation is needed. The amount can be rounded to 250 or 240 for simplicity. However, rounding to 250 is straightforward, especially for percentage calculations.

Now, 20% of $250 is $50, allowing us to estimate the new cost: $250 - $50 = $200. This estimate is fairly close to the actual cost of $195.99, showing that rounding and approximate calculations can lead to reasonable results.

Example 2: Estimate a sensible answer to 54,893 x 29.

Solution: To approach this problem efficiently, we round 54,893 to the nearest ten thousand: 50,000. Similarly, 29 rounds to 30. The rounded values allow for quicker calculations: 50,000 x 30 = 1,500,000.

Rounding

Rounding to the Nearest Integer

Focus on the digit immediately following the decimal point:

If this digit is between 0 and 4, drop the decimal part, leaving the integer part as it is.

If this digit is between 5 and 9, drop the decimal part and add 1 to the integer.

Example 1: round 12.7 to the nearest integer

Solution: The digit immediately after the decimal point is 7, so the decimal part is dropped, and 1 is added to the integer part, making the number 13.

Rounding to the Nearest Ten

Focus on the ones digit:

If the digit is between 0 and 4, change it to 0, keeping the other digits the same.

If the digit is between 5 and 9, change it to 0 and add 1 to the tens digit.

Example 2: round 347 to the nearest ten

Solution: The ones digit is 7, so it changes to 0, and 1 is added to the tens digit, making the number 350.

Rounding to the Nearest Hundred

Focus on the tens digit:

If the digit is between 0 and 4, change it and the ones digit to 0.

If the digit is between 5 and 9, change it and the ones digit to 0 and add 1 to the hundreds digit.

Example 3: round 2836 to the nearest hundred

Solution: The tens digit is 3, which changes to 0, along with the ones digit, making the number 2800.

You must get the idea now! If you are asked to round to the nearest thousand, million, billion, trillion, and so on, just follow similar procedures as the above.

Inequalities

Inequalities provide a way to compare two values or expressions that are not necessarily equal. Common symbols used to express inequalities include:

$<$ (less than): Indicates that the value on the left is smaller than the value on the right.

$>$ (greater than): Indicates that the value on the left is larger than the value on the right.

$\leq$ (less than or equal to): Indicates that the value on the left is smaller than or equal to the value on the right.

$\geq$ (greater than or equal to): Indicates that the value on the left is larger than or equal to the value on the right.

Below is a summary of inequalities symbols, their meanings, and how they appear on the number line.

Symbol	Meaning	On the Number Line
$<$	Less than	Open circle
$>$	Greater than	Open circle
$\leq$	Less than or equal to	Closed circle
$\geq$	Greater than or equal to	Closed circle

Example 1: Comparing Numbers Directly

Solution: Compare 45 and 38:

Since $45 > 38$, we write the inequality as: $45 > 38$.

Example 2: Inequality Expressions

For more complex scenarios, the AFOQT test may require evaluating and simplifying expressions. Consider the expressions $2x + 10$ and $4x - 8$. Question: when the first expression is larger than the second? To find out, we write an inequality:

$$2x + 10 > 4x - 8.$$

Rearranging by subtracting 2x and adding 8 from both sides:

$$10 + 8 > 4x - 2x$$

$$18 > 2x.$$

Dividing both sides by 2:

$$9 > x.$$

This example shows how inequalities can help find ranges of values for variables.

Example 3: Inequalities in Real-Life Context

In practical applications, inequalities may compare numerical data. For example, a store offers a discount on products that cost at least $50. A customer has a coupon for 15% off any qualifying item. Question: should the customer qualify to use the coupon, how much will he/she save at a minimum?

We write an inequality to show how much the customer would save: $Savings \geq 0.15 * 50$.

This inequality shows that the customer would save at least $7.50 with the coupon.

This is illustrated on the following number line.

Below are more examples of inequalities illustrated on a number line.

Example 4: $-5 < X \leq 12$

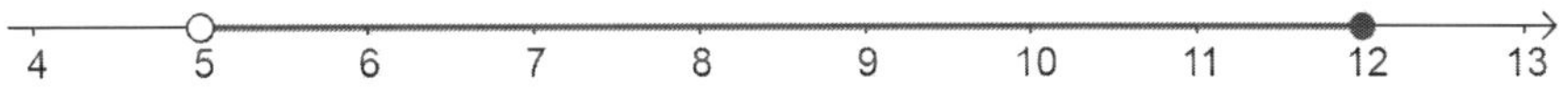

Example 5: $X \leq -1 \; or \; X > 3$

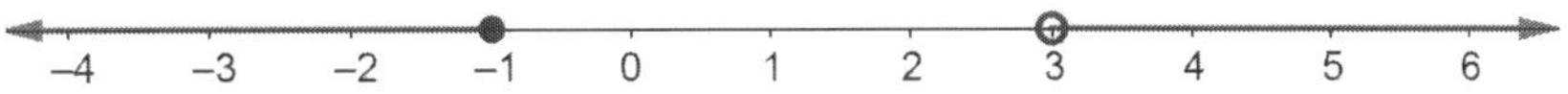

FRACTION OPERATIONS

In this section, we will review arithmetic operations with fractions, including addition, subtraction, multiplication, and division of fractions and mixed numbers, estimation and rounding.

Basic Fraction Concepts

Fraction: A way to represent parts of a whole, a fraction consists of a numerator and a denominator separated by a fraction bar. For example: In the fraction $\frac{3}{4}$, 3 parts of a whole that is divided into 4 equal parts are considered. The top part of a fraction is the **numerator**, indicating how many parts of the whole are being considered. The bottom part of a fraction is the **denominator**, indicating into how many parts the whole is divided. In the above example, 3 is the numerator, and 4 is the denominator. The numbers 3 and 4 are also called the **terms** of the fraction. The term on top of the fraction bar is the numerator, and the term on the bottom is the denominator.

Proper Fractions: Fractions where the numerator is less than the denominator. For example: $\frac{3}{5}$ is a proper fraction because the numerator, 3, is less than the denominator, 5.

Improper Fractions: Fractions where the numerator is greater than or equal to the denominator. For example: $\frac{7}{5}$ is a proper fraction because the numerator, 7, is bigger than the denominator, 5.

Reciprocals: Two numbers whose product is 1. Essentially, you flip the numerator and the denominator. For example: $\frac{7}{11}$ is the reciprocal $\frac{11}{7}$.

Zero as Numerator: Indicates that the fraction represents zero, as nothing is taken from the whole.
For example: $\frac{0}{9} = 0$.

Zero as Denominator: This is undefined in mathematics because you cannot divide by zero.

Mixed Number: A whole number combined with a proper fraction. A mixed number is a way to express a number that includes both a whole part and a fraction part. It combines a whole number (like 1, 2, 3, and so on) with a proper fraction (where the top number, or numerator, is smaller than the bottom number, or denominator), e.g., $2\frac{1}{2}$.

Converting an Improper Fraction to a Mixed Number

Converting an improper fraction to a mixed number involves turning a fraction where the numerator (top number) is larger than the denominator (bottom number) into a number that shows how many whole parts there are, along with a proper fraction. Let's go through how to do this with an example.

Example: Convert $\frac{22}{7}$ to a Mixed Number.

Solution: This can be achieved by the following steps.

1. Divide the Numerator by the Denominator: Divide 22 by 7. When you divide 22 by 7, the quotient (the whole number part of the division) is 3, because 7 goes into 22 three times completely.
2. Calculate the Remainder: To find out how much is left over, multiply the whole number part (3) by the denominator (7), which equals 21. Subtract this product from the original numerator: $22 - 21 = 1$.
3. The Whole Number: The quotient from your division (3) is the whole number part of the mixed number.
4. The Fraction Part: The remainder (1) is the new numerator, while the denominator remains the same (7), so the fraction part is $\frac{1}{7}$.
5. Write the Mixed Number: Combine the whole number and the fraction part to form the mixed number, which is $3\frac{1}{7}$.

This conversion demonstrates how an improper fraction can be expressed more intuitively as a mixed number, making it easier to visualize and understand.

Converting a Mixed Number to an Improper Fraction

Converting a mixed number to an improper fraction involves combining the whole number and the fractional parts into a single fraction, where the numerator is larger than the denominator.

Example: Convert $3\frac{2}{5}$ to an improper fraction.

Solution: This can be achieved by the following steps.

1. Multiply the Whole Number by the Denominator: Multiply the whole number part of the mixed number (3) by the denominator of the fraction part (5). This calculation represents the total number of fifths in the three whole parts: $3 \times 5 = 15$
2. Add the Numerator of the Fraction Part: Add the numerator of the fraction part (2) to the result from the first step. This step accounts for all the parts: $15 + 2 = 17$
3. Write the Improper Fraction: The sum from the second step (17) becomes the new numerator, and the original denominator (5) stays the same. So, the improper fraction is: $\frac{17}{5}$.

This improper fraction, $\frac{17}{5}$, represents the total number of fifths in $3\frac{2}{5}$. This method ensures that all parts of the mixed number are accounted for, translating it back into a single fraction format. This can be particularly useful in calculations involving multiple fractions or when needing a consistent format for mathematical operations.

Equivalent Fractions

Equivalent fractions are fractions that, although they have different numerators and denominators, actually represent the same value or portion of a whole. This concept is crucial for simplifying fractions, comparing them, and performing operations like addition and subtraction when the fractions involved have different denominators.

How to Find Equivalent Fractions: To find equivalent fractions, you multiply or divide the numerator and the denominator of a fraction by the same non-zero number. Here are the steps for each method.

Multiplication method

1. Choose a number (other than zero) to multiply both the numerator and the denominator.
2. This number is known as the 'scale factor'.

Example: Find two equivalent fractions for $\frac{1}{4}$.

Solution 1: Multiply both the numerator and the denominator by 2:

$$\frac{1 \times 2}{4 \times 2} = \frac{2}{8}$$

Solution 2: Multiply both the numerator and the denominator by 3:

$$\frac{1 \times 3}{4 \times 3} = \frac{3}{12}$$

Division method

If both the numerator and the denominator can be evenly divided by the same number (this number is known as the 'common divisor'), you can simplify the fraction by dividing the numerator and denominator with the common divisor.

Example: Find an equivalent fractions for $\frac{6}{9}$.

Solution: Since both 6 and 9 are divisible by 3, divide both by 3:

$$\frac{6 \div 3}{9 \div 3} = \frac{2}{3}$$

These examples illustrate how you can either increase or decrease the terms of a fraction while keeping its value unchanged.

The Fundamental Property of Fractions

The Fundamental Property of Fractions, also known as the Multiplicative Property of Equality, states that multiplying the numerator and the denominator of a fraction by the same non-zero number does not change the value of the fraction.

If a, b, and c are numbers, and $b \neq 0$ and $c \neq 0$, then: $\frac{a}{b} = \frac{a \cdot c}{b \cdot c}$

Both the multiplication method and the division method of finding equivalent fractions utilize the Fundamental Property of Fractions.

Greatest Common Divisor (GCD)

The Greatest Common Divisor (GCD), also known as the Greatest Common Factor (GCF), is the largest number that can evenly divide both the numerator and the denominator of a fraction without leaving a remainder. Finding the GCD is crucial for simplifying fractions to their lowest terms.

Steps to Find the GCD:

1. List the Factors of Each Number: Begin by listing all the factors (numbers that divide without leaving a remainder) for both the numerator and the denominator.
2. Identify the Common Factors: Compare the lists of factors for both the numerator and the denominator and identify the numbers that appear in both lists.
3. Select the Largest Common Factor: The largest number in the list of common factors is the GCD.

Example: Find the GCD of 18 and 24.

Solution:

- Factors of 18: 1, 2, 3, 6, 9, 18
- Factors of 24: 1, 2, 3, 4, 6, 8, 12, 24

Common Factors are: 1, 2, 3, 6

Greatest Common Factor is: 6

Thus, the GCD of 18 and 24 is 6.

Cancellation and Lowest Terms

Cancellation and simplifying fractions to their lowest terms are are used to simplify calculations and make fractions easier to understand and compare.

Cancellation involves dividing both the numerator and the denominator of a fraction by the same non-zero number. This process reduces the fraction to a simpler form, often making it easier to work with in calculations. A fraction is in its **lowest terms** (or simplest form) when the numerator and the denominator have no common factors other than 1. This means the fraction cannot be simplified any further.

Steps for Simplifying Fractions:

1. **Identify a Common Factor**: Find a number that divides evenly into both the numerator and the denominator.
2. **Divide Both Terms**: Divide the numerator and the denominator by this common factor.
3. **Repeat if Necessary**: Continue this process with the new fraction until no further common factors can be found.

Example 1: Simplifying $\frac{18}{24}$.

Solution: First, identify the **greatest common divisor (GCD)** for 18 and 24, which is 6.

Then, divide both the numerator and the denominator by 6:

$$\frac{18 \div 6}{24 \div 6} = \frac{3}{4}$$

Now, $\frac{3}{4}$ is in its lowest terms because the only common factor between 3 and 4 is 1.

Example 2: Simplifying $\frac{40}{60}$.

Solution: The GCD for 40 and 60 is 20. Divide both the numerator and the denominator by 20:

$$\frac{40 \div 20}{60 \div 20} = \frac{2}{3}$$

$\frac{2}{3}$ is in its lowest terms, as 2 and 3 are coprime (they have no common factors other than 1).

These examples show how cancellation reduces fractions to their simplest form, facilitating easier manipulation and comparison of fractional values.

Multiplication of Fractions

Multiplying fractions is straightforward once you understand the basic rule: multiply the numerators together to get the new numerator, and multiply the denominators together to get the new denominator.

Steps for Multiplying Fractions:

1. Multiply the Numerators: Take the numerator of each fraction and multiply them together.
2. Multiply the Denominators: Take the denominator of each fraction and multiply them together.

3. Simplify the Resulting Fraction: If possible, simplify the new fraction by dividing both the numerator and the denominator by their greatest common divisor (GCD).

Example 1: Multiply $\frac{1}{4}$ and $\frac{3}{5}$.

Solution: $\frac{1}{4} \times \frac{3}{5} = \frac{1 \times 3}{4 \times 5} = \frac{3}{20}$

Example 2: Multiply $\frac{7}{8}$ and $\frac{12}{15}$

Solution: $\frac{7}{8} \times \frac{12}{15} = \frac{7 \times 12}{8 \times 15} = \frac{84}{120}$

Simplify $\frac{84}{120}$ by finding the GCD of 84 and 120, which is 12. Divide both the numerator and the denominator by 12:

$$\frac{84 \div 12}{120 \div 12} = \frac{7}{10}$$

This example shows how multiplication of fractions can also involve simplification to reduce the fraction to its lowest terms.

Pre-cancelling when Multiplying Fractions

Pre-cancelling (also known as cross-cancelling) when multiplying fractions is a technique that simplifies the multiplication process by reducing the fractions before actually performing the multiplication. This method involves cancelling common factors between the numerators and denominators of the fractions involved in the multiplication. It makes calculations easier and helps to avoid dealing with unnecessarily large numbers.

Steps for Pre-cancelling:

1. **Identify Common Factors**: Look for any common factors that the numerator of one fraction has with the denominator of the other fraction.
2. **Cancel the Common Factors**: Divide the common factors out before multiplying.
3. **Multiply the Remaining Numbers**: Multiply the simplified numerators and denominators to get the final answer.

Example: Multiply $\frac{14}{45}$ by $\frac{27}{28}$

Solution: Here's how you can pre-cancel. From the first fraction $\frac{14}{45}$, and the second fraction $\frac{27}{28}$:

Notice that 14 in the numerator of the first fraction and 28 in the denominator of the second fraction share a common factor of 14.

Notice that 27 in the numerator of the second fraction and 45 in the denominator of the first fraction share a common factor of 9.

Pre-cancel the common factors:

$\frac{14}{28}$ reduces to $\frac{1}{2}$ (14 divided by 14 is 1, 28 divided by 14 is 2).

$\frac{27}{45}$ reduces to $\frac{3}{5}$ (27 divided by 9 is 3, 45 divided by 9 is 5).

Now multiply the simplified fractions:

$$\frac{1}{2} \times \frac{3}{5} = \frac{1 \times 3}{2 \times 5} = \frac{3}{10}$$

The entire process of solving this problem can be written this way:

$$\frac{14}{45} \times \frac{27}{28} = \frac{14}{28} \times \frac{27}{45} = \frac{1}{2} \times \frac{9 \times 3}{9 \times 5} = \frac{1}{2} \times \frac{3}{5} = \frac{1 \times 3}{2 \times 5} = \frac{3}{10}$$

This example shows how pre-cancelling simplifies the process of multiplying fractions, making the multiplication straightforward and reducing the numbers involved.

Division of Fractions

Division of fractions involves reversing the process of multiplication by using the reciprocal of the divisor. The **reciprocal** of a fraction is obtained by swapping its numerator and denominator. This method, often summarized by the phrase "invert and multiply," simplifies the process of dividing fractions.

Steps for Dividing Fractions:

1. **Find the Reciprocal**: Take the reciprocal of the fraction that you are dividing by (the divisor).
2. **Multiply the Fractions**: Multiply the first fraction (the dividend) by the reciprocal of the second fraction.

Example 1: Divide $\frac{3}{7}$ by $\frac{6}{5}$

Solution: $\frac{3}{7} \div \frac{6}{5} = \frac{3}{7} \times \frac{5}{6} = \frac{3\times5}{7\times6} = \frac{15}{42}$

As illustrated, the division of fractions is handled through the "invert and multiply" technique, transforming a division problem into a multiplication problem that is often simpler to solve.

Dividing fractions where one or both numbers are mixed fractions involves converting those mixed fractions to improper fractions first, and then following the "invert and multiply" method. Let's go through this with a detailed example.

Example 2: Divide $2\frac{1}{3}$ by $1\frac{1}{2}$

Solution: Convert Mixed Fractions to Improper Fractions first:

$$2\frac{1}{3} = \frac{2 \times 3 + 1}{3} = \frac{6+1}{3} = \frac{7}{3}$$

$$1\frac{1}{2} = \frac{1 \times 2 + 1}{2} = \frac{2+1}{2} = \frac{3}{2}$$

Invert the Divisor ($1\frac{1}{2}$) and Multiply:

$$\frac{7}{3} \div \frac{3}{2} = \frac{7}{3} \times \frac{2}{3} = \frac{7 \times 2}{3 \times 3} = \frac{14}{9}$$

Here the entire process:

$$2\frac{1}{3} \div 1\frac{1}{2} = \frac{7}{3} \div \frac{3}{2} = \frac{7}{3} \times \frac{2}{3} = \frac{7 \times 2}{3 \times 3} = \frac{14}{9}$$

Adding and Subtracting Fractions

The process of adding and subtracting fractions varies slightly depending on whether the fractions have the same denominator or different denominators.

Adding and Subtracting Fractions with the Same Denominator

When fractions have the same denominator (the bottom number of the fraction), the process is straightforward:

1. **Keep the Denominator**: The denominator remains the same.
2. **Add/Subtract the Numerators**: Simply add or subtract the numerators (the top numbers of the fractions) as indicated.

Example 1: Add $\frac{3}{7}$ and $\frac{2}{7}$

Solution: $\frac{3}{7} + \frac{2}{7} = \frac{3+2}{7} = \frac{5}{7}$

Example 2: Subtracting $\frac{5}{7}$ from $\frac{6}{7}$

Solution: $\frac{6}{7} - \frac{5}{7} = \frac{6-5}{7} = \frac{1}{7}$

Adding and Subtracting Fractions with Different Denominators

When fractions have different denominators, you must first find a common denominator before you can add or subtract them. This often involves finding the **Least Common Denominator** (LCD), which is the smallest number that both denominators can divide into without a remainder.

1. **Find the Least Common Denominator (LCD):** Determine the smallest common multiple of the denominators.
2. **Adjust the Fractions**: Convert each fraction to an equivalent fraction with the LCD as the new denominator.
3. **Add/Subtract the Adjusted Numerators**: With the same denominators, add or subtract the numerators.

Example 1: adding $\frac{1}{4}$ and $\frac{1}{6}$

Solution: Observe that the LCD of 4 and 6 is 12. Convert $\frac{1}{4}$ to $\frac{3}{12}$ and convert $\frac{1}{6}$ to $\frac{2}{12}$

Now, add the adjusted fractions:

$$\frac{1}{4}+\frac{1}{6}=\frac{3}{12}+\frac{2}{12}=\frac{3+2}{12}=\frac{5}{12}$$

Combined operations with fractions and mixed numbers

Combined operations with fractions and mixed numbers involve performing multiple arithmetic operations—such as addition, subtraction, multiplication, and division—on numbers in fractional and mixed number forms.

Steps for Combined Operations:

1. Convert Mixed Numbers: If the problem involves mixed numbers, convert them to improper fractions first. This simplifies the process of combining them with other fractions.
2. Find a Common Denominator: For addition and subtraction, ensure all fractions involved have a common denominator. This may involve converting each fraction to an equivalent form.
3. Perform Operations: Apply the relevant arithmetic operations. If the operation is addition or subtraction, combine the numerators as appropriate. For multiplication or division, follow the standard rules for fractions.
4. Simplify: Always simplify the resulting fraction to its lowest terms. This may involve finding the greatest common divisor (GCD) and reducing the fraction.

Example: Calculate $1\frac{2}{3}-\frac{3}{4}\times\frac{5}{6}+\frac{1}{2}$

Solution: Step 1: Convert Mixed Numbers

Convert $1\frac{2}{3}$ to an improper fraction:

$$1\frac{2}{3}=\frac{1\cdot 3+2}{3}=\frac{5}{3}$$

Step 2: Perform Multiplication First by following PEMDAS rules.

Calculate $\frac{3}{4}\times\frac{5}{6}$:

$$\frac{3}{4}\times\frac{5}{6}=\frac{3\cdot 5}{4\cdot 6}=\frac{15}{24}=\frac{5}{8}$$

Step 3: Perform Subtraction

Subtract $\frac{5}{8}$ from $\frac{5}{3}$ (first find a common denominator, which is 24):

$$\frac{5}{3}=\frac{40}{24},\quad \frac{5}{8}=\frac{15}{24}$$

$$\frac{40}{24}-\frac{15}{24}=\frac{25}{24}$$

Step 4: Add $\frac{1}{2}$ (convert $\frac{1}{2}$ to $\frac{12}{24}$):

$$\frac{25}{24}+\frac{12}{24}=\frac{37}{24}$$

Step 5: Simplify or Convert to Mixed Number

The result $\frac{37}{24}$ can be expressed as a mixed number:

$$\frac{37}{24}=1\frac{13}{24}$$

Summary: When performing combined operations with fractions and mixed numbers, the key is to handle one operation at a time, simplify at each step, and always keep track of the order of operations to ensure accuracy.

DECIMAL OPERATIONS

Decimals are a way of expressing numbers that are not whole, using a base of ten. Here's an introduction to some fundamental concepts of decimals:

Decimal Point: The decimal point separates the whole number part from the fractional part of a number. It is denoted by a dot (.)

Writing Whole Numbers as Decimals: Any whole number can be written as a decimal by adding a decimal point and zeros. For example, the whole number 25 can be written as a decimal: 25=25.0. This shows that 25 is equivalent to 25 plus zero tenths.

Powers of Ten: Decimals are based on powers of ten. Each place to the right of the decimal point represents a negative power of ten. The first place to the right of the decimal point is the tenths place (10^{-1}), the next is the hundredths place (10^{-2}), and so on. Conversely, places to the left of the decimal point represent positive powers of ten, like tens (10^{1}), hundreds (10^{2}), etc.

Leading Zeros: In decimal fractions that fall strictly between -1 and 1, the leading zero digits between the decimal point and the first non-zero digit are essential for conveying the magnitude of a number and must not be omitted. For instance, for decimal 0.000357, the three zeros between decimal point and number 3 cannot be omitted.

The zero that appears immediately to the left of the decimal point, such as the 0 in 0.468, is sometimes dropped, although the decimal point must remain. Most of the time though, the zero to the left of the decimal point is not omitted. Keeping the zero to the left of the decimal point enhances clarity.

Trailing Zeros: Zeros after the last non-zero digit in a decimal number can affect the precision in a scientific or mathematical context but do not change the value of the number. For example: Compare 0.2500 and 0.25. The first number, 0.2500, suggests a precision measurement to the ten-thousandths place. This indicates a higher precision than 0.25, which suggests a precision measurement to the hundredths place. However, the two numbers are equal in value.

Decimal Place Values

Each position or place in a decimal number has a value based on powers of ten.

- Tenths (10^{-1}): This is the first place to the right of the decimal point. Each unit in this place is one-tenth of a whole.
- Hundredths (10^{-2}): The second place to the right of the decimal point, where each unit is one-hundredth of a whole.
- Thousandths (10^{-3}): The third place to the right of the decimal point, where each unit is one-thousandth of a whole.
- And so on, with ten-thousandths, hundred-thousandths, etc.

Example: Consider the decimal number 45.6789. Here's how each digit fits into the place value system:

- 4 is in the tens place (10^{1}),
- 5 is in the units or ones place (10^{0}),
- 6 is in the tenths place (10^{-1}),
- 7 is in the hundredths place (10^{-2}),
- 8 is in the thousandths place (10^{-3}),
- 9 is in the ten-thousandths place (10^{-4}).

Each position affects the value of the number. For instance, the 6 in 45.6789 contributes 6 tenths to the value, or 0.6. The 7 contributes 7 hundredths, or 0.07, and so on.

Multiplying and Dividing Decimals by Powers of 10

When you multiply a decimal by a power of 10, shift the decimal point to the right by as many places as there are zeros in the power of 10.

Example: Multiply 4.567 by 100 (which is 10^2): $4.567 \times 100 = 456.7$

The decimal point moves two places to the right.

When you divide by a power of 10, shift the decimal point to the left by as many places as there are zeros in the power of 10.

Example: Divide 3.25 by 1000 (which is 10^3): $3.25 \div 1000 = 0.00325$

The decimal point moves three places to the left.

If the number of zeros exceeds the number of digits before the decimal in the original number, you may need to add leading zeros. For example, multiplying 0.123 by 10^5 results in 12300, and dividing 0.00456 by 10^2 results in 0.0000456.

Rounding of Decimal Numbers

Rounding decimal numbers is used to make them easier to work with, particularly when precision is less critical. Rounding involves increasing or decreasing a number to a certain place value.

Steps for Rounding Decimals:

1. Identify the Place Value to round to: Determine the decimal place to which you want to round. This could be to the nearest tenth, hundredth, thousandth, etc.
2. Look at the Digit Immediately to the Right: This is the deciding digit.
3. If the deciding digit is 5 or higher, round up by adding one to the digit in the place you are rounding to and dropping all digits to the right.
4. If the deciding digit is less than 5, round down by keeping the digit in the place you are rounding to the same and dropping all digits to the right.

Example: Round 14.5379 to the nearest tenth, hundredth, and thousandth:

To the Nearest Tenth: The digit in the tenths place is 5. Look at the next digit (3). Since it's less than 5, the 5 stays the same. Hence, $14.5379 \approx 14.5$.

To the Nearest Hundredth: The digit in the hundredths place is 3. Look at the next digit (7). Since it's 5 or higher, round up the 3 to a 4. Hence, $14.5379 \approx 14.54$.

To the Nearest Thousandth: The digit in the thousandths place is 7. Look at the next digit (9). Since it's 5 or higher, round up the 7 to an 8. Hence, $14.5379 \approx 14.538$.

Adding and Subtracting Decimals

The key to successfully performing adding and subtracting decimals is ensuring the decimal points of all numbers involved are lined up correctly. Here's how to add and subtract decimals:

Steps for Adding Decimals

1. Align the Decimal Points: Write the numbers so that the decimal points are vertically aligned.
2. Fill in Missing Places: If the numbers have different numbers of digits after the decimal point, add zeros to the ends of the shorter decimals to make them equal in length.
3. Add as Whole Numbers: Ignore the decimal point and add the numbers as if they were whole numbers.
4. Place the Decimal Point: In the sum, place the decimal point directly below the other decimal points.

Example: Add 2.75 and 3.006

Solution: Write the numbers with aligned decimal points:

$$\begin{array}{r} 2.75 \\ +\quad 3.006 \\ \hline 5.756 \end{array}$$

Steps for Subtracting Decimals

1. Align the Decimal Points: As with addition, ensure the decimal points are vertically aligned.
2. Fill in Missing Places: Pad the number with fewer decimal places with zeros.
3. Subtract as Whole Numbers: Ignore the decimal point temporarily and subtract as if they were whole numbers.
4. Place the Decimal Point: Ensure the decimal point in the result lines up with the decimal points above.

Example: Subtract 7.82 from 10.5:

Solution: Write the numbers with aligned decimal points and pad with zeros if necessary:

$$\begin{array}{r} 10.50 \\ -\quad 7.82 \\ \hline 2.68 \end{array}$$

Key Points in this example are:

Alignment: When aligning decimals, it is crucial to be accurate with the placement of decimal points to ensure correct calculations.

Zero Padding: Adding zeros to the ends of shorter decimals does not change their value but helps make the addition and subtraction operations straightforward.

Carry/Borrow: In subtraction, remember to borrow as you would with whole numbers when subtracting one digit from another that is smaller. Similarly, remember to carry over in addition when sums of digits exceed 9.

Multiplying and Dividing Decimals

When multiplying decimals, it's crucial to handle the decimal points correctly to ensure the product has the correct number of decimal places.

Steps for Multiplying Decimals:

1. **Ignore the Decimal Points**: Multiply the numbers as if they were whole numbers.
2. **Count Decimal Places**: Add up the total number of decimal places in both factors.
3. **Place the Decimal Point**: In the product, position the decimal point so that it has the combined number of decimal places from the factors.

Example: Multiply 3.2 by 2.5

Solution: Multiply as whole numbers: $32 \times 25 = 800$

Count decimal places: $1+1=2$

Position the decimal point: $3.2 \times 2.5 = 8.00$ or simply 8.

Steps for Dividing Decimals:

1. Make the Divisor a Whole Number: Shift the decimal point in the divisor right until it is a whole number, doing the same shift to the dividend.
2. Divide as Whole Numbers: Perform the division as you would with whole numbers.
3. Place the Decimal Point: Insert the decimal point in the quotient based on the initial shifts made.

Example: Divide 6.75 by 1.5

Solution: Steps for Dividing Decimals:

1. Make the Divisor a Whole Number: We'll adjust the divisor 1.5 to make it a whole number by shifting the decimal point to the right, so 1.5 becomes 15.
2. Adjust the Dividend Accordingly: Shift the decimal point in the dividend the same number of places as the divisor to maintain the balance. So, 6.75 becomes 67.5.
3. Divide as Whole Numbers: Perform the division on the adjusted numbers.
4. Position the Decimal Point: After the division, ensure the decimal point in the quotient is correctly placed based on the shifts made.
5. Now, perform the division:

$$\frac{67.5}{15} = 4.5$$

PERCENTAGE

Percents are a way to express a number as a fraction of 100. The word "percent" comes from the Latin phrase "per centum," which means "by the hundred." This makes percents very useful for describing proportions and comparisons.

For example, imagine you have a jar of 100 marbles, and 25 of them are red. You could say that 25% of the marbles are red. This percentage tells us how many marbles out of every 100 are red, making it easy to understand proportions even if the total number of marbles were to change.

Here's a simple mathematical example involving percents: Suppose you scored 45 out of 50 questions correct on a test. To find out the percentage of questions you got right, you divide the number of questions you answered correctly by the total number of questions, and then multiply by 100 to convert it to a percentage. The equation looks like this:

$$\text{Percentage} = \left(\frac{\text{Correct answers}}{\text{Total questions}}\right) \times 100$$

Plugging in the numbers:

$$\text{Percentage} = \left(\frac{45}{50}\right) \times 100$$

This calculation shows you got 90% of the questions correct.

Any problem involving percents can be expressed in the form "A is P percent of B." In this statement, one of the values A, B, or P is typically unknown. To handle such problems mathematically, we translate the statement into an equation:

$$A = \left(\frac{P}{100}\right) \times B$$

If we divide both sides of this equation by B, we derive the formula for P, the percentage:

$$\frac{A}{B} = \frac{P}{100}$$

This equation shows that the percentage P can be calculated by dividing A by B and then multiplying the result by 100. Let's explore this concept through three practical examples:

Example 1: Finding the Unknown Percent

Suppose you have savings of $500 and you learn that it is a part of your annual savings goal. If your total savings goal is $2000, what percent of your goal have you already saved?

Using the formula: $P = \left(\frac{A}{B}\right) \times 100$ to calculate the percent:

$$P = \left(\frac{500}{2000}\right) \times 100 = 25\%$$

So, you have saved 25% of your annual savings goal.

Example 2: Finding the Total (B)

Imagine you scored 92% on a test, and this percentage represents getting 46 questions correct. How many questions were on the test?

Rearrange the formula to solve for B: $B = \frac{A}{(P/100)}$, plug in the values:

$$B = \frac{46}{92\%} = \frac{46}{0.92} = 50$$

There were about 50 questions on the test.

Example 3: Finding the Part (A)

You want to buy a laptop that is on sale for 30% off its original price of $800. How much discount are you getting?

Apply the formula: $A = \left(\frac{P}{100}\right) \times B$

Calculate the discount: $A = 30\% \times 800 = 240$

You get a $240 discount on the laptop.

These examples demonstrate how versatile the percent formula is for solving various real-world problems involving percentages.

Percentage Increase and Decrease

Percentage increase and decrease are important concepts used to describe how much something grows or reduces in proportion over time.

Percentage increase is used to measure how much a quantity has grown relative to its original amount. It's calculated by finding the difference between the new value and the original value, dividing that difference by the original value, and then multiplying the result by 100 to convert it to a percentage. Here's the formula for calculating percentage increase:

$$\text{Percentage Increase} = \left(\frac{\text{New Value} - \text{Original Value}}{\text{Original Value}}\right) \times 100$$

Example: Suppose last year a store sold 150 units of a product, and this year the store sold 180 units. The percentage increase in sales is calculated as follows:

$$\text{Percentage Increase} = \left(\frac{180 - 150}{150}\right) \times 100 = 20\%$$

This means that sales increased by 20% from last year.

Percentage decrease is used to measure how much a quantity has reduced relative to its original amount. The formula is similar to that of percentage increase, but it starts with the original value being higher than the new value. Here's how to calculate percentage decrease:

$$\text{Percentage Decrease} = \left(\frac{\text{Original Value} - \text{New Value}}{\text{Original Value}}\right) \times 100$$

Example: If a car's value decreases from $20,000 to $15,000 over a year, the percentage decrease is:

$$\text{Percentage Decrease} = \left(\frac{20000 - 15000}{20000}\right) \times 100 = 25\%$$

This calculation shows that the car's value has decreased by 25%.

These calculations help us understand changes in terms of percentages, which are easier to compare than just absolute numbers. Percentage increases and decreases offer a clear and standardized method of measurement.

Convert between Percentages, Fractions, and Decimals

Converting between percentages, fractions, and decimals allows you to interpret and compare different forms of numerical expressions.

1. From Percentages to Fractions and Decimals

Converting Percentages to Fractions

To convert a percentage to a fraction, you simply place the percentage number over 100 and then simplify the fraction if possible.

Example: Convert 75% to a fraction.

Solution: First, write the percentage as a fraction: $\frac{75}{100}$. Then, simplify the fraction by dividing the numerator and the denominator by their greatest common divisor, which is 25 in this case:

$$\frac{75 \div 25}{100 \div 25} = \frac{3}{4}$$

Converting Percentages to Decimals

To convert a percentage to a decimal, divide the percentage by 100 or simply move the decimal point two places to the left.

Example: Convert 75% to a decimal.

Solution: Divide 75 by 100: $75 \div 100 = 0.75$

2. From Fractions and Decimals to Percentages

Converting Fractions to Percentages

To convert a fraction to a percentage, divide the numerator by the denominator to get a decimal, and then multiply by 100 to get the percentage.

Example: Convert 3/4 to a percentage.

Solution: First, divide the numerator by the denominator: $3 \div 4 = 0.75$

Then, multiply by 100 to convert to a percentage: $0.75 \times 100 = 75\%$

Converting Decimals to Percentages

To convert a decimal to a percentage, multiply the decimal by 100.

Example: Convert 0.75 to a percentage.

Solution: Multiply by 0.75 by 100: $0.75 \times 100 = 75\%$

3. Convert between Decimals and Fractions

Converting Decimals to Percentages

To convert a decimal to a percentage, multiply the decimal by 100. This shift of the decimal point two places to the right transforms the decimal into a percentage, as it essentially converts the decimal into a fraction with a denominator of 100.

Example: Convert 0.85 to a percentage.

Solution: Multiply the decimal by 100 to get the percentage: $0.85 \times 100 = 85\%$

This means 0.85 is equivalent to 85%.

Converting Decimals to Fractions

To convert a decimal to a fraction, the steps are as follows:

1. Write down the decimal divided by 1 (e.g., 0.85/1).
2. Multiply both the numerator (the top number) and the denominator (the bottom number) by 10 for every number after the decimal point. This step is necessary to eliminate the decimal point.
3. Simplify the resulting fraction by dividing both the numerator and the denominator by their greatest common divisor.

Example: Convert 0.85 to a fraction.

Solution: Express the decimal as a fraction: $\frac{0.85}{1}$. Since there are two digits after the decimal, multiply both the numerator and the denominator by 100 (10 raised to the power of 2):

$$\frac{0.85 \times 100}{1 \times 100} = \frac{85}{100}$$

Simplify the fraction:

$$\frac{85 \div 5}{100 \div 5} = \frac{17}{20}$$

Thus, the decimal 0.85 is equivalent to the fraction $\frac{17}{20}$.

NUMBER COMPARISONS AND EQUIVALENTS

One type of AFOQT test question is comparing values between fractions, decimals, and percentages.

Comparing Decimals

The key to comparing decimal values is understanding the place value of each digit. The digits to the left of the decimal point represent whole numbers, while the digits to the right represent fractions of a whole (tenths, hundredths, etc.).

Steps for Comparing Decimals:

1. Align Decimal Points: Make sure both numbers are written with their decimal points in the same position. If one number has fewer decimal places than the other, add zeros to make the numbers have the same number of decimal places.
2. Compare Whole Numbers: Look at the digits to the left of the decimal point. The number with the larger whole number is greater.
3. Compare Decimal Places: If the whole numbers are the same, compare each decimal place from left to right. The number with the larger digit in the first differing decimal place is greater.

Example: Find the largest among the following decimal numbers: 0.756, 0.765, 0.76, and 0.75.

Solution: Align the decimal points and pad with zeros if necessary to ensure that all have the same number of decimal places:

- 0.756
- 0.765
- 0.760 (added a zero)
- 0.750 (added a zero)

Compare from left to right: The whole number part is the same for all numbers: 0. The tenths place (first digit after the decimal point) is also the same: 7. The hundredths place is where differences appear:

- 0.756 has a 5
- 0.765 has a 6
- 0.760 has a 6
- 0.750 has a 5

The thousandths place further distinguishes the numbers with 0.765 having the largest value (5 compared to 0).

Hence, the answer is that 0.765 is the largest number among the group of numbers given.

Comparing Fractions

Comparing fractions with the same denominator is straightforward, so let's go directly to comparing Fractions with the different denominators. There are two methods: 1.) Cross Multiplication method, and 2.) Common Denominator Method.

Example 1: Compare $\frac{7}{9}$ and $\frac{8}{11}$ using Cross Multiplication Method.

Solution: Observe that $7 \times 11 = 77$, $8 \times 9 = 72$. Since $77 > 72$, we conclude that:

$$\frac{7}{9} > \frac{8}{11}$$

Example 2: Compare $\frac{7}{9}$ and $\frac{8}{11}$ using Common Denominator Method.

Solution: Convert $\frac{7}{9}$ and $\frac{8}{11}$ to fractions with a common denominator: the LCD of 9 and 11 is 99. So, convert $\frac{7}{9}$ to $\frac{77}{99}$ (by multiplying both numerator and denominator by 11). Convert $\frac{8}{11}$to $\frac{72}{99}$ (by multiplying both numerator and denominator by 9).

Since $\frac{77}{99} > \frac{72}{99}$, we conclude that $\frac{7}{9} > \frac{8}{11}$.

Comparing Decimals, Fractions, and Percents

Some AFOQT questions ask you to compare a mixture of Decimals, Fractions, and Percents. The two primary strategies are converting each value to a decimal or a fraction, depending on which method is easier. Let's explore both methods.

Convert All to Decimals

This method involves converting every value into a decimal, which allows for a direct comparison between the numbers.

Steps:

- Fractions to Decimals: Divide the numerator by the denominator.
- Percentages to Decimals: Divide the percentage by 100.

Example: Compare these values: $\frac{3}{4}$, 0.65, and 80%.

Solution: Convert the fraction to a decimal: $\frac{3}{4}$=0.75.

Convert the percentage to a decimal: 80%=0.80

Now, compare all values as decimals: 0.75, 0.65, 0.80.

Conclusion: The correct order from smallest to largest is 0.65, 0.75, and 0.80. Hence, the original values can be order from smallest to largest as 0.65, $\frac{3}{4}$, 80%.

Convert All to Fractions

This method involves converting all values to fractions before making comparisons. This method is preferrable if converting all values to decimals is cumbersome. Consider the following example.

Example: Rank the value of $\frac{5}{11}$, $\frac{7}{13}$, and 45%, from smallest to largest.

Solution: It is obvious converting $\frac{5}{11}$ and $\frac{7}{13}$ into decimals is not the easiest task if you are doing long division manually. At the same time, it is obvious 45% can readily be converted into a fraction. Based on these observations, we can determine that the best step forward is probably to compare the three values as fractions using the cross-multiplication method.

1. Convert 45% to a fraction: $45\% = \frac{45}{100} = \frac{9}{20}$. So, the three values to compare become: $\frac{5}{11}$, $\frac{7}{13}$, and $\frac{9}{20}$.

2. Compare $\frac{5}{11}$ and $\frac{7}{13}$. Cross-multiply to find: $5 \times 13 = 65, 7 \times 11 = 77$.

   Since 65 < 77, we know $\frac{5}{11} < \frac{7}{13}$.

3. Compare $\frac{7}{13}$ and $\frac{9}{20}$. Cross-multiply to find: $7 \times 20 = 140, 9 \times 13 = 117$.

   Since 140 > 117, $\frac{7}{13} > \frac{9}{20}$.

4. Now that we know $\frac{7}{13}$ is the largest value, we still need to compare $\frac{5}{11}$ and $\frac{9}{20}$. Cross-multiply to find: $7 \times 20 = 140, 9 \times 13 = 117$. Since 140 > 117, $\frac{7}{13} > \frac{9}{20}$.

5. Hence, we conclude that: $\frac{9}{20} < \frac{5}{11} < \frac{7}{13}$.

An astute test-take may also be able to determine $\frac{7}{13}$ to be bigger than $\frac{5}{11}$ and 45% by noticing that $\frac{7}{13}$ is larger than 0.5, while the other two items are smaller than 0.5. This will help save two cross-multiplication comparisons.

In conclusion, when faced with AFOQT comparisons questions between decimals, fractions, and percents, it is advisable to first assess quickly which method above is easier and then solve the problem accordingly.

PRE-ALGEBRA CONCEPTS

AFOQT evaluates your ability to interpret numerical data and apply mathematical concepts in real-world situations. You'll handle rates, ratios, proportions, unit conversions, and so on, all crucial for practical problem-solving in daily life and non-STEM academic fields.

We will also cover the following Pre-Algebra concepts in this section: rational numbers, exponents, radicals, fractional exponents, and scientific notation.

Real Numbers

Real numbers form the comprehensive set of numbers used in mathematics, including rational and irrational ones. They represent all points on the number line and include various number types, from the simplest counting numbers to complex fractional and irrational forms.

Types of Real Numbers:

Natural Numbers: These are also known as counting numbers, which are the numbers used for counting objects. They start from 1 and continue infinitely: 1, 2, 3, 4, 5,...

Integers: This set includes all natural numbers, their negatives, and zero. Unlike natural numbers, integers cover both positive and negative values: ...,−3, −2, −1, 0, 1, 2, 3,...

Rational Numbers: Any number that can be expressed as a ratio (or fraction) of two integers is a rational number. This set includes all integers and fractions where the denominator is not zero. Examples include: 34, −2, $-\frac{5}{7}$, 0.75, $\frac{49}{57}$, and so on.

Irrational Numbers: These are numbers that cannot be expressed as simple fractions. They have non-terminating, non-repeating decimal representations. Famous examples include:

- $\sqrt{2}$ (the square root of 2)
- π (the ratio of a circle's circumference to its diameter)

To recap, real numbers encompass all rational and irrational numbers. This vast collection allows for accurate representation and measurement of distances, quantities, and other numerical concepts.

Absolute Value

The absolute value of a number is its distance from zero on the number line, regardless of direction. It is always a non-negative value. The notation for absolute value is two vertical bars surrounding the number, like this: $|x|$.

Examples:

- $|5| = 5$ because 5 is 5 units away from zero.
- $|-8| = 8$ because -8 is 8 units away from zero.
- $|0| = 0$ because 0 is exactly at zero.

Absolute value is handled similarly to parentheses in the order of operations (PEMDAS):

- P: Parentheses (including absolute value bars)
- E: Exponents (including roots)
- MD: Multiplication and Division (left to right)
- AS: Addition and Subtraction (left to right)

This means that any operation inside the absolute value bars should be evaluated first, and the result's absolute value is taken afterward.

Examples:

- Simple Absolute Value: $|-4| = 4$
- Absolute Value with Expressions: $|5 - 8| = |-3| = 3$
- Combining Absolute Value with Other Operations:

$$3 \times |4 - 6| + 2 = 3 \times |-2| + 2 = 3 \times 2 + 2 = 6 + 2 = 8$$

- Absolute Value with Nested Operations: $|(2^3 - 10)| = |8 - 10| = |-2| = 2$

In all these cases, operations inside the absolute value bars are completed first, followed by taking the absolute value itself and then any other arithmetic operations outside the bars.

Rates

In arithmetic, a rate is a specific kind of ratio that compares two quantities with different units. It is a measure of one quantity relative to another, allowing us to understand how one variable changes with respect to another. Rates are commonly used to describe things like speed, price per unit, or productivity, etc.

Examples of Rates:

1. **Speed (Distance/Time):**
   - Speed is often expressed as a rate, such as miles per hour (mph) or kilometers per hour (km/h).

     Example: If a car travels 120 miles in 2 hours, its speed is: $\frac{120\ miles}{2\ hours} = 60\ mph$
2. **Price (Cost/Unit):**
   - Prices are rates that indicate how much something costs per unit.

     Example: If a 10-pound bag of apples costs $20, the cost per pound is:

$$\frac{20\ dollars}{10\ pounds} = 2\ dollars\ per\ pound$$

3. **Productivity (Output/Time):**
   - Productivity rates describe the output or result produced within a given period.

     Example: If a worker assembles 300 items in 8 hours, the productivity rate is:

$$\frac{300\ items}{8\ hours} = 37.5\ items\ per\ hour$$

Unit Conversion

When you know the rate between two units, you can easily convert from one to the other. This is particularly useful when dealing with measurements like speed, currency, and volume.

Example: Suppose you have a speed of 60 miles per hour (mph) and want to convert it to kilometers per hour (km/h). The conversion rate is: 1 mile=1.60934 kilometers.

To convert 60 mph to km/h, multiply by the conversion factor:

$$60\, mph \times 1.60934\, km/mile = 96.56\, km/h$$

Therefore, 60 mph is equivalent to approximately 96.56 km/h.

Unit Rate

A unit rate is a rate where the denominator is reduced to 1. For example, if the speed of a vehicle is 60 mph, the rate is a unit rate—"miles per one hour," or simply "miles per hour". This contrasts with "miles per every two hours", or "miles per every 10 minutes", which are also rates, but not unit rates.

Ratios and Proportional Relationships

A **ratio** is a comparison of two or more numbers, often representing how many times one quantity is contained within another. Ratios can be written in different ways, including:

- Fraction Form: $\frac{a}{b}$
- Colon Form: a:b
- Word Form: "a to b"

Example: 3/5 or 3:5 or "3 to 5" are all ratios expressed differently but are equal in value.

In practice, you might encounter questions involving simplifying ratios, finding equivalent ratios, or comparing different ratios.

A **proportional relationship** occurs when two quantities always have the same ratio or are directly proportional. In other words, as one quantity increases or decreases, the other changes at a consistent rate.

These relationships can often be expressed as: $y = kx$, where y and x are the variables in the relationship, k is the constant of proportionality.

Example: If a recipe calls for 3 cups of flour to make 24 cookies, you can determine the cups of flour needed for 48 cookies through a proportion:

$$\frac{3}{24} = \frac{x}{48}$$

Cross-multiplying gives:

$$x = \frac{3 \times 48}{24} = 6$$

This shows that 6 cups of flour are needed to make 48 cookies.

You may get asked these types of questions on the test:

- **Proportional Reasoning**: Questions may ask you to determine the missing value in a proportion or identify if two sets of values are proportional.
- **Rates and Unit Rates**: These could involve calculating unit prices, speeds, or other rates.
- **Scale Factors**: You might need to work with maps, diagrams, or blueprints involving scaling.

Exponents

Exponents represent repeated multiplication of a base number. Understanding different types of exponents and their applications is fundamental in algebra and scientific calculations. Here's an overview of various exponent concepts.

Positive Whole-Number Exponents

A positive whole-number exponent represents how many times a base number is multiplied by itself. It's a form of repeated multiplication that simplifies large calculations.

General Form: If a is the base and n is the exponent (a positive whole number), then:

$$a^n = a \times a \times a \times \dots (n \text{ times})$$

Examples:

- $2^3 = 2 \times 2 \times 2 = 8$
- $5^4 = 5 \times 5 \times 5 \times 5 = 625$
- $10^2 = 10 \times 10 = 100$

Properties of Positive Whole-Number Exponents:

1. **Multiplying with the Same Base:** When multiplying two expressions with the same base, add their exponents: $a^m \times a^n = a^{m+n}$

   Example: $2^3 \times 2^2 = 2^{3+2} = 2^5 = 32$

2. **Dividing with the Same Base:** When dividing two expressions with the same base, subtract their exponents: $a^m \div a^n = a^{m-n}$

   Example: $7^5 \div 7^2 = 7^{5-2} = 7^3 = 343$

3. **Power of a Power:** When raising an expression with an exponent to another power, multiply the exponents: $(a^m)^n = a^{m \times n}$

   Example: $(3^2)^3 = 3^{2\times3} = 3^6 = 729$

Zero Exponents

Any nonzero base raised to the power of zero equals 1. This rule simplifies calculations involving expressions with zero exponents.

General Rule: $a^0 = 1$, where $a \neq 0$

Example: $(-5)^0 = 1$

This rule applies because, by definition, an expression like $a^n \div a^n$ equals 1.

Meanwhile, $a^n \div a^n = a^{(n-n)} = a^0$. Hence, $a^0 = 1$.

Negative Exponents

A negative exponent represents the reciprocal of a base raised to the corresponding positive exponent. This concept flips the base to its reciprocal and changes the sign of the exponent to positive. Sounds confusing? The example below will clarify it.

General Rule: For any nonzero number a and positive integer n:

$$a^{-n} = \frac{1}{a^n}$$

Examples:

- $4^{-2} = \frac{1}{4^2} = \frac{1}{16}$

- $10^{-3} = \frac{1}{10^3} = \frac{1}{1000} = 0.001$
- $(2x)^{-3} = \frac{1}{(2x)^3} = \frac{1}{8x^3}$

Properties of Negative Exponents:

1. **Multiplying with the Same Base:** When multiplying two expressions with the same base, add the exponents even if one or both are negative: $a^m \times a^{-n} = a^{m-n}$

   Example: $5^3 \times 5^{-2} = 5^{3-2} = 5^1 = 5$

2. **Dividing with the Same Base:** When dividing two expressions with the same base, subtract the exponents: $\frac{a^m}{a^{-n}} = a^{m+n}$

   Example: $\frac{2^4}{2^{-2}} = 2^{4+2} = 2^6 = 64$

3. **Power of a Power:** When raising a base with a negative exponent to another power, multiply the exponents: $(a^{-m})^n = a^{-m \times n}$

   Example: $(3^{-2})^4 = 3^{-8} = \frac{1}{3^8}$

Negative exponents allow for easy representation of reciprocals and small values. They are commonly used in scientific notation and simplify algebraic expressions involving division and reciprocal relationships.

Radicals

Radicals represent roots of numbers and provide a way to express roots in a simplified form. The most common type is the square root, but radicals can represent other roots like cube roots, fourth roots, etc.

Examples:

- **Square Root** ($\sqrt{x}$): The square root of a number x is a value y such that $y^2 = x$. Example: $\sqrt{16} = 4$ because $4^2 = 16$.
- **Cube Root** ($\sqrt[3]{x}$): The cube root of a number x is a value y such that $y^3 = x$.

  Example: $\sqrt[3]{27} = 3$ because $3^3 = 27$.
- **General Roots**: Other roots follow the same pattern, $\sqrt[n]{x}$ represents a radical where n is called the **root** of the radical.

  Example: $\sqrt[4]{81} = 3$

Properties of Radicals:

1. **Product Rule:** The product of two radicals can be combined into a single radical: $\sqrt[n]{a} \times \sqrt[n]{b} = \sqrt[n]{a \times b}$

   Example: $\sqrt{4} \times \sqrt{9} = \sqrt{36} = 6$

2. **Quotient Rule:** The quotient of two radicals can also be combined: $\sqrt[n]{\frac{a}{b}} = \frac{\sqrt[n]{a}}{\sqrt[n]{b}}$.

   Example: $\frac{\sqrt{25}}{\sqrt{4}} = \sqrt{\frac{25}{4}} = \frac{5}{2}$

Fractional Exponents

Fractional exponents are another way to represent roots or radicals. Instead of using the radical symbol, a root is expressed as an exponent in fractional form. This notation provides a concise way to represent both roots and powers.

General Form:

If a is the base and $\frac{m}{n}$ is the fractional exponent: $a^{\frac{m}{n}} = \sqrt[n]{a^m}$

where: n is the root (index), m is the power to which the base is raised before taking the root.

Examples:

- **Square Root** (Fractional Exponent as $\frac{1}{2}$): The square root of a number a is expressed using a fractional exponent as: $a^{\frac{1}{2}} = \sqrt{a}$.

  Example: $25^{\frac{1}{2}} = \sqrt{25} = 5$

- **Cube Root** (Fractional Exponent as $\frac{1}{3}$): The cube root of a number a is expressed using a fractional exponent as: $a^{\frac{1}{3}} = \sqrt[3]{a}$

  Example: $27^{\frac{1}{3}} = \sqrt[3]{27} = 3$.

- **Combining Roots and Powers:** A fractional exponent like $\frac{3}{4}$ indicates that the base should first be raised to the power of 3, and then the fourth root is taken: $a^{\frac{3}{4}} = \sqrt[4]{a^3}$.

  Example: $16^{\frac{3}{4}} = \sqrt[4]{16^3} = \sqrt[4]{4096} = 8$

Powers of 10

Powers of 10 refer to multiplying the base 10 by itself a certain number of times. They are especially important because our number system is based on powers of 10. This notation helps represent very large or very small numbers conveniently.

General Form:

If n is a positive integer, a power of 10 is expressed as: $10^n = 10 \times 10 \times 10 \times \ldots$ (n times)

Examples: Positive Powers of 10

- $10^1 = 10$
- $10^3 = 10 \times 10 \times 10 = 1000$
- $10^6 = 1{,}000{,}000$ (one million)

A positive power of 10 shows how many zeros follow the number 1 in the standard form we write numbers, e.g., 1000, 1,000,000.

Examples: Negative Powers of 10

- $10^{-1} = \frac{1}{10} = 0.1$
- $10^{-2} = \frac{1}{10^2} = \frac{1}{100} = 0.01$
- $10^{-6} = 0.000001$ (one millionth)

A negative power of 10 shows how many decimal places the 1 is shifted to the left of the decimal point.

Scientific Notation

Powers of 10 are used extensively in scientific notation, a method to express very large or very small numbers. Scientific notation combines a coefficient and a power of 10: $a \times 10^n$, where a is a number between 1 and 10, n is an integer representing the power of 10.

Examples:

- The mass of Earth is approximately 5.97×10^{24} kilograms, a very large number.
- The mass of a hydrogen atom is about 1.67×10^{-27} kilograms, a very small number.

Calculations involving scientific notation follow the same rules as ordinary exponents. Here's a guide to handling scientific notation in various arithmetic operations:

Multiplication: To multiply numbers in scientific notation, multiply the coefficients and add the exponents of the powers of 10.

Example: $(2.5 \times 10^3) \times (4 \times 10^2)$

Solution:

1. Multiply the coefficients: $2.5 \times 4 = 10$
2. Add the exponents: $10^3 \times 10^2 = 10^{3+2} = 10^5$
3. Combine the results: $10 \times 10^5 = 1.0 \times 10^6$

Division: To divide numbers in scientific notation, divide the coefficients and subtract the exponents of the powers of 10.

Example: $\frac{4.8 \times 10^5}{2 \times 10^3}$

Solution:

1. Divide the coefficients: $\frac{4.8}{2} = 2.4$
2. Subtract the exponents: $10^5 \div 10^3 = 10^{5-3} = 10^2$
3. Combine the results: 2.4×10^2

Addition and Subtraction: When adding or subtracting numbers in scientific notation, make sure the exponents are the same before combining the coefficients.

Example: $(3.5 \times 10^4) + (2.3 \times 10^3)$

Solution:

1. Adjust the second term so the exponents match: $2.3 \times 10^3 = 0.23 \times 10^4$
2. Add the coefficients: $3.5 + 0.23 = 3.73$
3. Combine with the power of 10: 3.73×10^4

Summary of the Exponent Operation Rules

Now that we have spent so much time on exponents, let's summarize all the fundamental rules of exponent operations in one table.

Rule	Example
$x^n x^m = x^{n+m}$	$7^3 \cdot 7^2 = 7^5$
$\frac{x^n}{x^m} = x^{n-m}$	$\frac{5^5}{5^3} = 5^{5-3} = 5^2 = 25$
$x^0 = 1 \; provided \; x \neq 0$	$39^0 = 1$
$(x \cdot y)^n = x^n \cdot y^n$	$(3 \cdot 4)^2 = 3^2 \cdot 4^2 = 9 \cdot 16 = 144$
$\left(\frac{x}{y}\right)^n = \frac{x^n}{y^n}$	$\left(\frac{1}{3}\right)^2 = \frac{1^2}{3^2} = \frac{1}{9}$
$(x^n)^m = x^{n \cdot m}$	$(2^3)^4 = 2^{3 \cdot 4} = 2^{12}$
$x^{-n} = \frac{1}{x^n} \; provided \; X \neq 0$	$10^{-3} = \frac{1}{10^3}$

§2. ALGEBRAIC REASONING

Algebra is a branch of mathematics that focuses on the manipulation of symbols and variables to represent and solve equations and expressions. It extends the basic principles of arithmetic by using letters (variables) to stand in for numbers. This allows for the generalization of mathematical concepts and relationships. Key concepts of Algebra include the following.

Variables: Symbols (typically letters) that represent unknown or general values. Examples include x and y.

Coefficients: Numerical values that multiply the variables. For instance, in the expression $3x$, 3 is the coefficient, meaning the variable x is multiplied by 3.

Terms: A term is a single mathematical expression involving a number (coefficient), a variable, or both, separated by addition or subtraction. Examples of terms include: $4x, -5y, 7$.

Expressions: Combinations of terms involving variables, numbers, and arithmetic operations (addition, subtraction, multiplication, and division). Example: $3x + 2y - 5$. This expression includes three terms: $3x, 2y, -5$. Notice that a minus sign is always included with the term that it immediately precedes. In this case, the third term is -5. To make it easier, the forgoing expression can be alternatively written as: $3x + 2y + (-5)$.

Equations: Statements that two expressions are equal, often containing one or more unknowns (variables). Example: $2x + 3 = 7$.

Inequalities: Statements that compare two expressions using inequality symbols such as greater than (>), less than (<), greater than or equal to (≥), or less than or equal to (≤). Example: $4x - 1 < 9$

Functions: Relationships between two sets of variables, usually expressed as a rule or equation. Example: $f(x) = x^2 - 4$.

EVALUATING ALGEBRAIC EXPRESSIONS

Evaluating algebraic expressions means finding the value of an expression by substituting specific values for the variables involved and then performing the necessary calculations.

Steps for Evaluating Algebraic Expressions:

1. **Identify the Expression:** The expression is a combination of variables, coefficients, constants, and arithmetic operators like addition, subtraction, multiplication, and division.
2. **Substitute Values:** Replace the variables with specific numerical values provided in the problem.
3. **Perform the Calculations:** Follow the order of operations (PEMDAS): Parentheses, Exponents, Multiplication and Division (from left to right), Addition and Subtraction (from left to right).

Example: Given $x = 2$ and $y = 4$, evaluating $3x + 5y - 7$.

Solution: Substitute the values of x and y into the above expression:

$$3x + 5y - 7 = 3 \cdot 2 + 5 \cdot 4 - 7 = 6 + 20 - 7 = 19$$

OPERATIONS OF ALGEBRAIC EXPRESSIONS

In algebra, **like terms** are terms that contain the same variables raised to the same power. The coefficients (numbers in front of variables) can be different. For instance, in the expression:

$3x^2 + 4x - 5 + 7x^2 - 2x + 8$, like terms are as follows:

- $3x^2$ and $7x^2$ are like terms because both contain the variable x raised to the power of 2.
- $4x$ and $-2x$ are like terms because both contain the variable x raised to the power of 1.
- -5 and 8 are like terms because they are constants (terms without variables).

Addition and Subtraction

To add or subtract algebraic expressions, combine like terms.

Example: Simplify $3x^2 + 4x - 5 + 7x^2 - 2x + 8$

Solution:

1. Combine like terms involving x^2: $3x^2 + 7x^2 = 10x^2$
2. Combine like terms involving x: $4x - 2x = 2x$
3. Combine the constants: $-5 + 8 = 3$
4. Final Result: $10x^2 + 2x + 3$

Multiplication

To multiply algebraic expressions, apply the distributive property and combine like terms.

Example: Multiply the following: (x+3)(x−2)

Solution:

1. Distribute x over x−2: $x \cdot (x - 2) = x^2 - 2x$
2. Distribute 3 over $x - 2$: $3 \cdot (x - 2) = 3x - 6$
3. Add the results: $x^2 - 2x + 3x - 6 = x^2 + x - 6$

The above process is also called the **FOIL** method of multiplying two binomials. The acronym "FOIL" stands for: First, Outer, Inner, Last.

- First: Multiply the first terms of each binomial.
- Outer: Multiply the outer terms of each binomial.
- Inner: Multiply the inner terms of each binomial.
- Last: Multiply the last terms of each binomial.

In some cases, an algebraic expression is complicated by parentheses, and simplifying it requires removing the parentheses. To achieve this, distribute the term that directly precedes the parentheses by multiplying it with each term inside. Let's tackle an example where a polynomial is multiplied by a binomial:

Example: Multiply the following: $(2x^2 - 3x + 4)(3x + 5)$

Solution: Distribute each term in the polynomial $(2x^2 - 3x + 4)$ over the terms in the binomial $(3x + 5)$.

1. **First Term:** Distribute $2x^2$

   $2x^2 \cdot 3x + 2x^2 \cdot 5 = 6x^3 + 10x^2$

2. **Second Term:** Distribute $-3x$

   $-3x \cdot 3x + (-3x) \cdot 5 = -9x^2 - 15x$

3. **Third Term:** Distribute 4

   $4 \cdot 3x + 4 \cdot 5 = 12x + 20$

4. **Combine the Results:**

   $6x^3 + 10x^2 - 9x^2 - 15x + 12x + 20$

5. **Simplify by Combining Like Terms:**

   $6x^3 + (10x^2 - 9x^2) + (-15x + 12x) + 20 = 6x^3 + x^2 - 3x + 20$

Division

For division, expressions are often divided through factorization or by reducing fractions.

Example: Divide the following: $\frac{2x^2-8}{2x}$

Solution:

1. Factor out 2 from the numerator: $2x^2 - 8 = 2(x^2 - 4)$
2. Recognize that $x^2 - 4$ is a difference of squares: $x^2 - 4 = (x + 2)(x - 2)$
3. Substitute back to rewrite the original expression: $\frac{2(x+2)(x-2)}{2x}$
4. Cancel out the common factor of 2 and reduce: $\frac{(x+2)(x-2)}{x}$

GCF Factoring

The greatest common factor (GCF) of a set of terms is the largest expression that divides each term evenly. Factoring out the GCF is the most basic type of polynomial factoring and simplifies expressions by grouping common factors.

Steps to Factor Out the GCF:

1. **Identify the GCF**: Determine the largest factor shared by all terms in the polynomial.
2. **Factor Out the GCF**: Divide each term by the GCF, leaving a simpler polynomial inside parentheses.
3. **Rewrite**: Multiply the GCF by the simplified polynomial inside the parentheses.

Example 1: Factor out the GCF from: $12x^3 - 18x^2 + 24x$

Solution:

1. Identify the GCF: The GCF of 12, 18, and 24 is 6. The common variable is x, and the smallest power among all terms is x. Thus, the GCF is $6x$.
2. Factor Out the GCF: Divide each term by $6x$:

   $12x^3 \div 6x = 2x^2, \quad 18x^2 \div 6x = 3x, \quad 24x \div 6x = 4$
3. Rewrite the Expression: $12x^3 - 18x^2 + 24x = 6x(2x^2 - 3x + 4)$

Example 2: GCF Factoring to Simplify Division: $\frac{15\ ^3+20x^2}{5x}$

Solution:

1. Factor Out the GCF (Numerator): The GCF of the numerator, $15x^3 + 20x^2$, is $5x^2$. Thus, factor it out: $15x^3 + 20x^2 = 5x^2(3x + 4)$
2. Rewrite the Expression: $\frac{15x^3+20x^2}{5x} = \frac{5x^2(3x+4)}{5x}$
3. Simplify the Division: Cancel out the common factor of 5x: $\frac{5x^2}{5x} = x$,

   leaving: $x(3x + 4) = 3x^2 + 4x$.

   Hence, $\frac{15x^3+20x^2}{5x} = \frac{5x^2(3x+4)}{5x} = x(3x + 4) = 3x^2 + 4x$.

FACTORING

Factoring is a fundamental algebraic process used to simplify expressions, solve equations, and understand polynomial functions. Factoring involves breaking down a complex expression into simpler, multiplied components known as factors. These factors, when multiplied together, reconstruct the original expression. Factoring can simplify calculations and reveal useful properties and relationships within algebraic expressions.

Greatest Common Factor (GCF) and Factoring by Grouping

Mastering the techniques of finding the greatest common factor (GCF) and factoring by grouping are essential for simplifying and solving algebraic expressions. This section will guide you through both processes, helping you break down more complex polynomial expressions.

Finding the Greatest Common Factor (GCF)

The GCF of two or more expressions is the largest expression that divides each of the original expressions without leaving a remainder. Finding the GCF is often the first step in factoring polynomials effectively.

Steps to Find the GCF:

1. **List Factors**: List all factors (or use prime factorization) of each term in the expression.
2. **Identify Common Factors**: Identify the largest factor that is common to all terms.

Example: Finding the GCF of $12x^2$ and $18x$

Solution:

1. Prime Factorization:

$$12x^2 = 2 \times 2 \times 3 \times x \times x$$

$$18x = 2 \times 3 \times 3 \times x$$

2. Common Factors: The common prime factors are 2, 3, and x.
3. Greatest Common Factor: The GCF is $6x$ (since $2 \times 3 \times x = 6x$).

Factoring by Grouping

Factoring by grouping involves rearranging and grouping terms in a polynomial to simplify factoring, especially useful for polynomials with four or more terms.

Steps to Factor by Grouping:

1. **Group the Terms**: Arrange the polynomial into groups that have a common factor.
2. **Factor Out the GCF from Each Group**: Apply the GCF method within each group.
3. **Factor Out the Common Binomial Factor**: Look for and factor out the common binomial factor from the grouped terms.

Example: Factoring $x^3 + 3x^2 + 2x + 6$ by Grouping

Solution:

1. Group the Terms: $x^3 + 3x^2$ and $2x + 6$
2. Factor Out the GCF from Each Group:
   - From $x^3 + 3x^2$, factor out x^2, resulting in $x^2(x + 3)$
   - From $2x + 6$, factor out 2, resulting in $2(x + 3)$
3. Factor Out the Common Binomial Factor: The common factor is $x + 3$, hence:

$$\begin{aligned} x^3 + 3x^2 + 2x + 6 &= x^2(x+3) + 2(x+3) \\ &= (x^2 + 2)(x + 3) \end{aligned}$$

By understanding and applying these methods, you can efficiently simplify and solve a variety of polynomial expressions.

Factoring Quadratics $x^2 + bx + c$

Quadratic expressions in the standard form $x^2 + bx + c$ are a common sight in algebra. These are polynomials where the highest exponent is 2, and they do not have a coefficient in front of x^2. The goal of factoring such quadratics is to express them as the product of two binomials. This section will guide you through the process of factoring these types of quadratics using a straightforward method that involves finding two numbers.

The key to factoring a quadratic expression of the form $x^2 + bx + c$ is to find two numbers that multiply to c (the constant term) and add up to b (the coefficient of x). These numbers will be used to create the binomials.

Steps to Factor $x^2 + bx + c$:

1. **Identify the Numbers**: Find two numbers that multiply to give the constant term c, and at the same time, add up to the coefficient b.
2. **Write the Binomials**: Use these numbers to write two binomials. The binomials will have the form $(x + m)(x + n)$, where m and n are the numbers identified in the first step.

Example: Factor $x^2 + 5x + 6$

1. Identify the Numbers: We need two numbers that multiply to 6 (the constant term) and add to 5 (the coefficient of x). After checking possible pairs, we find that 2 and 3 fit the requirements because $2 \times 3 = 6$ and $2 + 3 = 5$.
2. Write the Binomials: Using the numbers 2 and 3, we can write the quadratic as $(x + 2)(x + 3)$.

Check the Factorization: To verify, expand $(x + 2)(x + 3)$ using the distributive property (also known as the FOIL method for binomials):

$$(x + 2)(x + 3) = x^2 + 3x + 2x + 6 = x^2 + 5x + 6$$

The expansion matches the original expression, confirming our factorization is correct.

Factoring Quadratics $ax^2 + bx + c$

Factoring quadratic expressions where the leading coefficient (a) is not 1, such as in the general form $ax^2 + bx + c$, can be more challenging than factoring when $a = 1$. This section will introduce a method known as the "ac method" or "splitting the middle term," which is effective for tackling these more complex quadratics.

The ac method involves manipulating the middle term of the quadratic by finding two numbers that multiply to ac (the product of the coefficients of x^2 and the constant term) and add up to b (the coefficient of x). These numbers are then used to split the middle term, allowing the expression to be factored by grouping.

Steps to Factor $ax^2 + bx + c$:

1. **Multiply and Find**: Calculate ac and find two numbers that multiply to ac and add up to b.
2. **Split the Middle Term**: Use these numbers to split the middle term into two terms.

$$ax^2 + bx + c = ax^2 + nx + mx + c$$

3. **Group and Factor**: Group the terms in pairs and factor out the greatest common factor from each group: $(ax^2 + nx) + (mx + c)$
4. **Factor Out the Common Binomial Factor**: Extract the common binomial factor from the grouped terms to complete the factorization.

Example: Factor $6x^2 + 11x + 4$

Solution:

1. Multiply and Find: Calculate $ac = 6 \times 4 = 24$.
2. Find two numbers that multiply to 24 and add to 11. These numbers are 8 and 3.
3. Split the Middle Term: Rewrite the middle term using 8 and 3: $6x^2 + 8x + 3x + 4$.
4. Group and Factor: Group the terms: $(6x^2 + 8x) + (3x + 4)$.
5. Factor out the GCF from each group: $2x(3x + 4) + 1(3x + 4)$.
6. Factor Out the Common Binomial Factor: The common factor is $(3x + 4)$, so the expression factors as $(2x + 1)(3x + 4)$.

$$\begin{aligned} 6x^2 + 11x + 4 &= 2x(3x + 4) + 1(3x + 4) \\ &= (2x + 1)(3x + 4) \end{aligned}$$

By following these steps, you can successfully factor quadratic expressions with a leading coefficient greater than one.

Factoring Perfect Square Trinomials

Perfect square trinomials are a special form of quadratic expressions that result from squaring a binomial. These trinomials are always in the form of $ax^2 + 2abx + b^2$ or $ax^2 - 2abx + b^2$ where the square of the first term and the square of the last term are perfect squares, and the middle term is twice the product of the square roots of these squares. Factoring perfect square trinomials is straightforward once you recognize the pattern.

Steps to Factor $ax^2 + 2abx + b^2$ or $ax^2 - 2abx + b^2$

To factor a perfect square trinomial, you need to identify whether it can be written as the square of a binomial. Here are the steps to follow:

1. **Identify the Squares:** Ensure that the first and last terms are perfect squares.
2. **Check the Middle Term:** Verify that the middle term is twice the product of the roots of the first and last terms.
3. **Write the Binomial:** Express the trinomial as the square of a binomial.

Example: Factor $x^2 + 6x + 9$

Solution:

1. Identify the Squares: The first term x^2 is the square of x, and the last term 9 is the square of 3.
2. Check the Middle Term: The middle term $6x$ should be twice the product of x and 3, which it is: $2 \cdot 3 \cdot x = 6x$.
3. Write the Binomial: Since all conditions are satisfied, the trinomial can be written as the square of a binomial: $(x + 3)^2$.

$$x^2 + 6x + 9 = (x + 3)^2$$

Factoring Differences of Squares $a^2 - b^2$

The difference of squares is a common algebraic pattern that is simple to factor once recognized. This pattern applies to expressions where two terms are squared and subtracted from each other, taking the general form $a^2 - b^2$. Factoring differences of squares relies on a fundamental algebraic identity, making it a quick and effective process.

Steps to Factor $a^2 - b^2$

To factor a difference of squares, you utilize the identity $a^2 - b^2 = (a - b)(a + b)$. Here are the steps to factor such expressions:

1. **Identify the Squares**: Ensure both terms are perfect squares.
2. **Apply the Difference of Squares Formula**: Write the expression as the product of two binomials, one representing the sum of the square roots and the other the difference of the square roots.

Example: Factor $x^2 - 16$

Solution:

1. Identify the Squares: The first term, x^2, is the square of x, and the last term, 16, is the square of 4.
2. Apply the Difference of Squares Formula: Since x^2 and 16 are perfect squares, apply the formula $a^2 - b^2 = (a - b)(a + b)$ with $a = x$ and $b = 4$.

$$x^2 - 16 = (x - 4)(x + 4)$$

This straightforward factorization shows how the difference of squares identity simplifies the process, quickly breaking down the expression into a product of binomials.

Factoring the Sum or Difference of Cubes

Factoring the sum or difference of cubes involves breaking down expressions that are either the sum or the subtraction of two cubed terms. These expressions take the forms $a^3 + b^3$ and $a^3 - b^3$, respectively. The factoring of these forms is based on specific algebraic identities, which allow for simplification into products of binomials and trinomials.

Factoring $a^3 + b^3$ and $a^3 - b^3$

To factor the sum or difference of cubes, you use the following identities:

- For the sum of cubes: $a^3 + b^3 = (a + b)(a^2 - ab + b^2)$
- For the difference of cubes: $a^3 - b^3 = (a - b)(a^2 + ab + b^2)$

The mnemonic "SOAP" is a helpful way to remember the signs used in the formulas for factoring the sum or difference of cubes. There are a total of three plus or minus signs to the right of the equal sign in the above two formulas. Here's how each letter in the mnemonic corresponds to these three plus or minus signs in the formulas:

1. **S - Same**: This stands for the first sign in the binomial part of the factorization formula, which is the same as the sign in the original cubic expression. For instance:
   - In $a^3 + b^3$, the sign between a and b in the binomial $(a + b)$ is positive.
   - In $a^3 - b^3$, the sign between a and b in the binomial $(a - b)$ is negative.
2. **O - Opposite**: This represents the first sign in the trinomial part of the factorization formula, which is the opposite of the sign in the binomial. For example:
   - In $a^3 + b^3 = (a + b)(a^2 - ab + b^2)$, the sign after a^2 is negative, which is the opposite of the positive sign in $(a + b)$.
   - In $a^3 - b^3 = (a - b)(a^2 + ab + b^2)$, the sign after a^2 is positive, opposite the negative sign in $(a - b)$.
3. **AP - Always Positive**: This refers to the last sign in the trinomial part of the factorization formula, which is always positive, regardless of whether the expression is a sum or a difference of cubes. This means:
   - Both $a^2 - ab + b^2$in the sum of cubes and $a^2 + ab + b^2$ in the difference of cubes end with a positive term $(+b^2)$.

Example: Factor $x^3 - 27$

Solution:

1. Identify the Cubes: The term x^3 is the cube of x, and 27 is the cube of 3.
2. Apply the Difference of Cubes Formula: Since the expression involves subtraction, use the difference of cubes formula: $a^3 - b^3 = (a - b)(a^2 + ab + b^2)$ with $a = x$ and $b = 3$.

$$x^3 - 27 = (x - 3)(x^2 + 3x + 9)$$

General Strategy of Factoring Polynomials

Factoring polynomials efficiently requires a systematic approach, ensuring each polynomial is broken down into its simplest form. Below is a structured method to guide you through the factoring process:

Step 1—Identify the Greatest Common Factor (GCF): Check if there is a GCF in all terms of the polynomial. If one exists, factor it out first. This simplifies the polynomial, reducing the complexity of further factoring steps.

Step 2—Analyze the Structure of the Polynomial: Determine the type of polynomial and apply the appropriate factoring technique:

<u>Binomial Factors:</u>

Sum of Squares: Note that sums of squares generally do not factor over the real numbers.

Sum of Cubes: Apply the sum of cubes formula: $a^3 + b^3 = (a + b)(a^2 - ab + b^2)$

Difference of Squares: Factor as the product of conjugates using $a^2 - b^2 = (a - b)(a + b)$

Difference of Cubes: Use the difference of cubes formula: $a^3 - b^3 = (a - b)(a^2 + ab + b^2)$

<u>Trinomial Factors:</u>

Simple Form $(x^2 + bx + c)$: Undo the FOIL process by finding two numbers that multiply to c and add to b.

Complex Form ($ax^2 + bx + c$): Check if a and c are perfect squares and fit the trinomial square pattern where the first term, ax^2, and the last term, c, must both be perfect squares, and the middle term, bx, must be twice the product of the square roots of ax^2 and c.

Otherwise, use trial and error or the "ac" method, finding two numbers that multiply to ac and add to b, then group and factor.

Polynomials with More than Three Terms:

Grouping Method: Group terms to create factorable chunks, typically aiming for common factors or binomial factors within the groups.

Step 3—Verification: Check if the polynomial is factored completely. Ensure that no further factorization is possible. Multiply the factors to verify that they combine to form the original polynomial.

The following table summarizes the factoring methods we have covered, and the general strategy of factoring polynomials.

Table: General Strategy of Factoring Polynomials

Category	Type	Formula/Method
GCF		Always check for and factor out the GCF first.
Binomial	Difference of Squares	$a^2 - b^2 = (a - b)(a + b)$
	Sum of Squares	Sums of squares do not factor (over real numbers)
	Sum of Cubes	$a^3 + b^3 = (a + b)(a^2 - ab + b^2)$
	Difference of Cubes	$a^3 - b^3 = (a - b)(a^2 + ab + b^2)$
Trinomial	Simple: $x^2 + bx + c$	Look for two numbers that multiply to c and add to b
	Complex: $ax^2 + bx + c$	Use the 'ac' method
More than 3 terms	Grouping	Group terms to factor by common factors

FUNCTIONS

A function is a relation that uniquely associates each element of a given set, called the domain, with an element in another set, known as the range. In simpler terms, a function takes an input, applies a specific rule to it, and then produces exactly one output. Below are a couple of examples:

- Numerical Example: Consider the function $f(x) = x + 3$. If the input x is 2, then the output $f(2)$ is $2 + 3 = 5$.
- Real-life Example: Consider the relationship between the distance traveled and the amount of fuel used by a vehicle. This function, often referred to as the fuel efficiency function, calculates the fuel consumption based on the distance traveled.

Function Notation

Function notation is a concise way to represent functions. The notation $f(x)$ denotes a function named f evaluated at an input x. Here, f represents the function itself, and x is the variable or input to the function. The expression $f(x)$ represents the output of the function.

Example: In the function $f(x) = x^2 - 4x + 4$,

- $f(0) = 0^2 - 4 \times 0 + 4 = 4$
- $f(2) = 2^2 - 4 \times 2 + 4 = 0$

Different Representations of Functions

Functions can be represented in various forms, each providing different insights:

1. **Verbal Descriptions**: Describing the relationship in words, such as "add three to any number."
2. **Equations**: A mathematical statement like $f(x) = 2x + 1$ that defines the output for each input.
3. **Tables**: Listing input values alongside their corresponding output values.
4. **Graphs**: Visual representations, where the input values are plotted on the x-axis, and the output values are plotted on the y-axis.

Evaluating Linear and Quadratic Functions

Linear Functions

Linear functions are functions of the form $f(x) = mx + b$, where m is the slope and b is the y-intercept. Evaluating a linear function means substituting a value for x and calculating the corresponding y value.

Example: For the function $f(x) = 3x + 2$, find $f(4)$.

To find f(4), substitute x=4: $f(4) = 3(4) + 2 = 14$

Quadratic Functions

Quadratic functions are expressed as $f(x) = ax^2 + bx + c$, with a, b, and c being constants. Evaluating a quadratic function involves substituting the x value into the equation and calculating the result.

Example: For $f(x) = x2 - 5x + 6$, find $f(3)$.

To evaluate $f(3)$, substitute $x = 3$: $f(3) = 3^2 - 5 \cdot 3 + 6 = 0$

Radical Functions

Radical functions are functions that involve a variable within a radical symbol, most commonly a square root. These functions are characterized by the presence of roots, such as square roots, cube roots, or higher-order roots. The general form of a radical function is: $f(x) = \sqrt[n]{x}$, where n is the degree of the root.

Domain of Radical Functions

The domain of a radical function includes all the values of x for which the expression under the radical is defined. This depends on the type of radical:

1. **Square Root Functions**: For a function of the form $f(x) = x$ or $f(x) = \sqrt{g(x)}$ the expression under the square root must be non-negative. Therefore, the domain is found by solving: $g(x) \geq 0$. For example, for$f(x) = \sqrt{x-2}$, the domain is determined by $x - 2 \geq 0$, so $x \geq 2$.
2. **Higher-Order Root Functions**: For functions involving even roots (such as fourth roots), the domain also requires the expression under the radical to be non-negative. For odd roots (such as cube roots), the expression under the radical can be any real number, so the domain is all real numbers.

Range of Radical Functions

The range of a radical function depends on the values that the function can output:

1. **Square Root Functions**: Since the square root function $\sqrt{x}$ produces non-negative values, the range of $f(x) = \sqrt{x}$ is $y \geq 0$. For $f(x) = \sqrt{g(x)}$, the range is determined by evaluating the function over its domain.
2. **Higher-Order Root Functions**: The range of functions involving even roots is also non-negative. For odd roots, the range is all real numbers because these functions can produce both positive and negative values.

Examples

1. Square Root Function: $f(x) = \sqrt{x-1}$
   - Domain: $x - 1 \geq 0 \Rightarrow x \geq 1$
   - Range: $y \geq 0$

2. Cube Root Function: $f(x) = \sqrt[3]{x+2}$

- Domain: All real numbers (since cube roots are defined for all real numbers)
- Range: All real numbers

Determining the domain and range helps in graphing the function accurately and understanding its behavior within the specified intervals.

Simplifying Radical Expressions

Simplifying radical expressions involves rewriting them in their simplest form. This process includes reducing the radicand (the expression under the radical) and rationalizing the denominator if necessary.

Steps to Simplify Radical Expressions

1. **Factor the Radicand**: Break down the expression inside the radical into its prime factors or simpler expressions.
   - For example, $\sqrt{50}$ can be factored as $\sqrt{25 \cdot 2}$
2. **Simplify the Radical**: Apply the property $\sqrt{a \cdot b} = \sqrt{a} \cdot \sqrt{b}$ to separate the factors.
   - Simplify each part: $\sqrt{25 \cdot 2} = \sqrt{25} \cdot \sqrt{2} = 5\sqrt{2}$.
3. **Combine Like Terms**: If you have multiple radical terms, combine like terms by ensuring they have the same radicand.
   - For example, $3\sqrt{2} + 5\sqrt{2} = 8\sqrt{2}$
4. **Rationalize the Denominator**: If a radical expression is in the denominator, multiply the numerator and the denominator by a term that will eliminate the radical in the denominator.
   - For example, to simplify $\frac{1}{\sqrt{3}}$ multiply by $\frac{\sqrt{3}}{\sqrt{3}}$ to get $\frac{\sqrt{3}}{3}$.

Examples:

1. Simplifying a Single Radical: Simplify $\sqrt{72}$
   - Factor $72: \ 72 = 36 \times 2$
   - Simplify: $\sqrt{72} = \sqrt{36 \times 2} = \sqrt{36} \cdot \sqrt{2} = 6\sqrt{2}$.
2. Combining Like Radicals: Simplify $2\sqrt{3} + 4\sqrt{3}$
   - Combine like terms: $2\sqrt{3} + 4\sqrt{3} = (2+4)\sqrt{3} = 6\sqrt{3}$.
3. Rationalizing the Denominator: Simplify $\frac{5}{\sqrt{7}}$
   - Multiply by $\frac{\sqrt{7}}{\sqrt{7}}$: $\frac{5}{\sqrt{7}} \cdot \frac{\sqrt{7}}{\sqrt{7}} = \frac{5\sqrt{7}}{7}$.
4. Simplifying a Radical with a Variable: Simplify $\sqrt{18x^3}$
   - Factor: $18x^3 = 9 \times 2 \times x^2 \times x$
   - Simplify: $\sqrt{18x^3} = \sqrt{9 \times 2 \times x^2 \times x} = \sqrt{9} \cdot \sqrt{2} \cdot \sqrt{x^2} \cdot \sqrt{x} = 3x\sqrt{2x}$

Tips for Simplifying Radical Expressions

- **Perfect Squares**: Look for perfect squares within the radicand to simplify.
- **Prime Factorization**: Use prime factorization to break down complex radicands.
- **Rationalization**: Always rationalize the denominator if a radical is present.

Graphing Radical Functions

Below are the two most common radical functions along with their corresponding graphs. Although there are countless variations of radical functions, understanding the principles of function transformations allows us to determine their approximate shapes and locations based on these two graphs below. Radical functions with even roots all resemble a transformed (translated, dilated, or flipped) graph of $f(x) = \sqrt{x}$, while those with odd roots resemble a transformed graph of $y = \sqrt[3]{x}$.

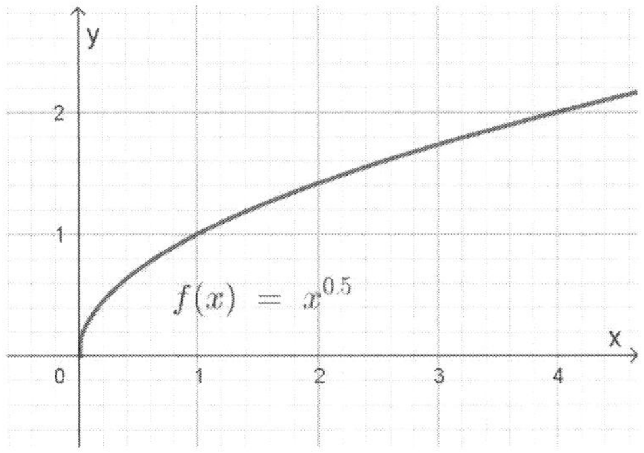

$f(x) = \sqrt{x}$

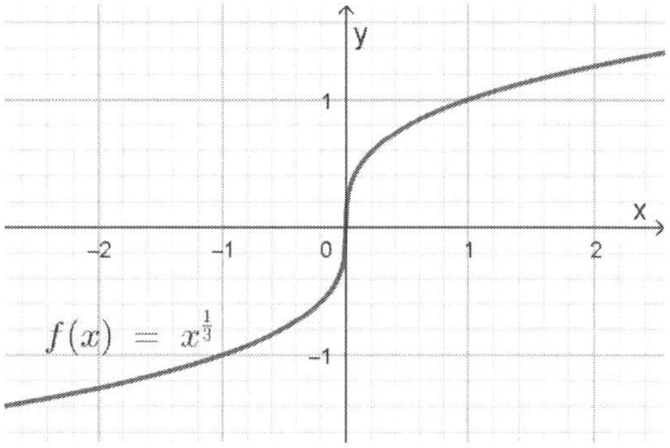

$y = \sqrt[3]{x}$

LINEAR EQUATIONS

A one-variable linear equation is an equation with a single unknown variable, typically denoted by x, and it forms a straight line when graphed on the coordinate plane. The standard form of a one-variable linear equation is: $ax + b = 0$ where: a and b are constants (with $a \neq 0$), and x is the unknown variable.

Solve One-Variable Linear Equations

To solve a linear equation means finding the value of the unknown variable that makes the equation true.

Steps to Solve:

1. **Isolate the Variable**: Rearrange the equation to express the unknown variable (x) on one side of the equation.
2. **Simplify**: Combine like terms or reduce fractions as needed.
3. **Solve for the Variable**: Perform arithmetic operations to obtain the value of the unknown variable.

Example 1—Simple Equation: Solve the equation: $2x + 5 = 13$

Solution:

1. Isolate $2x$: Subtract 5 from both sides.

$$2x + 5 - 5 = 13 - 5$$
$$2x = 8$$

2. Solve for x: Divide both sides by 2.

$$x = \frac{8}{2} = 4$$

Example 2—Fractional Coefficients: Solve the equation: $\frac{3x}{5} - 2 = 1$

Solution:

1. Isolate the Fractional Term: Add 2 to both sides.

$$\frac{3x}{5} - 2 + 2 = 1 + 2$$
$$\frac{3x}{5} = 3$$

2. Solve for x: Multiply both sides by 5 to clear the fraction.

$$3x = 3 \cdot 5$$
$$3x = 15$$

3. Divide both sides by 3.

$$x = \frac{15}{3} = 5$$

Example 3— Cross-Multiplication: Solve the equation: $\frac{2x-1}{3} = \frac{x+4}{5}$

Solution:

1. **Cross-Multiply**: Multiply the numerator on one side by the denominator on the other, and vice versa:

$$5 \cdot (2x - 1) = 3 \cdot (x + 4)$$

2. **Simplify Both Sides**: Perform the multiplication:

$$10x - 5 = 3x + 12$$

3. **Isolate the Variable**: To solve for x, get all the x-terms on one side of the equation by subtracting $3x$ from both sides:

$$\begin{aligned} 10x - 3x - 5 &= 12 \\ 7x - 5 &= 12 \end{aligned}$$

4. **Solve for x**: Add 5 to both sides to isolate $7x$, and then divide by 7:

$$\begin{aligned} 7x - 5 + 5 &= 12 + 5 \\ 7x &= 17 \\ x &= \frac{17}{7} \end{aligned}$$

Example 4— Multiplying LCD: Solve the equation: $\frac{x}{4} + \frac{3}{2} = \frac{x-1}{3}$

Solution:

1. **Identify the LCD:** The denominators in the equation are 4, 2, and 3. The least common denominator (LCD) for these is 12.
2. **Multiply Each Term by the LCD:** Multiply both sides of the equation by 12 to clear the fractions.

$$\begin{aligned} 12 \cdot \left(\frac{x}{4} + \frac{3}{2}\right) &= 12 \cdot \frac{x-1}{3} \\ 12 \cdot \frac{x}{4} + 12 \cdot \frac{3}{2} &= 4(x-1) \\ 3x + 18 &= 4x - 4 \end{aligned}$$

3. **Solve for x:**

$$\begin{aligned} 3x - 4x &= -18 - 4 \\ -x &= -22 \\ x &= 22 \end{aligned}$$

Solving Equations in Terms of Other Variables

Sometimes, an equation involves multiple variables, and the goal is to solve for one variable in terms of the others. This allows us to express that variable as a function of the remaining variables. The steps to solve equations in terms of other variables are not any different from what we have already discussed above.

Example: Solve for x in terms of y in the equation: $2x + 3y = 12$

Solution:

$$\begin{aligned} 2x &= 12 - 3y \\ x &= \frac{12 - 3y}{2} \end{aligned}$$

Solving Inequalities

Linear inequalities are similar to linear equations, but instead of using an equal sign (=), they involve inequality signs like <, >, ≤, or ≥. The goal is to find the range of values that satisfy the given inequality.

Interval notation is a method used to represent the set of all solutions to an inequality. It describes the range of numbers that satisfy a condition or a series of conditions. The notation uses parentheses () and square brackets [] to show whether endpoints are included or excluded from the interval.

- **Parentheses** (): Used to exclude endpoints. For instance, (a, b) means all values between a and b, but not including a or b.
- **Square Brackets** []: Used to include endpoints. For example, $[a, b]$ includes both a and b in the range.
- **Infinity** (∞) and **Negative Infinity** ($-\infty$): Represent unbounded intervals. Infinity symbols are always used with parentheses because infinity itself isn't a specific number that can be included.

Examples: Inequality solution and corresponding interval:

- $x > 3$ is equivalent to $(3, \infty)$
- $2 \leq x < 7$ is equivalent to $[2,7)$
- $x \leq -5$ is equivalent to $(-\infty, -5]$

Using interval notation provides a concise and standardized way to describe ranges of values that satisfy various conditions in inequalities.

Basic Steps for Solving Linear Inequalities

1. **Isolate the Variable**: Rearrange the inequality to isolate the unknown variable on one side, while keeping the other terms on the opposite side. This often involves addition, subtraction, multiplication, or division.
2. **Simplify if Needed**: Combine like terms or reduce fractions.
3. **Consider the Direction of the Inequality**: If you multiply or divide both sides of the inequality by a negative number, remember to reverse the direction of the inequality sign.
4. **Write the Solution**: Express the solution in interval notation or using inequality symbols.

Example 1—Simple Inequality: Solve the inequality: $3x - 7 < 5$

Solution:

$$3x - 7 + 7 < 5 + 7$$
$$3x < 12$$
$$x < \frac{12}{3} = 4$$
$$x < 4$$

The solution is all values of x less than 4. Interval Notation: $(-\infty,4)$.

Here is how the solution looks like on a number line:

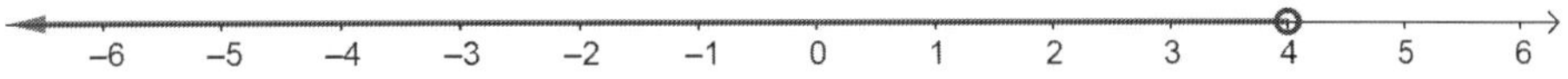

Example 2—Reversing the Inequality: Solve the inequality: $-2x + 8 \geq 4$

Solution:

$$-2x + 8 - 8 \geq 4 - 8$$
$$-2x \geq -4$$

Divide both sides by -2 to solve for x. Remember to reverse the direction of inequality because we are dividing both sides with a negative number.

$$x \leq \frac{-4}{-2} = 2$$
$$x \leq 2$$

The solution is all values of x less than or equal to 2. Interval Notation: $(-\infty, 2]$.

Here is how the solution looks like on a number line:

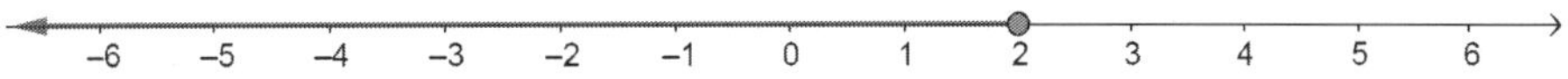

Solving Systems of Equations

A system of two equations consists of two equations with two variables, typically represented as x and y. The goal is to find a solution that satisfies both equations simultaneously.

Methods for Solving Systems of Equations

1. Substitution Method: This method involves solving one equation for one variable and substituting that expression into the other equation to find the second variable.

Example: Solve the following system:

$$\begin{aligned} x + 2y &= 8 \quad (1) \\ 3x - y &= 1 \quad (2) \end{aligned}$$

Solution:

1. Solve equation (1) for x:

$$x = 8 - 2y$$

2. Substitute $x = 8 - 2y$ into equation (2):

$$\begin{aligned} 3(8 - 2y) - y &= 1 \\ 24 - 6y - y &= 1 \\ 24 - 7y &= 1 \\ -7y &= -23 \\ y &= \frac{-23}{-7} = \frac{23}{7} \end{aligned}$$

3. Substitute back to find x:

$$\begin{aligned} x &= 8 - 2 \cdot \frac{23}{7} \\ x &= 8 - \frac{46}{7} = \frac{56}{7} - \frac{46}{7} = \frac{10}{7} \end{aligned}$$

So, the solution is: $\left(\frac{10}{7}, \frac{23}{7}\right)$.

2. Elimination Method: This method involves adding or subtracting the equations after multiplying one or both by suitable factors to cancel out one of the variables.

Example: Solve the following system:

$$\begin{aligned} 2x + 3y &= 11 \quad (1) \\ 4x - y &= 5 \quad (2) \end{aligned}$$

Solution:

1. Multiply equation (2) by 3 to match the coefficients of y in equation (1):

$$\begin{aligned} 3(4x - y) &= 3(5) \\ 12x - 3y &= 15 \end{aligned}$$

2. Add equations (1) and the transformed (2) to eliminate y:

$$\begin{aligned} (2x + 3y) + (12x - 3y) &= 11 + 15 \\ 14x &= 26 \\ x &= \frac{26}{14} = \frac{13}{7} \end{aligned}$$

3. Substitute $x = \frac{13}{7}$ into equation (1) to find y:

$$\begin{aligned} 2 \cdot \frac{13}{7} + 3y &= 11 \\ \frac{26}{7} + 3y &= 11 \\ 3y &= 11 - \frac{26}{7} = \frac{77}{7} - \frac{26}{7} = \frac{51}{7} \end{aligned}$$

$$y = \frac{\frac{51}{7}}{3} = \frac{51}{21} = \frac{17}{7}$$

So, the solution is: $\left(\frac{13}{7}, \frac{17}{7}\right)$.

Quadratic Equations

Quadratic equations are characterized by their standard form $ax^2 + bx + c = 0$, where a, b, and c are coefficients and $a \neq 0$. These equations often appear in various scientific, engineering, and mathematical contexts, representing parabolic shapes when graphed. This section introduces methods to solve quadratic equations, including factoring, using the quadratic formula, and completing the square.

Method 1: Factoring

Factoring involves expressing the quadratic equation in a product form $(px + q)(rx + s) = 0$, where $p, q, r,$ and s are numbers that satisfy the equation. This method works best when the quadratic can be easily decomposed into factors.

Steps:

1. **Write the equation in standard form**: Ensure the equation is in the form $ax^2 + bx + c = 0$.
2. **Factor the quadratic**: Look for two numbers that multiply to ac (the product of the coefficient of x^2 and the constant term) and add up to b (the coefficient of x). Break down the middle term using the identified pairs and factor by grouping.
3. **Set each factor to zero**: Solve $px + q = 0$ and $rx + s = 0$ for x.

Example: Solve $x^2 - 5x + 6 = 0$ by factoring.

Solution:

1. Factor pairs of 6 that add up to -5 are -2 and -3.
2. Factor the quadratic: $(x - 2)(x - 3) = 0$.
3. Set each factor to zero: $x - 2 = 0$ or $x - 3 = 0$, so $x = 2$ or $x = 3$.

Method 2: Quadratic Formula

The quadratic formula is a universal method that can solve any quadratic equation, regardless of its coefficients. The formula is derived from the process of completing the square.

Formula:

$$x = \frac{-b \pm \sqrt{b^2 - 4ac}}{2a}$$

Steps:

1. **Identify coefficients a, b, and c.**
2. **Substitute into the quadratic formula**: Plug the values of a, b, and c into the formula.
3. **Calculate the discriminant**: $b2 - 4ac$. The nature of the roots depends on the discriminant (real and distinct, real and equal, or complex).
4. **Solve for x**: Compute the values using plus and minus versions of the formula.

Example: Solve $2x^2 - 4x - 6 = 0$ using the quadratic formula.

Solution:

1. Substitute $a = 2, b = -4, c = -6$ into the formula.
2. Compute the discriminant: $(-4)^2 - 4 \times 2 \times (-6) = 16 + 48 = 64$.

3. Solve for x: $x = \frac{-(-4) \pm \sqrt{64}}{2 \times 2} = \frac{4 \pm 8}{4}$. So, $x = 3$ or $x = -1$.

Method 3: Completing the Square

Completing the square involves rewriting the quadratic equation in a way that it forms a perfect square trinomial $(dx + e)^2 + f = 0$, which can then be solved by taking square roots.

Solution:

1. **Isolate the quadratic and linear terms**: Move the constant term to the other side of the equation.
2. **Divide through by** a if $a \neq 1$.
3. **Complete the square**: Add and subtract the square of half the coefficient of x inside the equation.
4. **Factor and solve for** x.

Example: Solve $x^2 - 6x + 5 = 0$ by completing the square.

Solution:

1. Move the constant term: $x^2 - 6x = -5$.
2. Half the coefficient of x is -3, square it to get 9, add and subtract 9: $x^2 - 6x + 9 = 4$.
3. Factor and solve: $(x - 3)^2 = 4$, so $x - 3 = \pm 2$, thus $x = 5$ or $x = 1$.

Which Method to Choose

Choosing the right method to solve a quadratic equation—whether to factor, complete the square, or use the quadratic formula—depends largely on the equation's complexity and the specific coefficients involved.

Factoring is the fastest and most straightforward for equations with simple, small integer coefficients and obvious roots. However, not all quadratics neatly factor, especially those with large or complex coefficients. In these cases, completing the square can be useful for deriving the vertex form and understanding the equation's geometric properties, although it can be algebraically demanding.

The quadratic formula is the most universally applicable method, capable of solving any quadratic equation, but it can be computationally intensive.

Simplifying complex quadratic equations is always a good idea before attempting to solve it. Steps such as factoring out the greatest common divisor, rationalizing fractions, or methodically testing factor combinations can make solving the equation easier regardless of the chosen method.

SETTING UP ALGEBRA WORD PROBLEMS

AFOQT algebra word problems require translating real-world situations into mathematical expressions and equations. Understanding how to identify key information and represent it mathematically is crucial to solving them effectively.

Steps to Set Up and Solve Algebra Word Problems

1. Read and Understand the Problem: Identify the important information given in the problem and determine the unknowns to be found.
2. Assign Variables to the Unknowns: Choose symbols, typically letters like x, y, etc., to represent unknown quantities.
3. Write an Equation or System of Equations: Translate the relationships and conditions described in the word problem into mathematical expressions or equations. Ensure that each equation accurately represents the conditions in the problem.
4. Solve the Equation(s): Use algebraic methods like substitution or elimination (for systems) to find the value(s) of the unknown variable(s).
5. Interpret and Verify: Interpret the solution back into the context of the problem to ensure it makes sense. Verify by substituting back into the original equations if necessary.

Translating English Sentences into Math Equations

Different phrases and words can hint at mathematical operations. Here are some common phrases and their mathematical equivalents:

- Addition: "Sum of," "more than," "increased by," "plus."

  Example: "Five more than x" translates to $x + 5$.

- Subtraction: "Difference," "less than," "decreased by," "minus."

  Example: "Ten less than x" translates to $x - 10$.

- Multiplication: "Product of," "times," "of."

  Example: "Twice x" translates to $2x$.

- Division: "Quotient of," "divided by," "per."

  Example: "Half of x" translates to $\frac{x}{2}$.

- Equals: "Is," "are," "will be," "gives."

  Example: "The result is ten" translates to $= 10$.

Example 1— Setting Up a Proportion Problem: A juice recipe requires 4 oranges to make 500 ml of juice. If you need to make 750 ml of juice, how many oranges do you need?

Solution: Assign Variables: Let x represent the number of oranges needed to make 750 ml of juice.

1. Write the Proportion: Set up a ratio comparing oranges to juice in the original recipe: $\frac{4}{500}$
2. Set up a similar ratio for the desired quantity: $\frac{x}{750}$
3. Create an Equation: Establish the proportion by setting the two ratios equal:

$$\frac{4}{500} = \frac{x}{750}$$
$$4 \cdot 750 = 500 \cdot x$$
$$3000 = 500x$$
$$x = \frac{3000}{500} = 6$$

4. Interpret the Solution: You will need 6 oranges to make 750 ml of juice.

Example 2— Setting Up an Equation with One Unknown: In five years, John will be twice as old as he was three years ago. How old is John now?

Solution: Assign Variables: Let x represent John's current age.

1. Write an Equation: In five years, John's age will be $x + 5$. Three years ago, John's age was $x - 3$.
2. The problem says that in five years, John will be twice as old as he was three years ago. Therefore, the equation becomes:

$$x + 5 = 2(x - 3)$$
$$x + 5 = 2x - 6$$
$$5 = 2x - x - 6$$
$$5 = x - 6$$
$$x = 5 + 6 = 11$$

3. Interpret the Solution: John is currently 11 years old.

Example 3— Setting Up Systems of Equations: Sarah is twice as old as Jane. Five years ago, Sarah was three times as old as Jane was. How old are Sarah and Jane now?

Solution: Assign Variables: Let x represent Sarah's current age. Let y represent Jane's current age.

1. Write Equations: From "Sarah is twice as old as Jane," the equation becomes: $x = 2y$.

   From "Five years ago, Sarah was three times as old as Jane," the equation becomes:

   $$x - 5 = 3(y - 5)$$

2. Solve the System: Substitute $x = 2y$ into the second equation:

   $$\begin{aligned} 2y - 5 &= 3(y - 5) \\ 2y - 5 &= 3y - 15 \\ 2y - 3y &= -15 + 5 \\ -y &= -10 \\ y &= 10 \end{aligned}$$

3. Find x by substituting back into the equation $x = 2y$:

   $$x = 2 \cdot 10 = 20$$

4. Interpret the Solution: Sarah is 20 years old, and Jane is 10 years old.

5. Verify: Five years ago, Sarah was $20 - 5 = 15$ and Jane was $10 - 5 = 5$. Thus, Sarah was three times as old as Jane five years ago, confirming the solution is correct.

LINEAR APPLICATIONS AND GRAPHS

Linear equations can be used to model relationships between two variables in many real-life situations. When graphed, these relationships form straight lines, and their general form is:
$y = mx + c$, where y is the dependent variable, x is the independent variable, m is the slope (rate of change), and c is the y-intercept (the value of y when $x = 0$).

Here, it's crucial to distinguish between the input variable (often represented as x) and the output variable (often represented as y). Here's a clear understanding of each:

Input Variable (x): The input variable is the independent variable in a function or equation. It represents the values you can freely choose or manipulate. For instance, in a business context, x might represent the number of products manufactured. In science, x could represent time.

In a function $y = f(x)$, x is the value provided to the function.

Output Variable (y): The output variable is the dependent variable in a function or equation. Its value depends on the input variable and the relationship defined by the function or equation. For instance, in business, y might represent the total revenue earned. In science, y could represent the distance traveled after a certain period.

In the function $y = f(x)$, y is the value calculated after plugging in the input x.

Example—Travel Distance and Speed: Suppose you're driving a car at a constant speed of 60 mph (miles per hour). The distance (y) traveled depends on the amount of time (x) spent driving. The relationship between distance and time can be expressed using the equation: $y = 60x$.

- **Input (x)**: This represents the amount of time (in hours) spent driving.
- **Output (y)**: This represents the total distance traveled in miles.

For example: If you drive for 4 hours ($x = 4$), the distance traveled (y) is: $y = 60 \cdot 4 = 240$.

So, the output variable y (distance traveled) is 240 miles after driving for 4 hours at a speed of 60 mph. This example illustrates how the input (x representing time) directly influences the output (y representing distance traveled) based on a linear relationship.

Such a linear relationship between speed and travel distance can also be illustrated in a graph, as shown here.

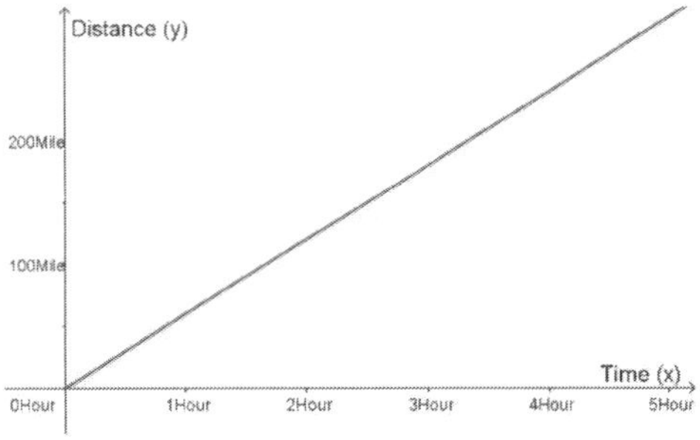

Linear Functions: Slope-Intercept Form ($y = mx + c$)

In linear equation $y = mx + c$, y is the output or dependent variable, x is the input or independent variable, m is the slope, which represents the rate of change, and c is the y-intercept, the value of y when $x = 0$. Let's examine the concept of slope and intercept further.

Slope (m): The slope is the ratio of the vertical change (rise) to the horizontal change (run) between two points on a line. A positive slope means the line is increasing, while a negative slope means the line is decreasing. If the slope is zero, the line is horizontal (constant function).

Intercept (c): The y-intercept is the point where the line crosses the y-axis, representing the value of y when $x = 0$.

Example—Continuing the Travel Example with an Intercept: In the previous travel example, we assumed that the starting point is zero miles (no intercept). However, let's consider that the car has already traveled 30 miles before starting to drive at a constant speed. The equation now changes to: $y = 60x + 30$, where $m = 60mph$ is the speed (slope), $c = 30$ miles is the initial distance already traveled (intercept).

In this case, if you drive for 4 hours ($x = 4$), the total distance traveled (y) is:

$$y = 60 \cdot 4 + 30 = 240 + 30 = 270$$

So, after driving for 4 hours, the total distance covered is 270 miles, including the 30 miles already covered. The relationship between time and distance can also been expressed graphically as follows.

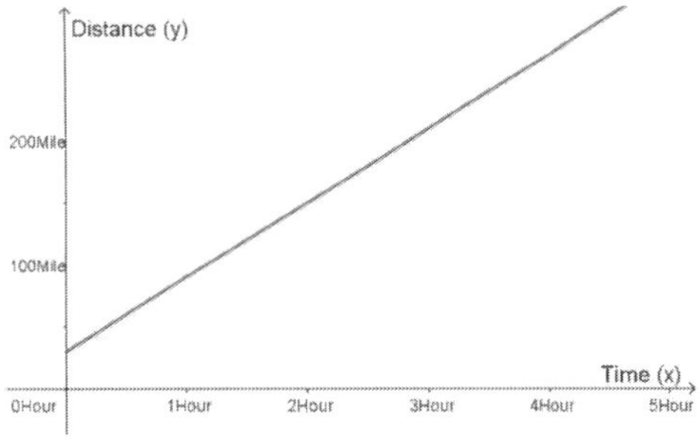

Compare the two figures and one will notice the second one has an intercept while the first one does not.

Graphing Linear Functions

Graphing linear functions involves plotting the relationship between an independent variable (x) and a dependent variable (y) on a two-dimensional plane.

Steps to Graph a Linear Equation

1. Identify the Slope and Intercept: The equation $y = mx + c$ provides the slope (m) and the y-intercept (c). The slope (m) measures the steepness of the line. The intercept (c) indicates the point where the line crosses the y-axis.
2. Plot the Y-Intercept: Mark the point on the y-axis where the line will pass. This is given by the y-intercept (c).
3. Use the Slope to Plot Another Point: The slope m is often written as a fraction $\Delta y/\Delta x$ (change in y over change in x). From the y-intercept, use the slope to find a second point on the graph: Move vertically by Δy (up if positive, down if negative). Move horizontally by Δx (right if positive, left if negative).
4. Draw the Line: Once you have two points, draw a straight line through them to complete the graph.

Example: Graphing the Equation $y = 2x + 3$

1. Identify Slope and Intercept: The slope (m) is 2, which means a vertical change of +2 for every horizontal change of +1. The y-intercept (c) is 3, meaning the line crosses the y-axis at $y = 3$.
2. Plot the Y-Intercept: Mark the point (0,3) on the y-axis.
3. Use the Slope to Find Another Point: Start at (0,3). Move up by 2 units (due to the slope's numerator) and right by 1 unit (due to the slope's denominator), reaching the point (1,5).
4. Draw the Line: Draw a line passing through the points (0,3) and (1,5).

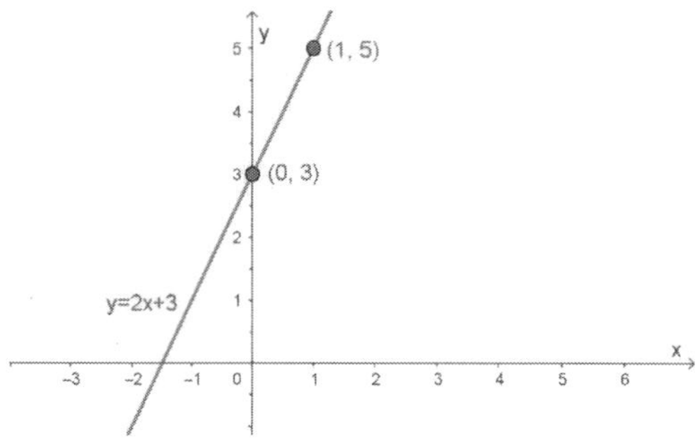

Note that the y-intercept marks the starting point of the line on the y-axis, and the slope controls the line's angle. Since the line will have a positive slope (2), it slants upward from left to right.

Writing Equations in Point-Slope Form

The point-slope form is particularly useful when you have a known point on the line and the slope. The equation in point-slope form is given by: $y - y_1 = m(x - x_1)$, where m is the slope of the line, (x_1, y_1) is a known point on the line.

Steps to Write an Equation in Point-Slope Form

1. Identify the Slope (m) and a Point (x_1, y_1): The slope indicates the rate of change or steepness of the line. A known point on the line helps anchor the line's position.
2. Substitute the Values into the Point-Slope Formula: Plug in the known slope (m) and the coordinates of the known point (x_1, y_1) into the formula.
3. Simplify or Leave in Point-Slope Form: You can either keep the equation in point-slope form or rearrange it to another form like slope-intercept ($y = mx + c$).

Example: Write the equation of a line with a slope of 3 passing through the point (2,−1).

Solution: Identify the Slope (m) and Point (x_1, y_1): $m = 3,\ (x_1, y_1) = (2, -1)$. Substitute into the Formula:

$$y - (-1) = 3(x - 2)$$
$$y + 1 = 3(x - 2)$$

This is the equation of the line in point-slope form. If you prefer, you can convert this to slope-intercept form by expanding and simplifying.

Writing Equations from Two Known Points

When given two points, you can find the equation of the line that passes through them using the slope formula and the point-slope form. Here's a step-by-step process:

Steps to Write a Linear Equation from Two Known Points

1. Find the Slope (m): Use the slope formula to calculate the slope between the two points: $m = \frac{y_2 - y_1}{x_2 - x_1}$. This formula calculates the rate of change between the two given points, (x_1, y_1) and (x_2, y_2).
2. Substitute the Slope and One Point into Point-Slope Form: $y - y_1 = m(x - x_1)$
3. Simplify or Convert to Slope-Intercept Form: You can keep the equation in point-slope form or convert it to slope-intercept form ($y = mx + c$) by expanding and simplifying.

Example: Find the equation of the line that passes through the points (1,2) and (3,−4).

Solution:

1. Calculate the Slope: $m = \frac{-4-2}{3-1} = \frac{-6}{2} = -3$
2. Substitute into Point-Slope Form: Use the point-slope form formula $y - y_1 = m(x - x_1)$ with $m = -3$ and the point (1,2): $y - 2 = -3(x - 1)$
3. Simplify: Expand and simplify to convert to slope-intercept form:

$$\begin{aligned} y - 2 &= -3x + 3 \\ y &= -3x + 5 \end{aligned}$$

This equation in slope-intercept form ($y = -3x + 5$) is the equation of the line that passes through the points (1,2) and (3,–4).

Parallel and Perpendicular Lines

In a two-dimensional coordinate system, lines can be parallel or perpendicular to each other based on the relationship of their slopes.

Parallel Lines

Parallel lines have the same slope but different y-intercepts. If two lines have the same slope (m), they are parallel and never intersect.

Example: Suppose the equation of an existing line is: $y = 4x + 2$. Any line parallel to this one will have the same slope, $m = 4$, but a different y-intercept (c).

Examples of a parallel lines include: $y = 4x - 3$, $y = 4x + 18$, $y = 4x - \frac{11}{17}$

Perpendicular Lines

Perpendicular lines intersect at right angles (90°). If the slope of one line is m, then the slope of any line perpendicular to it is the negative reciprocal, $-\frac{1}{m}$.

Example 1: Suppose the equation of an existing line is: $y = \frac{3}{4}x + 5$. The slope of this line is $m = \frac{3}{4}$. The slope of any line perpendicular to it will be the negative reciprocal: $m_\perp = -\frac{4}{3}$.

An example of a perpendicular line equation (using slope $m_\perp$ and an arbitrary y-intercept):

$$y = -\frac{4}{3}x + 1$$

By analyzing or adjusting the slopes of linear equations, you can quickly determine if lines are parallel or perpendicular to each other.

Example 2: Find a line that is perpendicular to the equation $y = 2x + 3$ and passes through the point (1, 5).

Solution:

1. Identify the Slope of the Given Line: The given line is: $y = 2x + 3$. The slope (m) is 2.
2. Find the Perpendicular Slope: The slope of a line perpendicular to this one is the negative reciprocal. If the original slope is m=2, the perpendicular slope ($m_\perp$) is:
$$m_\perp = -0.5$$
3. Use the Point-Slope Formula: We'll use the point-slope formula to find the equation of the perpendicular line: $y - y_1 = m(x - x_1)$. Here, $(x_1, y_1) = (1,5)$ and $m = -0.5$.

   Substitute into the Point-Slope Formula: $y - 5 = -0.5(x - 1)$
4. Simplify the Equation:

$$\begin{aligned} y - 5 &= -0.5x + 0.5 \\ y &= -0.5x + 5.5 \end{aligned}$$

Hence, the equation of the line perpendicular to the equation $y = 2x + 3$ and passes through the point (1, 5) is: $y = -0.5x + 5.5$.

§3. GEOMETRIC AND SPATIAL REASONING

In the AFOQT Mathematics Test, you won't need to tackle geometric proofs, and the exam only covers a well-defined list of practical geometry topics, which we will cover below.

PERIMETER, CIRCUMFERENCE, AREA, AND VOLUME

The perimeter and circumference formula of common geometric shapes is summarized in the following table. The AFOQT test will usually give you the needed formula except the simplest.

Shape	Formula	Description
Rectangle	Perimeter of a rectangle $P = 2(l + w)$	l and w are length and width
Square	Perimeter of a square $P = 4s$	s is the length of one side
Triangle	Perimeter of a triangle $P = a + b + c$	a, b, c are the length of the three sides
Circle	Circumference of a circle $C = 2\pi r$ (or $C = \pi d$)	r is the radius, d is the diameter

Area

Here's a table summarizing the formulas for calculating the area of common geometric shapes.

Shape	Formula	Description
Triangle	$A = \frac{1}{2} bh$	b is the base, h is the height
Square	$A = s^2$	s is the length of one side
Rectangle	$A = lw$	l and w are the length and width
Parallelogram	$A = bh$	b is the base, h is the height
Trapezoid	$A = \frac{1}{2}(a + b)h$	a and b are the bases, h is the height
Circle	$A = \pi r^2$	r is the radius

Volume

Shape	Formula	Description
Rectangular Prism	$V = lwh$	$l, w,$ and h are the length, width, and height of the prism
Right Cylinder	$V = \pi r^2 h$	r is the radius of the base, h is the height of the cylinder
Sphere	$V = \frac{4}{3}\pi r^3$	r is the radius of the sphere

PYTHAGOREAN THEOREM

The Pythagorean Theorem is a fundamental principle in geometry that relates the lengths of the sides of a right triangle.

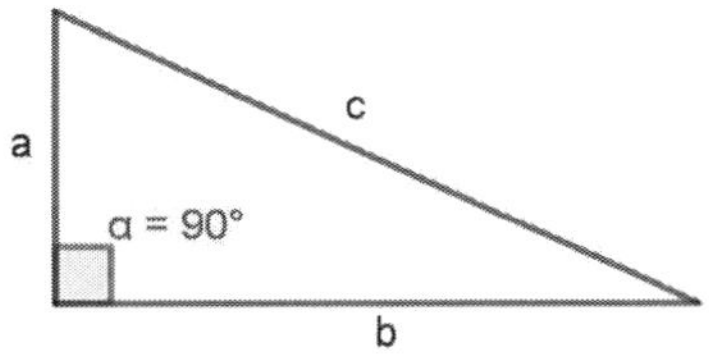

According to the theorem, the square of the length of the hypotenuse (the side opposite the right angle) is equal to the sum of the squares of the lengths of the other two sides. This relationship is usually expressed with the formula:

$$c^2 = a^2 + b^2$$

Here, c represents the length of the hypotenuse, and a and b represent the lengths of the other two sides. The Pythagorean Theorem is frequently used in trigonometry to solve problems involving right triangles. Let's go through a couple of examples where we apply the Pythagorean Theorem.

Example 1—Finding the Hypotenuse: Suppose you have a right triangle with legs of lengths 3 cm and 4 cm. You want to find the length of the hypotenuse.

According to the Pythagorean Theorem: $c^2 = a^2 + b^2$

Plugging in the values:

$$\begin{aligned} c^2 &= 3^2 + 4^2 \\ c^2 &= 9 + 16 \\ c^2 &= 25 \\ c &= \sqrt{25} \\ c &= 5\ cm \end{aligned}$$

So, the hypotenuse is 5 cm long.

Example 2—Finding a Leg of the Triangle: Imagine you know the hypotenuse of a right triangle is 10 cm and one of the legs is 8 cm. You need to find the length of the other leg.

Using the Pythagorean Theorem: $c^2 = a^2 + b^2$

Let b be the unknown side, and rearrange the equation: $b^2 = c^2 - a^2$

Substitute the known values:

$$\begin{aligned} b^2 &= 10^2 - 8^2 \\ b^2 &= 100 - 64 \\ b^2 &= 36 \\ b &= 6 \text{ cm} \end{aligned}$$

The unknown leg measures 6 cm.

DISTANCE FORMULA

The distance formula is used to calculate the distance between two points in a coordinate system. It is derived from the Pythagorean Theorem, which we just discussed. The formula provides the distance between two points (x_1, y_1) and (x_2, y_2) in a 2-dimensional Cartesian plane. The distance d between two points (x_1, y_1) and (x_2, y_2) is given by:

$$d = \sqrt{(x_2 - x_1)^2 + (y_2 - y_1)^2}$$

The expression $(x_2 - x_1)^2 + (y_2 - y_1)^2$ calculates the sum of the squares of the differences in the x and y coordinates. This is analogous to finding the square of the lengths of the two legs in a right triangle, where the line segment between the two points forms the hypotenuse.

Example: Calculating Distance Between Two Points

Let's calculate the distance between two points, $A(1,2)$ and $B(4,6)$.

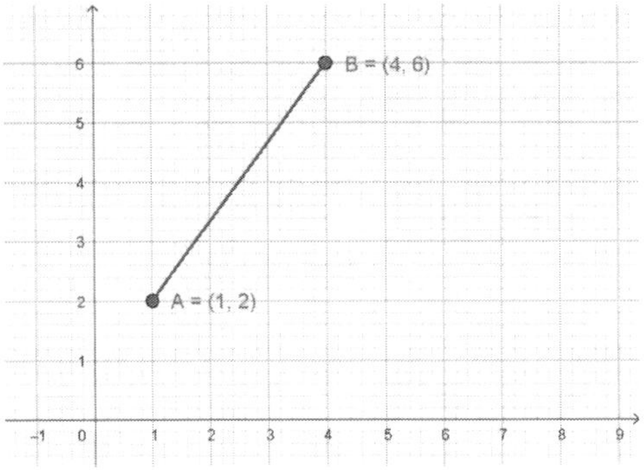

Using the distance formula:

$$d = \sqrt{(x_2 - x_1)^2 + (y_2 - y_1)^2}$$

Plugging in the coordinates:

$$d = \sqrt{(4-1)^2 + (6-2)^2}$$
$$d = \sqrt{3^2 + 4^2}$$
$$d = \sqrt{9 + 16}$$
$$d = \sqrt{25}$$
$$d = 5 \text{ units}$$

Thus, the distance between points A and B is 5 units.

INTERSECTING LINE THEOREMS

Intersecting lines create a variety of angles and geometric relationships that are governed by several theorems. Here's an overview:

Vertical Angles Theorem

If two lines intersect, then the vertical angles are congruent.

Example:

Given lines AB and CD intersecting at point E :

$$\angle AEC = \angle BED$$

$$\angle AED = \angle BEC$$

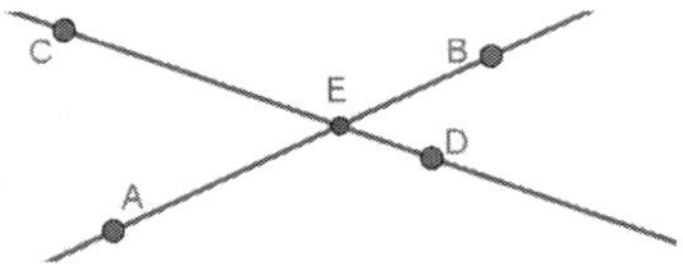

Linear Pair Theorem

If two angles form a linear pair, then they are supplementary (their measures add up to 180°).

Example:

Given lines AB and CD intersecting at point E :

$$\angle AEC + \angle CEB = 180°$$

$$\angle AED + \angle DEB = 180°$$

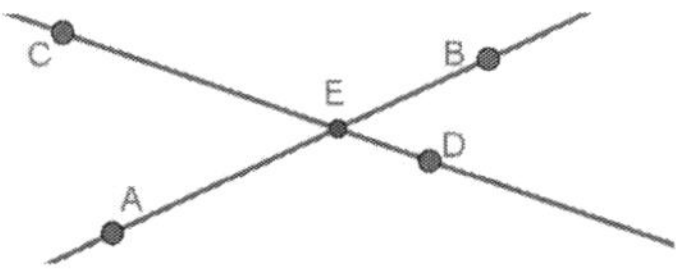

Corresponding Angles Postulate (with Parallel Lines)

If a transversal intersects two parallel lines, then each pair of corresponding angles is congruent.

Example:

Given parallel lines AB and CD intersected by transversal EF :

$$\angle AEF \cong \angle CDF$$

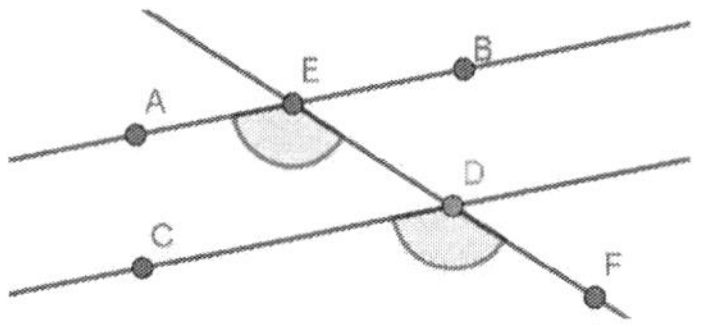

Alternate Interior Angles Theorem (with Parallel Lines)

If a transversal intersects two parallel lines, then each pair of alternate interior angles is congruent.

Example:

Given parallel lines AB and CD intersected by transversal EF :

$$\angle BED \cong \angle \mathrm{CDE}$$

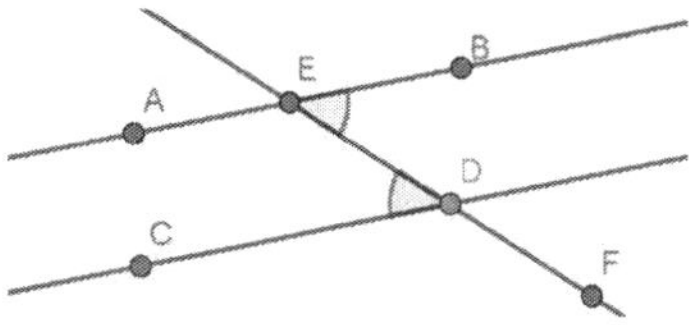

Same-Side Interior Angles Theorem (with Parallel Lines)

If a transversal intersects two parallel lines, then each pair of same-side interior angles is supplementary.

Example:

Given parallel lines AB and CD intersected by transversal EF :

$$\angle AEF + \angle \mathrm{CDE} = 180°$$

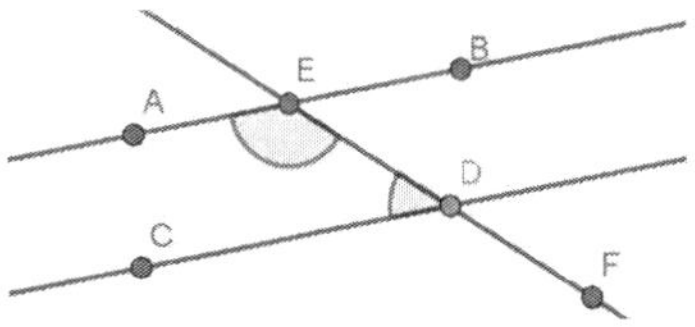

Alternate Exterior Angles Theorem (with Parallel Lines)

If a transversal intersects two parallel lines, then each pair of alternate exterior angles is congruent.

Example:

Given parallel lines AB and CD intersected by transversal EF :

$$\angle ABE \cong \angle DCF$$

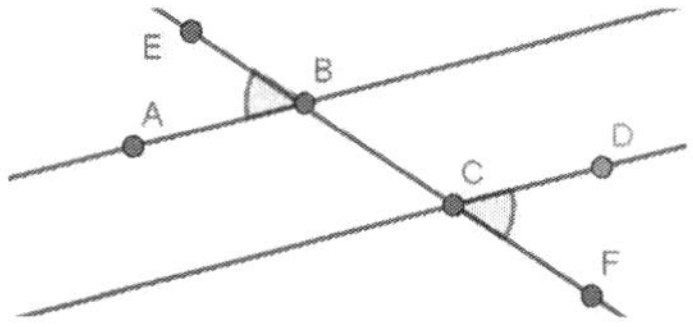

Now, let's put to these theorems to use in the following example.

Example: Given MN and ST are parallel lines, and $\angle a = 30°$, find all the rest of the angles.

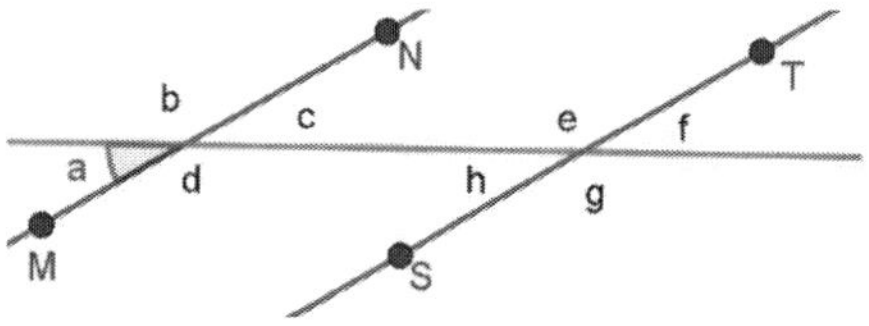

Solution:

Step 1: Identify the Vertical Angles

The vertical angle opposite $\angle a$ is $\angle c$ Hence, $\angle c = \angle a = 30°$.

Step 2: Identify the Linear Pairs

Angles a and d form a linear pair, so they add up to $180°$.

$$\angle d = 180° - \angle a = 180° - 30° = 150°$$

Since b and d are vertical angles: $\angle b = \angle d = 150°$

Step 3: Identify the Corresponding Angles

Since MN and ST are parallel, $\angle a$ corresponds to $\angle$ h. Hence, $\angle h = \angle a = 30°$.

Similarly, $\angle d$ *corresponds to* $\angle g$. Hence, $\angle g = \angle d = 150°$.
$\angle e$ *corresponds to* $\angle b$ Hence, $\angle e = \angle b = 150°$.
$\angle f$ *corresponds to* $\angle c$ Hence, $\angle f = \angle c = 30°$.

Example: Given angle $\angle a = 36°$, find $\angle \theta$

Solution:

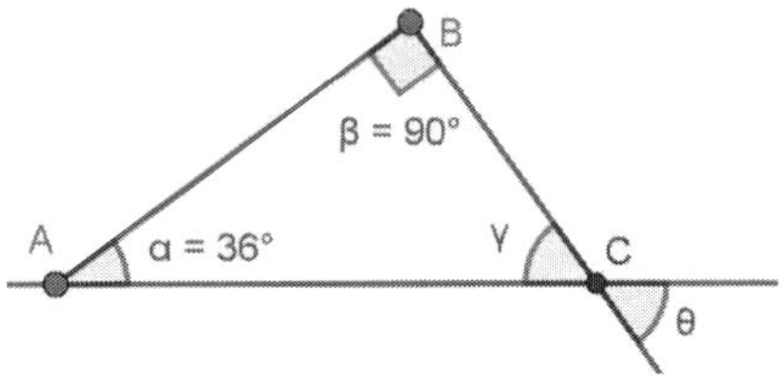

From the diagram:

1. Identify Angles in Triangle ΔABC : $\alpha = 36°, \beta = 90°$.
2. Triangle Sum Theorem: The sum of the angles in a triangle is $180°$.

$$\begin{aligned} \alpha + \beta + \gamma &= 180° \\ 36° + 90° + \gamma &= 180° \\ \gamma &= 180° - 36° - 90° \\ \gamma &= 54° \end{aligned}$$

3. Find $\angle \theta$: since $\angle \theta$ and $\angle \gamma$ are vertical angles. $\angle \theta = \angle \gamma = 54°$.

TRIANGLE CONGRUENCY THEOREMS

Triangle congruency describes a condition where two triangles are exactly identical in shape and size. Two triangles are congruent if all their corresponding sides are equal in length and all their corresponding angles are equal in measure. When two triangles are congruent, they can be perfectly overlaid on each other, with every corresponding side and angle matching precisely.

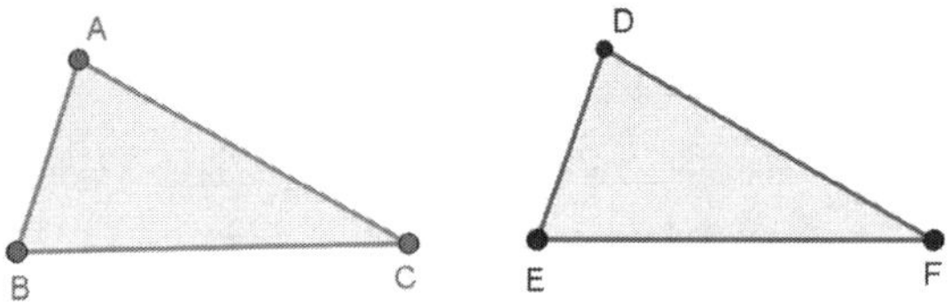

If two triangles $\triangle ABC$ and $\triangle DEF$ are congruent, this relationship is denoted as:
$\triangle ABC \cong \triangle DEF$. This means that $AB = DE,\ BC = EF,\ CA = FD$ and $\angle A = \angle D, \angle B = \angle E,\ \angle C = \angle F$.

SSS (Side-Side-Side) Congruency

Two triangles are congruent if all three sides of one triangle are equal to all three sides of another triangle. This is known as the SSS congruency criterion.

If $AB = DE,\quad BC = EF,\quad CA = FD$, then $\triangle ABC \cong \triangle DEF$.

SAS (Side-Angle-Side) Congruency

Two triangles are congruent if two sides and the included angle of one triangle are equal to two sides and the included angle of another triangle. This is known as the SAS congruency criterion.

If $AB = DE,\ AC = DF,\ \angle A = \angle D$, then $\triangle ABC \cong \triangle DEF$.

ASA (Angle-Side-Angle) Congruency

Two triangles are congruent if two angles and the included side of one triangle are equal to two angles and the included side of another triangle. This is known as the ASA congruency criterion.

If $\angle A = \angle D,\ \angle B = \angle E,\ AB = DE,$ then $\triangle ABC \cong \triangle DEF.$

AAS (Angle-Angle-Side) Congruency

Two triangles are congruent if two angles and a non-included side of one triangle are equal to two angles and the corresponding non-included side of another triangle. This is known as the AAS congruency criterion.

If $\angle A = \angle D, \angle B = \angle E, BC = EF,$ then $\triangle ABC \cong \triangle DEF.$

HL (Hypotenuse-Leg) Congruency (Right Triangles Only)

Two right triangles are congruent if the hypotenuse and one leg of one triangle are equal to the hypotenuse and one leg of another triangle. This is known as the HL congruency criterion.

If $AC = DF,\ \ BC = EF,$ and $\angle C = \angle F = 90°$, then $\triangle ABC \cong \triangle DEF.$

SSA (Side-Side-Angle) ≠ Congruency

There is one side and angle combination that does not prove triangle congruency: SSA. The SSA (Side-Side-Angle) condition fails to prove that a pair of triangles are congruent due to the possibility of multiple non-congruent triangles satisfying this condition. To understand why, let's look at the following scenario:

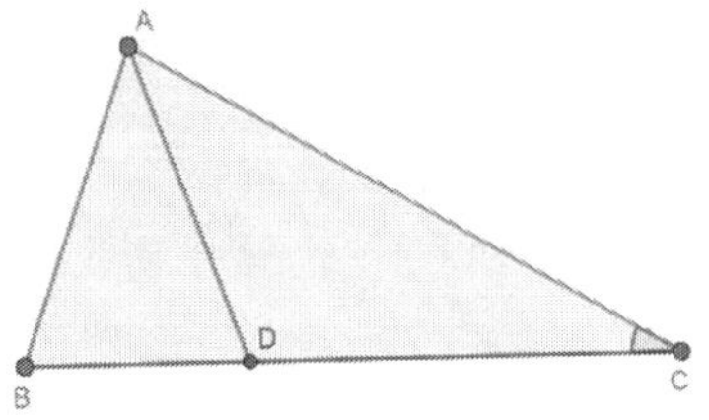

Consider triangles $\triangle ABC$ and $\triangle ACD$ as shown in the figure. Clearly, triangles $\triangle ABC$ and $\triangle ACD$ are not congruent. In fact, $\triangle ABC$ includes all of $\triangle ACD$ and then some. However, the two triangles share the same angle $\angle C$ and the same side AC. Also, they have one side where the length is equal: $AB = AD$.

The key here is the two sides that $\triangle ABC$ and $\triangle BCD$ have the same length ($AB = AD, AC = AC$) don't inscribe their shared angle, $\angle C$, leading to a SSA situation. If the two congruent sides inscribe the angle of equal value, leading to a SAS situation, then the two triangles will be congruent without any ambiguity.

AAA (Angle-Angle-Angle) ≠ Congruency

This should be self-evident. Two triangles with all three angles being equal can differ greatly in size. So, AAA does not necessarily guarantee two triangles are congruent.

Here is a tip on how to remember these theorems and exceptions: ALL combinations of three elements of A and S, be it AAS, ASA, SSA, SSS, will prove congruency, except AAA, and SSA. It is easy to remember AAA does not prove congruency—again, it is self-evident.

As for remembering SSA, the author has always remembered it as ASS instead. Should you associate it with anything other than a donkey though, it is on you, not me. The author's innocence is proven by a poem he wrote for this situation:

Two asses live on a farm,
Both tan and both tame.
One is healthy and one is lame,
Two asses are NOT always the same!

Triangle Similarity Theorems

Triangle similarity describes a condition where two triangles have the same shape but not necessarily the same size. Two triangles are similar if their corresponding angles are equal, and their corresponding sides are proportional. This means that one triangle can be obtained from the other by scaling (enlarging or reducing), possibly followed by a translation, rotation, or reflection.

If two triangles $\triangle ABC$ and $\triangle DEF$ are similar, this relationship is denoted as: $\triangle ABC \sim \triangle DEF$, which means that $\angle A = \angle D,\quad \angle B = \angle E,\quad \angle C = \angle F$ and $\frac{AB}{DE} = \frac{BC}{EF} = \frac{CA}{FD}$.

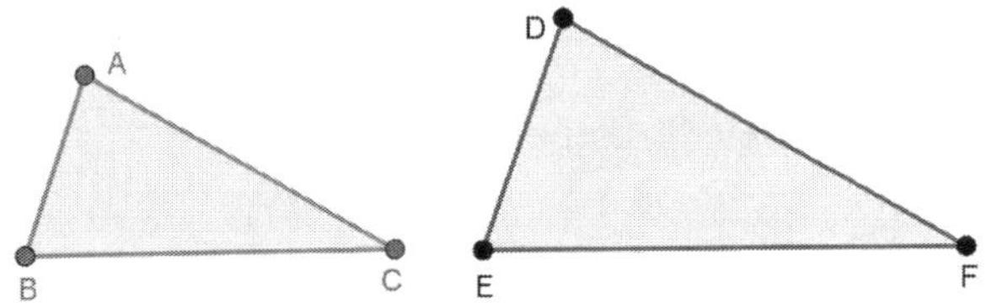

AA (Angle-Angle) Similarity

Two triangles are similar if two angles of one triangle are congruent to two angles of the other triangle. This is known as the AA similarity criterion. When two triangles are similar, their corresponding sides are proportional, and their corresponding angles are equal.

If $\angle A = \angle D$ and $\angle B = \angle E$, then $\triangle ABC \sim \triangle DEF$.

SSS (Side-Side-Side) Similarity

Two triangles are similar if the corresponding sides of one triangle are proportional to the corresponding sides of the other triangle. This is known as the SSS similarity criterion.

If $\frac{AB}{DE} = \frac{BC}{EF} = \frac{CA}{FD}$, then $\triangle ABC \sim \triangle DEF$.

SAS (Side-Angle-Side) Similarity

Two triangles are similar if one angle of one triangle is congruent to one angle of the other triangle, and the lengths of the sides inscribing these angles are proportional. This is known as the SAS similarity criterion.

If $\frac{AB}{DE} = \frac{AC}{DF}$ and $\angle A = \angle D$, then $\triangle ABC \sim \triangle DEF$.

Triangle Congruency and Similarity Questions

On the AFOQT Test, some questions may ask you to prove that a pair of triangles are congruent. These questions typically require you to be familiar with the four congruency theorems and apply them to the question to establish triangle congruency.

Example 1: For triangle ABC and triangle DEF, if $\angle A$ is congruent to $\angle D$, which of the following must be true in order to prove that triangles ABC and DEF are congruent?

A. $\angle B = \angle E$ and $AC = DF$

B. $BC = EF$ and $DF = AC$

C. $AB = DE$ and $BC = EF$

D. $\angle C = \angle F$ and $\angle B = \angle E$

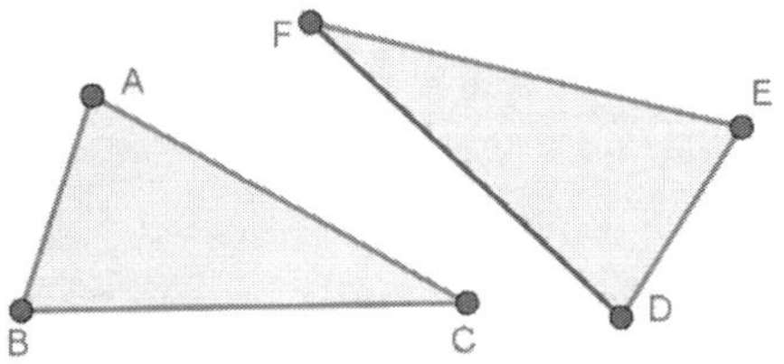

Answer: A

Explanation: Triangles are congruent if they meet the SSS, SAS, AAS, or ASA criterion. Here, choice A will satisfy AAS, so it is correct. Choice B and C constitute SSA situation, choice D constitutes AAA situation. Hence, B, C, and D are not correct answers.

Other AFOQT questions give you some choices and may ask you to identify a pair of similar triangles. This type of question often involves some calculation, such as creating and solving a proportion, before you can decide.

Example 2: Triangle ABC and triangle DEF are shown below. If $\angle A$ is congruent to $\angle D$ and $AC = 4, DE = 3, DF = 6$, which of the following must be true for triangles ABC and DEF to be similar?

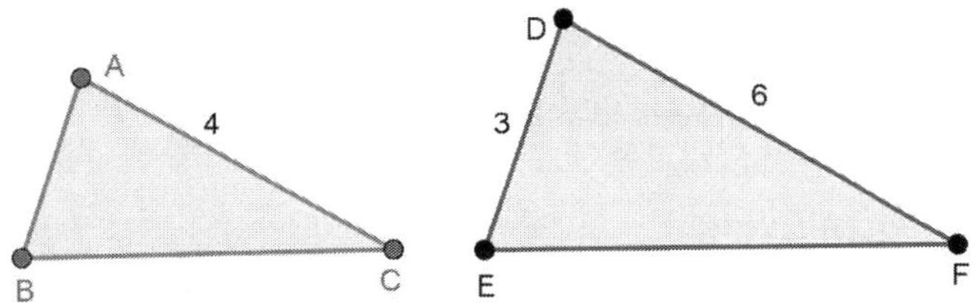

A. $AB = 1.5$

B. $AB = 2$

C. $AB = 2.5$

D. $AB = 3$

Answer: B

Explanation: Two triangles are similar if one angle of one triangle is congruent to one angle of the other triangle, and the lengths of the sides inscribing these angles are proportional. This is known as the SAS similarity criterion. So, to have $\triangle ABC \sim \triangle DEF$, it must be true that: $\frac{AB}{DE} = \frac{AC}{DF}$ and $\angle A = \angle D$. Thus, $\frac{AB}{3} = \frac{4}{6} \Rightarrow AB = \frac{3 \times 4}{6} = 2$.

Hence, B is the right choice.

§4. PROBABILISTIC AND STATISTICAL REASONING

Probability is a measure of how likely an event is to occur. It's a value between 0 and 1, where 0 means the event will not happen, and 1 means it will definitely happen. For example, if you toss a coin, the probability of it landing on heads is 0.5, meaning there's a 50% chance of it happening.

Key Probability Terms

- **Experiment:** A procedure that results in one or more outcomes. Tossing a coin is an example of an experiment.
- **Outcome:** A possible result of an experiment. For example, "heads" is an outcome of tossing a coin.
- **Sample Space:** The set of all possible outcomes. For example, for a coin toss, the sample space is {Heads,Tails}.
- **Event:** A subset of the sample space, representing one or more outcomes that we're interested in. If we're only interested in getting "heads," that outcome is an event.

CALCULATING PROBABILITY

Simple Probability: Simple probability refers to the likelihood of a single event occurring.

Formula: The probability of an event E is:

$$P(E) = \frac{Number\ of\ favorable\ outcomes}{Total\ number\ of\ possible\ outcomes}$$

Example: If you roll a six-sided die, the probability of rolling a 4 is:

$$P(Rolling\ a\ 4) = \frac{1}{6}$$

Compound Probability: Compound probability involves finding the likelihood of two or more events occurring together. For independent events (events that don't affect each other), multiply their probabilities to get Compound Probability.

Example: If you roll a die and flip a coin, the probability of rolling a 3 <u>and</u> flipping heads is:

$$P(Rolling\ a\ 3) \cdot P(Flipping\ Heads) = \frac{1}{6} \cdot \frac{1}{2} = \frac{1}{12}$$

Conditional Probability: Conditional probability is the likelihood of an event occurring given that another event has already happened.

Example: If 80% of students in a class pass math and 60% of those who pass also pass science, the probability of a student passing science given that they passed math is:

$$P(\,Pass\ Science\ |\ Pass\ Math\,) = \frac{0.6}{0.8} = 0.75$$

Table-Based Probability Questions

On AFOQT, probability questions are sometimes based on facts presented in a table. So, let's use the following example to review the three types of probabilities.

Example: Let's consider a table summarizing the distribution of animals in a nature reserve based on two traits: whether they are mammals or reptiles and whether they are nocturnal or diurnal.

Table of Animal Distribution

	Nocturnal	Diurnal	**Total**
Mammal	15	30	45
Reptile	10	25	35
Total	**25**	**55**	**80**

Simple Probability: Find the probability that a randomly chosen animal is nocturnal.

From the table, 25 animals are nocturnal out of a total of 80.

$$P(Nocturnal) = \frac{25}{80} = 0.3125$$

Compound Probability: Find the probability that a randomly chosen animal is both a mammal and nocturnal.

From the table, 15 animals are both mammals and nocturnal.

$$P(Mammal\ and\ Nocturnal) = \frac{15}{80} = 0.1875$$

Conditional Probability: Find the probability that an animal is a reptile given that it is diurnal.

From the table, 25 out of the 55 diurnal animals are reptiles.

$$P(\ Reptile \mid Diurnal\ ) = \frac{25}{55} \approx 0.455$$

DESCRIPTIVE STATISTICS

Descriptive statistics involve methods for summarizing and organizing data, providing simple insights into the patterns and characteristics of a dataset. Instead of analyzing every individual data point, descriptive statistics condense information into meaningful measures to help understand trends, variability, and distribution.

Measures of Central Tendency

Mean, median, and mode are measures of central tendency, used to describe the central point or typical value within a dataset. Each measure provides different insights into the nature of the data.

Mean (Average): The mean is the sum of all data values divided by the total number of values. It is influenced by every value in the dataset.

$$\text{mean} = \frac{x_1 + x_2 + \cdots + x_n}{n}$$

where:

- $x_1, x_2, \ldots, x_n$ are the data values,
- n is the number of data points.

Example: Given the dataset {3,5,7,8,10}, what is the mean?

Solution: mean $= \frac{3+5+7+8+10}{5} = \frac{33}{5} = 6.6$

Median: The median is the middle value when all data values are arranged in ascending or descending order. If the number of data points is even, the median is the average of the two middle values.

Example: Given the dataset {3,5,7,8,10}, the median value is 7 because it is the third value in a dataset with five numbers.

If the dataset were {3,5,7,8}, the median would be: $\frac{5+7}{2} = \frac{12}{2} = 6$.

Mode: The mode is the most frequently occurring value(s) in a dataset. There can be one mode (unimodal), more than one mode (bimodal or multimodal), or no mode if no value repeats.

Example: Given the dataset {3,5,5,8,10}, the mode is 5 because it occurs most frequently (twice). If the dataset were {3,3,5,5,8,10}, the mode would be both 3 and 5, making the dataset bimodal.

Measures of Dispersion (Spread)

The range of data helps describe the spread of values within a dataset. Here are some basic concepts:

Maximum (Max): The maximum is the largest value in a dataset.

Example: In the dataset {2,4,6,8,10}, the maximum value is: $\text{Max} = 10$.

Minimum (Min): The minimum is the smallest value in a dataset.

Example: In the dataset {2,4,6,8,10}, the minimum value is: $\text{Min} = 2$.

Range: The range is the difference between the maximum and minimum values in a dataset.

$\text{Range} = \text{Max} - \text{Min}$

Example: In the dataset {2,4,6,8,10}: $\text{Range} = 10 - 2 = 8$

Quantile: A quantile divides the data into intervals containing an equal number of data points.

Common quantiles include:

- Quartile: Divides data into four equal parts.
- Percentile: Divides data into 100 equal parts.

Example: Given the dataset {3,5,7,9,11,13,15}, let's find the first quartile (Q_1) and the third quartile (Q_3).

Solution: The following table illustrates how we analyze the data.

3	5	7	9	11	13	15
Minimum	First Quartile		Median		Third Quartile	Maximum

← Lower half → ← Upper half →

Hence, $Q_1 = 5$, $Q_3 = 13$.

In this example, the first quartile (Q_1) marks the value below which 25% of the data falls, and the third quartile (Q_3) marks the value below which 75% of the data falls. Together, these measures provide insight into the data's spread and distribution, helping identify potential outliers and the overall range of values.

Visual Representations

Graphical displays offer a visual way to analyze and interpret data. They reveal trends, patterns, and outliers that might not be evident through raw numbers alone. Here are three common types of data visualization tools appearing on AFOQT: histograms, box plots, and scatterplots.

Histogram

A histogram is a bar graph that displays the distribution of a dataset. The data is grouped into intervals (bins), and each bar represents the frequency (or count) of values within that interval.

Interpretation:

- **Height of Bars:** Indicates how many data points fall within each interval.
- **Shape:** The shape of the histogram provides insight into the data distribution (e.g., skewed, symmetric).
- **Outliers:** Bars that are separate from the main distribution can indicate outliers.

Example: A histogram depicting the scores of students on a test might show that most scores fall between 70 and 95, with an outlier below 50.

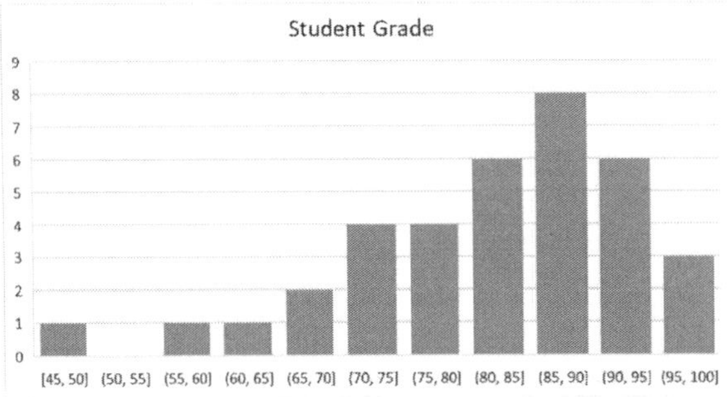

Box Plot (Box-and-Whisker Plot)

Definition: A box plot shows the data distribution by dividing it into quartiles and highlighting the median. It also displays potential outliers using whiskers.

Interpretation:

- **Box:** Represents the interquartile range (IQR) between the first quartile (Q_1) and third quartile (Q_3), which contains the middle 50% of the data.
- **Median Line:** Divides the box into two halves, showing the median of the data.
- **Whiskers:** In the usual convention, they extend from the box to the minimum and maximum values within a specified range.

Example: A box plot showing employee salaries might reveal that the median salary is near $82,000, while outliers earning above $100,000 form a distinct tail.

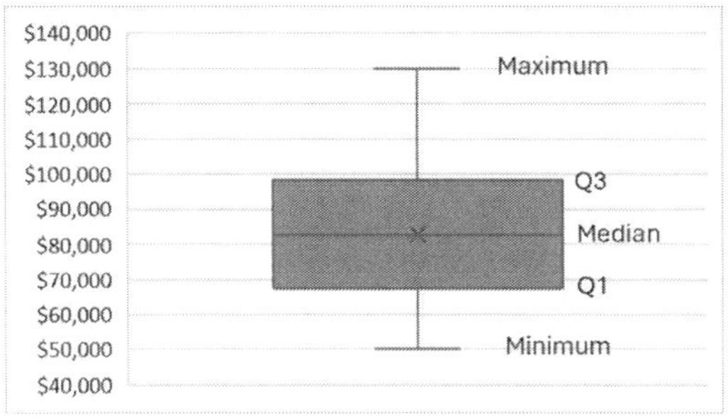

Scatter Plot

Definition: A scatter plot displays data points on a Cartesian plane to illustrate the relationship between two variables.

Interpretation:

- **Trend:** Points that form an upward or downward pattern indicate positive or negative correlations, respectively.
- **Clusters:** Groups of points in different areas may suggest different categories or clusters of data.
- **Outliers:** Points that lie far from the main cluster may represent outliers.

Example: A scatter plot comparing advertising expenditure and sales revenue might reveal that higher ad spending correlates with higher revenue.

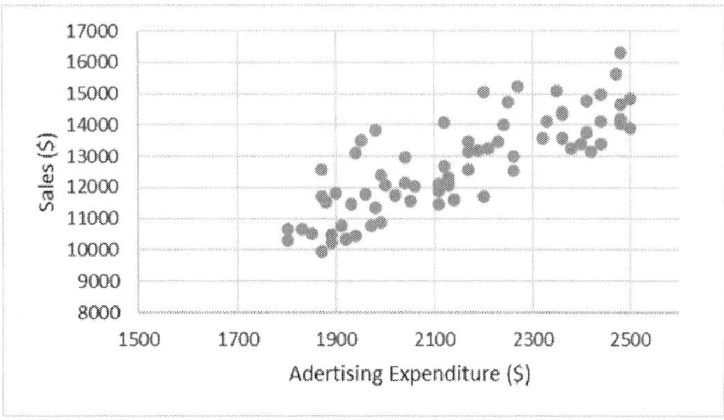

MATHEMATICS KNOWLEDGE PRACTICE SET 1

Select the correct answer from the choices given. For practice only. This practice set does not reflect the actual number of questions in the test.

1: $(3+5)^2 - 12/3 =$

A) 58
B) 62
C) 60
D) 64

2: What is the difference between 45.678 and 19.234, to the nearest integer?

A) 24
B) 26
C) 27
D) 28

3: 72 is 30% of what number?

A) 210
B) 230
C) 240
D) 220

4: Sara paid $45 for a jacket that was originally priced at $60. By what percent was the jacket discounted?

A) 15%
B) 20%
C) 25%
D) 30%

5: 0.85, $\frac{2}{3}$, and 80%.

Which of the following correctly orders these numbers from least to greatest?

A) $\frac{2}{3}$, 0.85, 80%
B) 80%, $\frac{2}{3}$, 0.85
C) 0.85, 80%, $\frac{2}{3}$
D) $\frac{2}{3}$, 80%, 0.85

6. Which of the following points (x, y) lies on the graph of $7x + 3y = 26$?

A) (1,3)
B) (2,5)
C) (4,3)
D) (2,4)

7. A cyclist traveled a distance of 45 kilometers in 3 hours. On average, how much distance did the cyclist cover each hour?

A) 12 km
B) 15 km
C) 18 km
D) 20 km

8. A circle has a diameter of 12 meters. What is the area, in square meters, of this circle? (The area of a circle with a radius of r , is equal to πr^2.)

A) 36π
B) 38π
C) 24π
D) 48π

9. The cost of renting a bicycle is $10 per hour. In addition, a flat helmet rental fee of $5 is charged. Which of the following represents the total cost, in dollars, of the bicycle and helmet rental for h hours?

A) $10h + 5$
B) $10h + 15$
C) $15h$
D) $10h - 5$

10. What is the value of $1.5 + 4 \times 3.2 - 10$?

A) 4.8
B) 5.3
C) 4.3
D) 6.3

11. The box plot below summarizes the calories of food a group of lab mice consume per day. What could be the range of these mice's calorie intake per day?

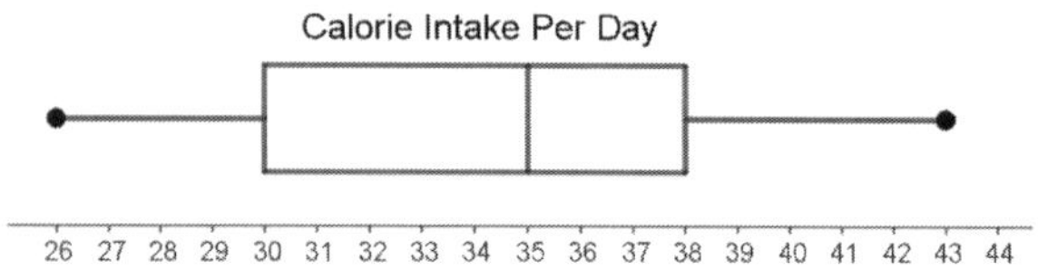

4.

A) 14
B) 15
C) 16
D) 17

12. There are 24 books on a shelf. 8 of the books are fiction, and the rest are non-fiction. What is the ratio of fiction books to non-fiction books on the shelf?

A) 1 to 2
B) 1 to 3
C) 1 to 1
D) 2 to 3

13. Which equation represents the line shown?

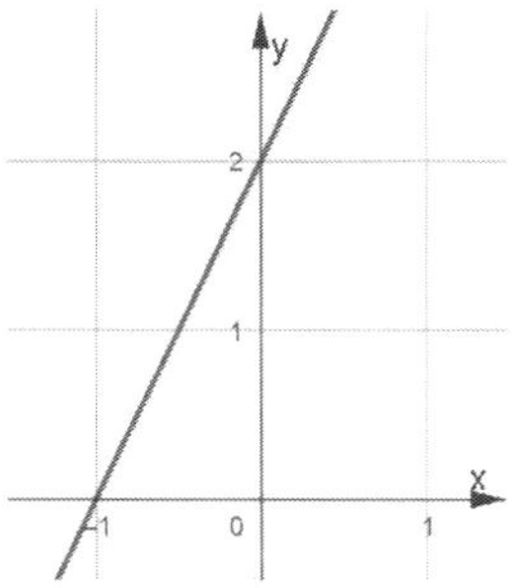

A) $y = -x + 2$
B) $y = x - 1$
C) $y = 2x + 2$
D) $y = 2x - 1$

14. Which of the following is equivalent to $\left(\frac{3}{4}\right)^{-2}$?

A) $\frac{9}{16}$
B) $-\frac{9}{16}$
C) $\frac{16}{9}$
D) $-\frac{16}{9}$

15. $1.25 \div 10^{-3} =$

A) 12.5
B) 125
C) 1250
D) 12500

Practice Set 1 Answers and Explanations

1. C) 60. Explanation: $(3+5)^2 - \frac{12}{3} = 64 - 4 = 60$

2. B) 26. Explanation: $45.678 - 19.234 \approx 26.444 \approx 26$

3. C) 240. Explanation: We know $0.30 \times N = 72$. Hence, $N = \frac{72}{0.30} = 240$.

4. C) 25%. Explanation: $\frac{60-45}{60} \times 100 = \frac{15}{60} \times 100 = 25\%$

5. D) $\frac{2}{3}$, 80%, 0.85. Explanation: Convert Each to Decimals: $\frac{2}{3} \approx 0.6667$, $80\% = 0.80$.

   Order the Decimals from Least to Greatest: 0.6667, 0.80, 0.85.

6. D) (2,4). Explanation: When faced with such problems, try each pair of solutions in the equation to see if it is true.

   Here, $7 \times 2 + 3 \times 4 = 14 + 12 = 26$. So D is correct.

7. B) 15 km. Explanation: Total distance = 45 km, Total time = 3 hours. $45/3 = 15$.

8. A) 36π. Diameter = 12 meters, thus, Radius $= \frac{\text{Diameter}}{2} = \frac{12}{2} = 6$ meters.

   Area $= \pi r^2 = \pi(6)^2 = \pi \times 36 = 36\pi$ square meters.

9. A) 10h + 5

10. C) 4.3. Explanation: $1.5 + 4 \times 3.2 - 10 = 1.5 + 12.8 - 10 = 1.5 + 2.8 = 4.3$

11. D) 17. Explanation: The range is the difference between the maximum and minimum values in a dataset.

    Range = Max – Min. So, $43 - 26 = 17$.

12. A) 1 to 2. Explanation: Total books = 24, Fiction books = 8, Non-fiction books $= 24 - 8 = 16$.

    Ratio of fiction books to non-fiction books $= \frac{8}{16} = \frac{1}{2}$.

13. C) $y = 2x + 2$. Explanation: From the graph, the line's slope is 2, while it intersects y-axis at (0,2). So, $y = 2x + 2$.

14. C) $\frac{16}{9}$. Explanation: $\left(\frac{3}{4}\right)^{-2} = \left(\frac{4}{3}\right)^2 = \frac{4^2}{3^2} = \frac{16}{9}$.

15. C) 1250. Explanation: $\frac{1.25}{10^{-3}} = 1.25 \times 10^3 = 1.25 \times 1000 = 1250$.

Mathematics Knowledge Practice Set 2

Select the correct answer from the choices given. For practice only. This practice set does not reflect the actual number of questions in the test.

1. Solve the following problem and choose the best $\frac{9.68}{2}$

A) 4.86
B) 4.84
C) 4.82
D) 4.87

2. To the nearest integer, what is the product of 15.7 and 4.8?

A) 74
B) 75
C) 76
D) 77

3. What is the sum of 23.85 and 2.155, to the nearest integer?

A) 26
B) 27
C) 25
D) 24

4. If 35% of a number N is 14, what is the value of N?

A) 35
B) 40
C) 45
D) 50

5. Which of the following has the least value?

A) 40% of 60
B) 60% of 40
C) 40% of 40
D) 60% of 60

6. Which of the following fractions is greater than $\frac{1}{3}$ and less than $\frac{2}{3}$?

A) $\frac{1}{6}$
B) $\frac{2}{5}$
C) $\frac{3}{4}$
D) $\frac{5}{6}$

7. Which of the following is equivalent to $\frac{3}{2} \times \frac{4}{5}$?

A) $\frac{7}{10}$
B) $\frac{6}{10}$
C) $\frac{3}{5}$
D) $\frac{6}{5}$

8. The bar graph represents the number bags of each type of fruit in a grocery store. Which fruit has the most number of bags in the store?

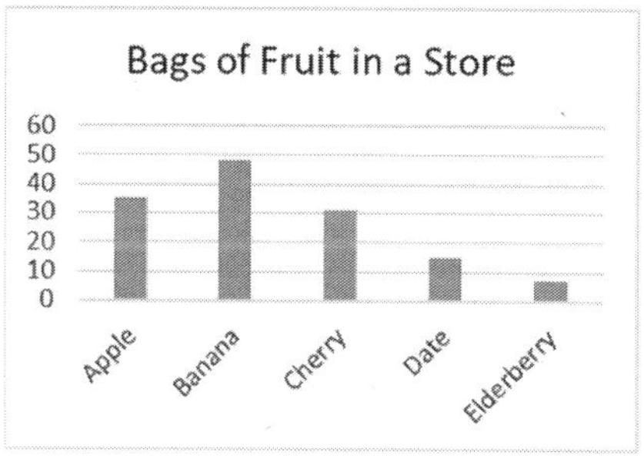

A) Apple
B) Banana
C) Cherry
D) Date

9. At a hardware store, nails cost \$0.05 each and screws cost \$0.10 each. Jenna spent less than \$5.00 on 20 nails and x screws. Which inequality represents this situation?

A) $\left(\frac{0.05+\ .10}{2}\right)x < 5.00$
B) $20(0.05) + 0.10x < 5.00$
C) $20(0.10) + 0.05x < 5.00$
D) $20(0.05 + 0.10)x < 5.00$

10. Based on the relationship between elevation and temperature established in the graph, what's the temperature at 5,000 ft?

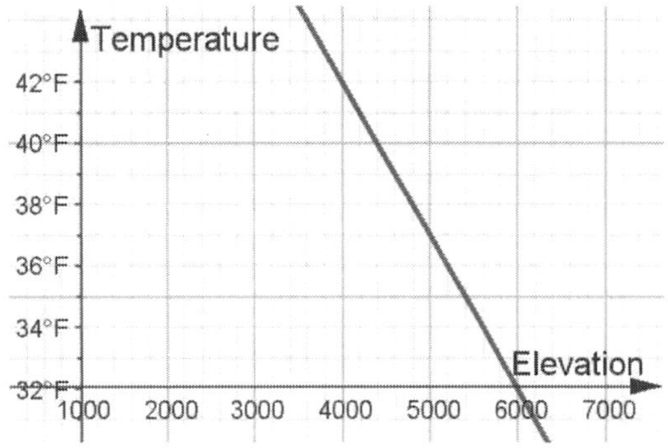

A) 35°
B) 36°
C) 37°
D) 38°

11. A rectangular field has an area of 1500 square feet. If the width of the field is 20 feet less than the length, what is the perimeter, in feet, of the field?

A) 140
B) 150
C) 160
D) 185

12. Which expression is equivalent to $5(d - 2^3) + 2d$?

A) $5d - 20$
B) $7d - 40$
C) $5d - 30$
D) $7d - 30$

13. In a triangle, one angle measures 40°, and another measures 60°. What is the measure of the third angle?

A) 80°
B) 100°
C) 70°
D) 60°

14. A right triangle has an area of 24 square units and one of its legs measures 6 units. What is the perimeter of the triangle?

A) 18 units
B) 24 units
C) 20 units
D) 22 units

15. Simplify the following polynomial expression:

$(3x^2 + 4x - 5) - (2x^2 - 3x + 6)$

A) $x^2 + 7x - 11$
B) $5x^2 + x + 1$
C) $x^2 + 7x - 1$
D) $x^2 + x - 11$

Practice Set 2 Answers and Explanations

1. B) 4.84. Explanation: $\frac{9.68}{2} = 4.84$

2. B) 75. Explanation: $15.7 \times 4.8 = 75.36 \approx 75$

3. A) 26. Explanation: $23.85 + 2.155 = 26.005 \approx 26$

4. B) 40. Explanation: We know $0.35 \times N = 14$. Thus, solving the equation: $N = \frac{14}{0.35} = 40$.

5. C) 40% of 40. Explanation: A. $0.40 \times 60 = 24$. B. $0.60 \times 40 = 24$. C. $0.40 \times 40 = 16$. D. $0.60 \times 60 = 36$.

6. B) $\frac{2}{5}$. Explanation: Convert the Fractions to Decimals: $\frac{1}{3} \approx 0.3333, \frac{2}{3} \approx 0.6667, \frac{1}{6} \approx 0.1667, \frac{2}{5} = 0.4, \frac{3}{4} = 0.75, \frac{5}{6} \approx 0.8333$. Identify the Fraction Between $\frac{1}{3}$ and $\frac{2}{3}$: $0.3333 < 0.4 < 0.6667$.

7. D) $\frac{6}{5}$. Explanation: $\frac{3}{2} \times \frac{4}{5} = \frac{3 \times 4}{2 \times 5} = \frac{12}{10} = \frac{6}{5}$.

8. B) Banana. Explanation: The bar associated with banana is the tallest.

9. B) $20(0.05) + 0.10x < 5.00$.

10. C) 37°. This can be observed in the graph: when elevation is at 5000, temperature is at 37°.

11. C) 160. Explanation: Let l = length of the field, w = width of the field. We know $w = l - 20$.

Hence, Area $= l \times w \Rightarrow 1500 = l \times (l - 20) \Rightarrow 1500 = l^2 - 20l \Rightarrow l^2 - 20l - 1500 = 0$.

Solve the quadratic equation: $l = \frac{-b \pm \sqrt{b^2 - 4ac}}{2a}$, where $a = 1,\ b = -20,\ c = -1500$.

So, $l = \frac{-(-20) \pm \sqrt{(-20)^2 - 4 \cdot 1 \cdot (-1500)}}{2 \cdot 1} = \frac{20 \pm \sqrt{400 + 6000}}{2} = \frac{20 \pm \sqrt{6400}}{2} = \frac{20 \pm 80}{2}$.

Hence, $l = \frac{20+80}{2}$ or $l = \frac{20-80}{2}$. Thus, $l = 50$ (since length cannot be negative).

$w = l - 20 \Rightarrow w = 50 - 20 = 30$. Perimeter $= 2(l + w) = 2(50 + 30) = 2 \times 80 = 160$ feet.

12. B) 7d – 40. Explanation: $5(d - 2^3) + 2d = 5(d - 8) + 2d \Rightarrow 5d - 40 + 2d \Rightarrow 7d - 40$.

13. A) 80°. Explanation: The sum of the internal angles of a triangle always equals 180°. Given that two of the angles are 40° and 60°, you can find the third angle by subtracting their sum from 180°: 180°−(40°+60°)=180°−100°=80°. Thus, the measure of the third angle is 80°.

14. B) 24 units. Explanation: The area formula for a triangle is: $\frac{1}{2} \times \text{base} \times \text{height} = 24$.

Given that one leg (which serves as the base or height) of the right triangle is 6 units, we can find the other leg using this equation. First, solve for the unknown height:

$\frac{1}{2} \times 6 \times \text{height} = 24 \Rightarrow 3 \times \text{height} = 24 \Rightarrow \text{height} = \frac{24}{3} = 8$ units

Now that we know both legs (6 units and 8 units), we can apply the Pythagorean theorem to find the hypotenuse:

hypotenuse $= \sqrt{6^2 + 8^2} = \sqrt{36 + 64} = \sqrt{100} = 10$ units

Finally, the perimeter of the triangle is the sum of all three sides: Perimeter $= 6 + 8 + 10 = 24$ units.

15. A) $x^2 + 7x - 11$. Explanation: simplify as follows.

$$(3x^2 + 4x - 5) - (2x^2 - 3x + 6) = (3x^2 - 2x^2) + (4x - (-3x)) + (-5 - 6) = x^2 + 7x - 11$$

Chapter 7. Arithmetic Reasoning

Introduction

The Arithmetic Reasoning subtest of the AFOQT focuses primarily on solving word problems. It measures your ability to understand and apply mathematical concepts to real-world scenarios. Rather than focusing on abstract calculations, this section requires you to interpret and solve problems based on the information provided in a practical context. These word problems often involve basic arithmetic, percentages, ratios, proportions, and occasionally some elementary algebra. These questions test not only a student's math skills but also their ability to comprehend English, requiring careful interpretation of problem descriptions to determine how to approach the solution.

Success in this section depends on your ability to break down the problem, identify the relevant mathematical operations, and apply logical reasoning to arrive at the correct solution. With practice, you'll become more comfortable navigating these types of questions and sharpening your problem-solving skills.

Basic steps when faced with word problems include:

Understand the Problem: Carefully read and comprehend the problem to identify key instructions and requirements.

Identify Key Information: Keep a mental note of relevant data, units, and mathematical operations required.

Plan and Solve: Translate the word problem into mathematical expressions, equations, or series of calculations, then solve step by step.

Below are examples of the type of word problems you may run into in the AFOQT Arithmetic Reasoning subtest.

Arithmetic Word Problems

Addition/Subtraction Problem

Example 1: After her gym session, Sarah decided to do some shopping. She purchased new workout gear for $35.75 and bought two protein bars from the counter, each costing $1.25. Afterward, she treated herself to a smoothie, which was $6.50. What was the total amount that Sarah spent on these three purchases?

Solution:

Workout Gear: 35.75

Protein Bars: $2 \times 1.25 = 2.50$

Smoothie: 6.50

Total $= 35.75 + 2.50 + 6.50 = 44.75$.

Sarah spent a total of $44.75 on her shopping trip.

Example 2: Charlotte, a baker, starts Monday with an inventory of 5,275 cookies at her bakery. During the day, she sells 600 cookies to customers and bakes more batches to keep up with the demand. By the end of the day, her bakery's cookie count has reached 5,700. How many cookies did Charlotte's bakery bake on Monday?

Solution: Initial Cookie Count (Monday Morning): 5,275

Cookies Sold During the Day: 600

Final Cookie Count (Monday Evening): 5,700

First, find out how many cookies should have been there if no new batches were baked:

Expected Count $= 5275 - 600 = 4675$

Now, calculate how many new cookies were baked to reach the final count of 5,700:

Cookies Baked $= 5700 - 4675 = 1025$

So, Charlotte's bakery baked 1,025 cookies on Monday to replenish the inventory after the sales.

Multiplication/Division Problem

Example 1: As part of a community outreach project, Carlos plans to distribute a 45-page brochure to each of the 20 organizations in his network. In addition, he wants to keep an extra 5% of the total number of brochures as a backup for any additional requests. How many pages will Carlos need to produce in total?

Solution: Step 1: Calculate the total number of brochures required, including the backup:

Total Brochures=20×1.05=21

Step 2: Calculate the total number of pages required: Total Pages=45×21=945.

Carlos will need to produce a total of 945 pages.

Example 2: A biologist needs to collect 528 leaf samples from various tree species in a forest research project. The project requires the biologist to divide the samples evenly among 24 research sites in the forest. How many leaf samples should be collected at each site?

Solution: Divide the total number of samples by the number of sites: Samples per Site $= \frac{528}{24} = 22$.

At each research site, one will need to collect 22 leaf samples.

Maximum/Minimum Problem

Example 1: A chemical storage tank can hold a maximum of 4,500 liters of liquid. Jordan needs to fill it with barrels of ethanol, each containing 400 liters. What is the maximum number of whole barrels that Jordan can fill the tank with?

Solution: Divide the tank's capacity by the volume per barrel:

Maximum Barrels $= \frac{4500}{400} = 11.25$.

Since the tank can only hold whole barrels, round down to the nearest whole number: Maximum Barrels=11. Jordan can fill the tank with a maximum of 11 barrels of ethanol.

Example 2: As part of a new diet, Benjamin plans to drink at least 2,000 milliliters of water each day. He uses a reusable bottle that holds 300 milliliters. How many full bottles does Benjamin need to drink each day to reach his goal?

Solution: Divide the daily water goal by the capacity of the bottle:

Bottles Required $= \frac{2000}{300} \approx 6.67$.

Since Benjamin can only drink whole bottles, round up to the nearest whole number: Bottles Required=7. Benjamin needs to drink at least 7 full bottles of water daily to meet his goal of 2,000 milliliters.

Parts of the Whole Problem

Example 1: In a local astronomy club, members voted on a new logo design. Design A received 36.2% of the votes, Design B garnered 29.4%, and Design C obtained 14.8%. The remaining votes were given to Design D. What percentage of the votes did Design D receive?

Solution: First, calculate the total percentage that Designs A, B, and C collectively obtained: 36.2+29.4+14.8=80.4.

To find the percentage of votes for Design D, subtract the total above from 100%: 100−80.4=19.6. Thus, Design D received 19.6% of the votes.

Example 2: A space research team is studying the different types of stars in a newly discovered star cluster. They determined that $\frac{1}{3}$ of the stars are red dwarfs and $\frac{1}{4}$ are white dwarfs. The remaining stars are either neutron stars or main-sequence stars. What fraction of the stars are neutron stars or main-sequence stars?

Solution: To find the fraction of neutron stars or main-sequence stars, first calculate the total fraction of stars that are either red dwarfs or white dwarfs.

Step 1: Find a common denominator to add the fractions $\frac{1}{3}$ and $\frac{1}{4}$. The least common denominator (LCD) is 12.

Step 2: Add the fractions: $\frac{1}{3} + \frac{1}{4} = \frac{4}{12} + \frac{3}{12} = \frac{7}{12}$. This sum represents the fraction of stars that are either red dwarfs or white dwarfs.

Step 3: Subtract this from 1 to find the fraction of stars that are neutron stars or main-sequence stars.

$$1 - \frac{7}{12} = \frac{12}{12} - \frac{7}{12} = \frac{5}{12}$$

So, $\frac{5}{12}$ of the stars are neutron stars or main-sequence stars.

ALGEBRA WORD PROBLEMS

Algebra Word Problems often require you to set up and solve equations based on real-world scenarios. If you're not already familiar with the process of setting up algebra word problems, please review the "Setting Up Algebra Word Problems" section from the previous chapter before diving into the examples below. Mastering the ability to translate word problems into algebraic expressions is key to solving these types of questions efficiently.

Example 1—Age Problem: Sarah is twice as old as her brother, Tom. In 4 years, the sum of their ages will be 44. How old are Sarah and Tom now?

Solution: Let Tom's current age be x. Since Sarah is twice as old as Tom, her age is $2x$.

In 4 years, Tom's age will be $x + 4$ and Sarah's age will be $2x + 4$. The problem states that the sum of their ages in 4 years will be 44. Therefore, we can set up the following equation: $(x + 4) + (2x + 4) = 44$.

Simplify the equation: $3x + 8 = 44$

Subtract 8 from both sides: $3x = 36$

Now, divide both sides by 3: $x = 12$

So, Tom is 12 years old, and Sarah is: $2x = 2 \times 12 = 24$

Thus, Tom is 12 years old, and Sarah is 24 years old.

Example 2—Distance Problem: A car rental company charges a base fee of $50 plus $0.20 per mile driven. If the total cost to rent a car for a day is $90, how many miles were driven?

Solution: Let the number of miles driven be x.

The total cost is given by the base fee plus the charge per mile, so the equation is: $50 + 0.20x = 90$

Subtract 50 from both sides: $0.20x = 40$

Now, divide both sides by 0.20: $x = \frac{40}{0.20} = 200$

Thus, 200 miles were driven.

GEOMETRY WORD PROBLEMS

Geometry Word Problems in the AFOQT Arithmetic Reasoning subtest test your ability to apply geometric principles to solve real-life problems. You'll encounter problems involving shapes, areas, volumes, perimeters, and other essential geometry concepts. It is crucial to recall formulas related to geometric shapes and be comfortable translating the given information into these formulas.

Example 1—Perimeter of a Rectangle: A rectangular garden has a length that is 5 feet longer than its width. The total perimeter of the garden is 50 feet. What are the dimensions of the garden?

Solution: Let the width of the garden be x The length is 5 feet longer than the width, so the length is $x + 5$.

The formula for the perimeter P of a rectangle is: $P = 2 \times (\text{length} + \text{width})$

Substitute the given values into the formula: $50 = 2 \times ((x + 5) + x)$. Solve the equation:

$$50 = 2 \times (2x + 5)$$
$$25 = 2x + 5$$

$$20 = 2x$$
$$x = 10$$

So, the width of the garden is 10 feet, and the length is: $x + 5 = 10 + 5 = 15$ feet

Thus, the dimensions of the garden are 10 feet by 15 feet.

Example 2—Area of a Triangle:

A triangular plot of land has a base of 12 meters and a height of 8 meters. What is the area of the triangle?

Solution: The formula for the area A of a triangle is: $A = \frac{1}{2} \times \text{base} \times \text{height}$

Substitute the given values into the formula: $A = \frac{1}{2} \times 12 \times 8$

Simplify the calculation: $A = \frac{1}{2} \times 96 = 48$ square meters. Thus, the area of the triangle is 48 square meters.

Example 3—Finding Angles within a Triangle: In a particular triangle, one angle is twice the size of the second angle, and the third angle is 20 degrees larger than the second angle. What are the measures of the three angles?

Solution: Let the second angle be x degrees.

The first angle is twice the size of the second angle, so the first angle is $2x$.

The third angle is 20 degrees larger than the second angle, so the third angle is $x + 20$.

We know that the sum of the three angles in a triangle is always 180 degrees, so we can set up the following equation:

$$2x + x + (x + 20) = 180$$
$$4x + 20 = 180$$
$$4x = 160$$
$$x = 40$$

So, the second angle is 40 degrees. Now, find the other two angles: The first angle is $2x = 2 \times 40 = 80$ degrees. The third angle is $x + 20 = 40 + 20 = 60$ degrees. Thus, the three angles of the triangle are 40 degrees, 80 degrees, and 60 degrees.

Arithmetic Reasoning vs Mathematics Knowledge Subtest

In this chapter, we explored the various types of word problems you may encounter on the AFOQT Arithmetic Reasoning subtest, including arithmetic, algebra, and geometry problems. Each type of problem requires a different approach, but they all emphasize the ability to apply mathematical concepts to real-world situations. While the Arithmetic Reasoning subtest focuses on word problems and logical reasoning, it's important to recognize the blurred line between this section and the Mathematics Knowledge subtest. Both assess your mathematical ability, but Arithmetic Reasoning emphasizes comprehension and problem-solving in everyday contexts, while the Mathematics Knowledge subtest leans more toward direct calculations and the understanding of formal mathematical principles.

ARITHMETIC REASONING PRACTICE SET

Select the correct answer from the choices given. For practice only. This practice set does not reflect the actual number of questions in the test.

1. A concert organizer charged $12 for each ticket. The organizer collected a total of $288 from ticket sales. How many tickets were sold?

A) 20
B) 22
C) 24
D) 26

2. The wholesale price of a chair is $80. The retail price of the chair is 15% more than the wholesale price. What is the retail price of the chair?

A) $90
B) $92
C) $94
D) $96

3. Lucas bought a 24-pack of water bottles. The pack was $\frac{1}{3}$ full before Lucas decided to refill it with new bottles. Each new bottle cost $0.75. How much did Lucas spend, in dollars, to refill the pack to its full capacity?

A) $12.00
B) $10.50
C) $9.75
D) $11.25

4. Liam bought some apples for $1 each and some pineapples for $4 each. He bought 2 more apples than pineapples and spent a total of $17. How many pineapples did Liam buy?

A) 1
B) 2
C) 3
D) 4

5. At 7 a.m., a hiker at Death Valley National Park was at an elevation of -200 feet relative to sea level. By 3 p.m. on the same day, the hiker reached an elevation of 1300 feet above sea level. What was the change in elevation, in feet, for the hiker from 7 a.m. to 3 p.m.?

A) -100 feet
B) 1200 feet
C) 1300 feet
D) 1500 feet

6. Sarah works at a library and a cafe. In a 30-day period, Sarah worked $\frac{1}{3}$ of the days at the library and did not work $\frac{1}{6}$ of the days. On the remaining days Sarah worked at the cafe. How many days did Sarah work at the cafe during the 30-day period?

A) 15
B) 10
C) 20
D) 5

7. If a climber ascends 25% of a 1,300 ft mountain in a day, how many feet does the climber ascend?

A) 325 ft
B) 25 ft
C) 375 ft
D) 97.5 ft

8. A conveyor belt is currently set to move at a speed of 5.921 meters per minute (MPM). The speed is increased to 6.088 MPM. By how much was the speed increased?

A) 0.167 MPM
B) 1.167 MPM
C) 1.833 MPM
D) 1.967 MPM

9. A faucet fills a bathtub at a rate of 2 gallons per minute. If the bathtub has a capacity of 160 gallons, how long will it take to fill the bathtub?

A) 10 minutes
B) 50 minutes
C) 80 minutes
D) 100 minutes

10. The volume of a cube is given by the formula $V = s^3$, where s is the length of a side of the cube. If the volume of a cube is 64 cubic inches, what is the length of each side?

A) 4 inches
B) 8 inches
C) 16 inches
D) 32 inches

11. Tim's total sales for the first four months of the year were $4500, $5200, $4800, and $5100. If he wants his average monthly sales to be $5000 after the fifth month, how much does he need to sell in the fifth month?

A) $5000
B) $5200
C) $5400
D) $4900

12. A salesman earns a commission of 7% of the total value of merchandise he sells. If he sells $48,000 worth of merchandise in one month, how much money will he earn in commission?

A) $1960
B) $2100
C) $3360
D) $3750

13. It costs $1.20 per square foot to paint a wall. How much will it cost to paint a 12-foot-by-20-foot wall?

A) $288.00
B) $240.00
C) $144.00
D) $200.00

14. A bakery has a monthly overhead of $4,500. It costs $1.50 to make each cupcake, and the cupcakes sell for $3.00 each. How many cupcakes must the bakery sell each month in order to make a profit?

A) 2,500
B) 3,000
C) 4,000
D) 3,500

15. Lily starts walking north at 3 miles per hour (mph) from her house. One hour later, her sister Emma realizes Lily forgot her water bottle and begins jogging after her at 6 miles per hour. How long will it take Emma to catch up with Lily?

A) 1 hour
B) 1.5 hours
C) 2 hours
D) 2.5 hours

Answers and Explanations

1. C) 24. Explanation: $\frac{288}{12} = 24$

2. B) $92. Explanation: $80 + 80 \times \frac{15}{100} = 80 + 12 = 92$

3. A) $12.00. Explanation: Calculate the Number of Bottles Needed: $24 \times \left(1 - \frac{1}{3}\right) = 24 \times \frac{2}{3} = 16$ bottles.

Then calculate the Cost to Refill the Pack: $16 \times 0.75 = 12.00$.

4. C) 3. Explanation: Let x be the number of pineapples. Then, $x + 2$ is the number of apples.

Hence, $1(x + 2) + 4x = 17 \Rightarrow x + 2 + 4x = 17 \Rightarrow 5x + 2 = 17 \Rightarrow 5x = 15 \Rightarrow x = 3$.

Thus, Liam bought 3 pineapples.

5. D) 1500 feet. Explanation: Change in elevation $= 1300 - (-200) = 1500$.

6. B) 10. Explanation: Total days $= 30$. Days at the library $= \frac{1}{3} \times 30 = 10$.

Days not worked $= \frac{1}{6} \times 30 = 5$. Remaining days $= 30 - 10 - 5 = 15$. Days at the cafe $= 15$.

$$30 - \left(\frac{1}{3} \times 30\right) - \left(\frac{1}{6} \times 30\right) = 10$$

7. A) 325 ft. Explanation: $0.25 \times 1300 = 325$ ft.

8. A) 0.167 MPM. Explanation: $6.088 - 5.921 = 0.167$ MPM

9. C) 80 minutes. Explanation: $\frac{160 \text{ gallons}}{2 \text{ gallons per minute}} = 80$ minutes.

10. A) 4 inches. Explanation: $s^3 = 64 \Rightarrow s = \sqrt[3]{64} = 4$.

11. C) $5400. Explanation: The total sales over five months for an average of $5000 would be: $5 \times 5000 = 25000$.

Now, add up the sales for the first four months: $4500 + 5200 + 4800 + 5100 = 19600$.

Subtract the current total from the required total to find the amount Tim needs to sell in the fifth month: $25000 - 19600 = 5400$. Thus, $5400 of sales is needed in the 5th month to reach an average of $5000.

12. C) $3360. Explanation: To find the commission the salesman will earn, we calculate 7% of $48,000:

$0.07 \times 48000 = 3360$. Thus, the salesman will earn $3360 in commission.

13. A) $288.00. Explanation: To find the cost of painting the wall, first calculate the area of the wall:

$$\text{Area} = 12 \times 20 = 240 \text{ square feet}$$

Now, multiply the area by the cost per square foot: $240 \times 1.20 = 288.00$.

14. B) 3,000. Explanation:

Each cupcake sells for $3.00, and it costs $1.50 to make, so the profit per cupcake is: 3.00 - 1.50 = 1.50.

Now, divide the bakery's monthly overhead by the profit per cupcake: $\frac{4500}{1.50} = 3000$

Thus, the bakery must sell 3,000 cupcakes each month to make a profit.

15. A) 1 hour. Explanation:

By the time Emma starts jogging, Lily has already been walking for 1 hour at 3 mph, so Lily is: $3 \times 1 = 3$ miles ahead. Emma is jogging 6 mph while Lily walks at 3 mph, so Emma is closing the gap at: $6 - 3 = 3$ mph.

To catch up with Lily, Emma must close the 3-mile gap. The time it will take her to do so is: $\frac{3 \text{ miles}}{3 \text{ mph}} = 1$ hour.

Chapter 8. Word Knowledge

Introduction

The Word Knowledge subtest of the AFOQT measures a candidate's ability to understand the meanings of words through definitions and synonyms. This section evaluates vocabulary skills, which are crucial not only for communication but also for succeeding in various military roles that involve reading, writing, intelligence gathering, and following instructions. A solid grasp of vocabulary is key to performing well in other AFOQT subtests as well, such as Paragraph Comprehension.

In the Word Knowledge subtest, you'll encounter questions where you're given a word and asked to select the best synonym or the closest meaning from four answer choices. The test aims to assess your understanding of word meanings based on context and your ability to decipher unfamiliar words.

Example Questions

Question 1: The word diligent most nearly means:

A) Careless
B) Hardworking
C) Lazy
D) Uncertain

Answer: B) Hardworking

Explanation: "Diligent" describes someone who is hardworking and consistently puts effort into their tasks. By eliminating opposites like "careless" and "lazy," you can narrow down to the correct answer.

Question 2: The weather was harsh during the hike.

A) mild
B) gentle
C) extreme
D) pleasant

Answer: C) extreme
Explanation: "Harsh" means severe or extreme, so the closest synonym is "extreme."

Question 3: The word most opposite in meaning to indifferent is:

A) uninterested
B) passionate
C) neutral
D) careless

Answer: B) passionate
Explanation: "Indifferent" means lacking interest or concern, so the antonym is "passionate," which indicates strong interest or intense emotion.

These examples show how the Word Knowledge subtest focuses on assessing your vocabulary through quick identification of word meanings. The key to success is building a strong vocabulary and developing strategies to approach unfamiliar words. In the next sections, we will cover specific techniques to decipher words and improve your overall performance on this subtest.

Building Blocks of Words

Understanding how to break down a word into its parts—prefixes, suffixes, and roots—is an essential skill for improving your vocabulary and performing well on the AFOQT Word Knowledge subtest. Many words in the English language are derived from Greek and Latin, and by learning the meaning of common word parts, you can often infer the meaning of unfamiliar words. This section will explain how to use these components to decode word meanings effectively.

Prefixes and Suffixes

By recognizing common prefixes (beginning parts of words) and suffixes (endings of words), you can often deduce the general meaning of words, even if you've never encountered them before. Both Greek and Latin origins heavily influence English, making these word components valuable tools for decoding language.

Prefixes

Prefixes appear at the start of a word and modify its meaning, often giving clues about direction, negation, or intensity. By learning some common prefixes, you can better guess the meaning of a word.

For example:

- Anti- (against): *Antibiotic* – a substance that works *against* bacteria.
- Pre- (before): *Preview* – a look *before* the main event.
- Re- (again): *Rewrite* – to write *again*.

Below is a table of **common prefixes**:

Prefix	Variations	Meaning	Examples
Anti-	Ant-	Against or opposite	Anti-inflammatory, antagonist
De-		Opposite	Decontaminate, deconstruct
Dis-		Not or opposite	Disagree, discontent, disable
En-	Em-	Cause to	Encourage, empower, embolden
Ex-		Out of, from, former	Export, ex-husband, exclude
In-	Im-, Il-, Ir-	Not	Incapable, impossible, illegitimate, irrational
Inter-		Between or among	Interact, international, interconnect
Mis-		Wrongly	Misunderstand, misplace, misuse
Non-		Not	Noncompliant, nonsense
Over-		Too much	Overcook, overestimate, overload
Post-		After	Postpone, post-war, postscript
Pre-		Before	Predict, pretest, preheat
Pro-		In favor of, forward	Proactive, promote, propel
Re-		Again or back	Rewrite, redo, replay
Semi-		Half, partly	Semicircle, semiannual, semiconscious
Sub-		Under	Submarine, substitute, subway
Super-		Above or over	Superhero, superimpose, supernatural
Trans-		Across or beyond	Transport, transmit, translate
Under-		Too little, below	Underestimate, underpaid, underground
Un-		Not	Unfair, unjust

Suffixes

Suffixes are added to the end of a word and often indicate the word's function in a sentence, such as whether it's a noun, verb, adjective, or adverb. They can also modify the meaning to reflect qualities, actions, or conditions.

For example:

- -able (capable of): *Manageable* – something that is capable of being managed.
- -ful (full of): *Hopeful* – full of hope.
- -ly (in the manner of): *Quickly* – done in a quick manner.

Below is a table of **common suffixes**:

Suffix	Variations	Meaning	Examples
-able	-ible	Capable of being	Readable, edible, manageable
-al		Relating to, characteristic of	Personal, natural, cultural
-ed		Past tense, having been	Walked, jumped, decorated
-en		To make or become	Strengthen, lengthen, darken
-er	-or	One who, more (comparative)	Teacher, actor, greater
-ful		Full of	Hopeful, thankful, beautiful
-ic	-ical	Pertaining to, related to	Artistic, musical, logical
-ing		Present participle, action	Running, cooking, swimming
-ion	-tion, -ation	The action or state of	Celebration, operation, promotion
-ist		A person who practices	Artist, scientist, biologist
-ity	-ty	State or quality	Clarity, responsibility, certainty
-ive	-ative, -itive	Having the nature of	Creative, sensitive, addictive
-ize	-ise	To make or become	Realize, organize, maximize
-less		Without	Hopeless, fearless, meaningless
-ly		In a manner	Quickly, softly, happily
-ment		Action or process	Development, engagement, movement
-ness		State of, quality of	Kindness, darkness, weakness
-ous	-eous, -ious	Full of, having qualities of	Curious, dangerous, mysterious
-ship		Position held, condition	Friendship, leadership, partnership
-sion	-tion	State or result of	Explosion, confusion, transition
-y		Characterized by, full of	Happy, sunny, tricky

By mastering common prefixes and suffixes, you can greatly improve your ability to infer the meaning of unfamiliar words on the AFOQT. Prefixes give you clues about the basic meaning or direction of the word, while suffixes often define the word's part of speech and refine its meaning.

Root Words

A root word is the core part of a word that carries the most essential meaning. Prefixes and suffixes are added to root words to modify their meaning, but the root remains central to understanding the word's fundamental idea. By learning common Greek and Latin roots, you'll be able to break down complex words and infer their meanings, even if you've never seen them before. Below are tables of commonly used Greek and Latin roots, their meanings, and examples.

a. Common Greek Root Words

Root	Meaning	Examples
Bio	Life	Biology, biography, biosphere
Geo	Earth	Geography, geology, geocentric
Therm	Heat	Thermometer, thermal, thermostat
Phon	Sound	Telephone, phonograph, symphony
Graph	Write or draw	Autograph, graphic, photography
Chron	Time	Chronology, chronic, synchronize
Hydr	Water	Hydration, hydroplane, hydroelectric
Path	Feeling or disease	Empathy, pathology, sympathy
Scope	See, view	Telescope, microscope, periscope
Tele	Distant	Telephone, television, telegraph

Root	Meaning	Examples
Psych	Mind	Psychology, psychiatrist, psyche
Auto	Self	Autonomy, autobiography, automatic
Macro	Large	Macroeconomics, macroscopic
Micro	Small	Microscope, microchip, microorganism
Mono	One	Monologue, monotheism, monopoly
Poly	Many	Polygon, polytheism, polygamy
Anti	Against	Antibiotic, antidote, anticlimax
Astro	Star, outer space	Astronomy, astronaut, asteroid
Photo	Light	Photograph, photosynthesis, photon
Meter	Measure	Thermometer, speedometer, barometer

b. Common Latin Root Words

Root	Meaning	Examples
Aud	Hear	Audible, audience, auditory
Bene	Good	Benefit, benevolent, benefactor
Cent	One hundred	Century, percent, centennial
Dict	Say, speak	Dictate, dictionary, predict
Form	Shape	Formation, reform, uniform
Ject	Throw	Eject, project, injection
Lum	Light	Illuminate, luminous, luminary
Mal	Bad	Malfunction, malnourished, malice
Mater	Mother	Maternal, maternity, matriarch
Mort	Death	Mortal, mortuary, immortality
Port	Carry	Transport, portable, import
Rupt	Break	Rupture, erupt, disrupt
Script	Write	Manuscript, prescription, inscription
Spect	See, look	Spectator, inspect, perspective
Struct	Build	Construct, structure, infrastructure
Tract	Pull, drag	Tractor, attract, contract
Vid/Vis	See	Video, vision, visible
Voc	Call	Vocal, invoke, advocate
Scrib	Write	Scribble, describe, inscription
Vit/Viv	Life	Vital, survive, revive

Applying Root Words to Decode Meanings

By recognizing these root words, you can better understand unfamiliar terms. For instance, if you encounter the word "auditory", knowing that "aud" means "hear" helps you quickly conclude that this word relates to hearing. Similarly, if you see "benevolent", you can break it down into "bene" (good) and "volent" (wishing), which tells you it means someone who is kind or wishing well.

Here is one more example: “thermometer”. The root "therm" relates to heat, and the root "Meter" means “measure” or “measuring device”. Hence, it makes sense that “thermometer” means a device that measures heat or temperature.

By learning and applying these common prefixes, suffixes, Greek and Latin root words, you can greatly improve your ability to decode unfamiliar words on the AFOQT Word Knowledge subtest. These roots not only help you determine the meaning of individual words but also strengthen your overall vocabulary, providing an edge in other parts of the AFOQT and beyond.

Inferring a Word's Meaning Through Context

One of the most effective strategies for deciphering unfamiliar words on the AFOQT Word Knowledge subtest is using context clues. These are hints or information within the sentence or passage that help you infer the meaning of an unknown word. There are several types of context clues that can assist you in determining the meaning without needing to know the exact definition. Let's explore the different kinds of context clues:

1. Definition Context Clues

Sometimes, the author provides a direct definition or explanation of the word right in the sentence or nearby. This is often signaled by phrases like "which means" or "that is."

Example: The artifact, which is an ancient object, was found in the ruins.

Explanation: The sentence directly defines the word "artifact" as an ancient object.

2. Synonym/Restatement Context Clues

In some instances, the unfamiliar word is explained or restated using a synonym, a word with the same or similar meaning, within the sentence.

Example: The student was meticulous, or very careful, in completing her project.

Explanation: The word "meticulous" is restated using its synonym "very careful," helping to clarify its meaning.

3. Antonym/Contrast Context Clues

Words with opposite meanings, known as antonyms, are sometimes used to infer the meaning of an unfamiliar word. Words like "but," "however," and "on the other hand" signal a contrast in meaning.

Example: Unlike her apathetic classmates, Jane was deeply concerned about the environment.

Explanation: By contrasting Jane's behavior with her classmates' indifference, you can infer that "apathetic" means showing little or no interest or concern.

4. Example Context Clues

Sometimes, the word is explained by providing examples. Words like "such as," "including," or "for example" often introduce these context clues.

Example: Large predators, such as lions, wolves, and sharks, are at the top of their respective food chains.

Explanation: By listing examples of "predators" like lions, wolves, and sharks, the sentence helps clarify the meaning of the word.

5. Cause and Effect Context Clues

In this type of clue, you infer the meaning of a word based on the cause or effect presented in the sentence.

Example: The river overflowed due to the torrential downpour, flooding the nearby homes.

Explanation: Since the downpour caused the river to overflow, you can infer that "torrential" means something very heavy or intense.

6. General Inference Context Clues

Sometimes, the meaning of a word can be inferred based on the general mood or atmosphere of the sentence or passage, even if no direct clues are given.

Example: The dark clouds and distant thunder gave the day a foreboding feeling.

Explanation: The overall mood suggests that "foreboding" relates to a sense of fear or something bad about to happen.

Practice Example: The professor's lecture was convoluted, making it difficult for the students to understand.

A) clear
B) complicated
C) engaging
D) simple

Answer: B) complicated
Explanation: The context indicates that the lecture was difficult to understand, which points to the word "convoluted" meaning complicated or intricate.

By using these different types of context clues—definition, synonym, antonym, example, cause and effect, and general inference—you can oftentimes figure out the meaning of unfamiliar words.

INFERRING MEANING BY IDENTIFYING CONNOTATION

Another effective strategy for deciphering unfamiliar words on the AFOQT Word Knowledge subtest is understanding whether the word carries a positive or negative connotation. Connotation refers to the implied or emotional meaning behind a word, beyond its literal definition. Words can evoke feelings or ideas that hint at whether they are used to describe something favorable or unfavorable. This technique can help you quickly eliminate wrong answer choices and guide you toward the correct one, even if you're not entirely familiar with the word itself.

1. Positive Connotation

Words with a positive connotation suggest something good, desirable, or favorable. They often evoke feelings of happiness, approval, or success. Even if you don't know the exact definition of a word, recognizing a positive connotation can help you choose the most appropriate synonym or antonym.

Example Word: Admirable
Connotation: Positive
Meaning: Worthy of admiration or praise.
Associated Words: Respectable, commendable.

If the word carries a positive connotation, you can rule out answer choices that have negative implications.

2. Negative Connotation

Words with a negative connotation convey something undesirable, harmful, or unpleasant. These words typically suggest disapproval, failure, or negativity. Understanding this can help you identify the correct answer by eliminating words with positive meanings.

Example Word: <u>Hostile</u>
Connotation: Negative
Meaning: Unfriendly or antagonistic.
Associated Words: Aggressive, belligerent.

If a word has a negative connotation, you can eliminate answer choices that describe positive or neutral ideas.

3. Neutral Connotation

Some words may have a neutral connotation, meaning they neither evoke strong positive nor negative feelings. These words often describe things factually, without implying approval or disapproval.

Example Word: <u>Academic</u>
Connotation: Neutral
Meaning: Relating to education or scholarly work.
Associated Words: Educational, scholarly.

Understanding connotation can help you interpret the word in its context and lead you to the correct answer.

Practice Example 1: The <u>turbulent</u> weather caused the flight to be delayed.

A) calm
B) stormy
C) glorious
D) pleasant

Answer: B) stormy
Explanation: "caused the flight to be delayed" has a negative connotation. Here, we derived the connotation not from the word being tested, but from the context of the sentence. It has to be a negative weather pattern that led to a delayed flight. Hence, you would eliminate positive or neutral answer choices like "calm", "glorious", or "pleasant" and lean toward an answer like "stormy".

Practice Example 2: The treaty was seen as a **pivotal** moment in the history of the conflict.

A) minor
B) insignificant
C) crucial
D) irrelevant

Answer: C) crucial
Explanation: The word "pivotal" has a positive and significant connotation, meaning something that is of great importance or critical. Words like "minor" and "insignificant" have opposite, less impactful meanings. The best answer is "crucial," which matches the importance implied by "pivotal."

Use This Strategy on the AFOQT

When approaching a question in the Word Knowledge subtest, look at the context in which the word is used. Ask yourself:

- Does this word suggest something good, bad, or neutral?
- Can I match the connotation to one of the answer choices?
- Can I eliminate choices that clearly do not fit the tone of the word?

This strategy works especially well when you're unsure about the exact meaning of a word but can sense whether it's being used in a positive, negative, or neutral way.

By practicing this approach, you can improve your ability to infer meanings on the AFOQT, even when you're unfamiliar with the specific word.

Inferring Meaning by Replacing with Answer Choices

A powerful strategy to use on the AFOQT Word Knowledge subtest is replacing the unfamiliar word with the answer choices. This method allows you to test each possible answer by fitting it into the sentence where the unfamiliar word appears, helping you determine which option makes the most sense in context. This approach is particularly helpful when you're unsure of a word's meaning but have a sense of the overall meaning of the sentence.

Example

Original Sentence: The scientist's hypothesis was <u>tenuous</u>, lacking sufficient evidence to be convincing.

A) strong
B) flimsy
C) irrefutable
D) elaborate

Step 1: Read the sentence carefully. The context suggests that the scientist's hypothesis is weak or not well-supported.

Step 2: Replace the word "tenuous" with each answer choice.

A) strong

"The scientist's hypothesis was strong, lacking sufficient evidence to be convincing."

(This doesn't make sense because "strong" contradicts the idea of lacking evidence.)

B) flimsy

"The scientist's hypothesis was flimsy, lacking sufficient evidence to be convincing."

(This fits well since "flimsy" implies weak or not convincing, matching the sentence.)

C) irrefutable

"The scientist's hypothesis was irrefutable, lacking sufficient evidence to be convincing."

(This doesn't fit because "irrefutable" means something that cannot be disproven, which doesn't match the idea of lacking evidence.)

D) elaborate

"The scientist's hypothesis was elaborate, lacking sufficient evidence to be convincing."

(While "elaborate" suggests something detailed or complex, it doesn't align with the idea of lacking evidence.)

Step 3: Eliminate options that don't make sense, which include "strong," "irrefutable," and "elaborate."

Step 4: Choose the best-fitting answer. In this case, B) flimsy is the correct choice.

Why This Strategy Works

This method works because it allows you to focus on the overall meaning of the sentence, which often provides enough information to guide you toward the right answer. Even if you don't know the meaning of the unfamiliar word, the context in which it is used and the logical flow of the sentence can help you find the correct answer.

Practice Example: The king's authority was <u>unassailable</u>; no one could question or challenge it.

A) vulnerable
B) invincible
C) questionable
D) uncertain

Answer: B) invincible
Explanation: The sentence indicates that the king's authority was beyond challenge, which aligns with the meaning of "invincible."

The replacement strategy is a simple yet effective tool for figuring out unfamiliar words. By substituting each answer choice into the sentence, you can quickly eliminate a few incorrect options and narrow down to the ones that best fit the context. This method helps you make logical, informed guesses, even when you are unfamiliar with the word itself.

Guessing a Choice Through Parts of Speech

Understanding a word's part of speech—whether it's a noun, verb, adjective, or adverb—you can often eliminate incorrect answer choices and make a better-informed guess.

The Four Key Parts of Speech

1. Nouns: Words that represent a person, place, thing, or idea.

   Example: The house was built last year.
   In this sentence, "house" is a noun, as it refers to a place.

2. Verbs: Words that describe an action, occurrence, or state of being.

   Example: She runs every morning.
   Here, "runs" is the verb, indicating an action.

3. Adjectives: Words that describe or modify nouns.

   Example: The bright sun shone all day.
   "Bright" is an adjective because it describes the noun "sun."

4. Adverbs: Words that modify verbs, adjectives, or other adverbs, often describing how something is done.

   Example: She ran quickly to catch the bus.
   "Quickly" is an adverb because it modifies the verb "ran."

Example

Original Sentence: The students were <u>assiduous</u>, always making sure to complete their homework on time.

The word "assiduous" most nearly means:

A) slow
B) careful
C) intelligence
D) lazy

Step 1: Identify the part of speech. The word "assiduous" is used to describe the students, which makes it an adjective.

Step 2: Eliminate options that don't match the part of speech. After "The students were", the sentence needs an adjective. Choice A, B, D are adjectives, but C) intelligence is a noun, which can be eliminated based on part of speech.

Step 3: Replace the word "assiduous" with choice A, B, D to see which fits best.

- "Slow" doesn't fit the context of completing homework on time.
- "Careful" fits, as it suggests paying attention to the task, which matches the idea of doing homework properly.
- "Lazy" contradicts the context of the sentence, which implies hard work and dedication.

Answer: B) careful
Explanation: "Assiduous" is an adjective meaning careful or hardworking, which aligns with the context of students consistently completing their homework on time.

Practice Example: The diplomat's response was circumspect, carefully avoiding any controversial remarks.

A) reckless
B) cautious
C) spontaneous
D) prudence

Answer: B) cautious
Explanation: We can immediately eliminate D) Prudence, which is a noun, not an adjective, and doesn't fit the sentence grammatically. "Circumspect" means being cautious or wary, especially in speech or action. In this context, the word "cautious" matches the diplomat's careful approach to avoiding controversy.

GUESSING A CHOICE THROUGH THE ELIMINATION METHOD

When you are unsure of the correct answer, you can improve your chances of selecting the right one by eliminating obviously incorrect choices. This method is especially useful when the word is completely unfamiliar, as it allows you to rule out options that don't fit, thus increasing your odds of guessing correctly.

Steps to Use the Elimination Method:

1. Read the Sentence Carefully: Start by understanding the context of the sentence. Even if you don't know the meaning of the word, the surrounding words and overall sentence structure can give you clues.

2. Eliminate Clearly Wrong Choices: Begin by eliminating any answer choices that are clearly incorrect based on context or part of speech. For example, if the word you're analyzing is an adjective, eliminate any options that are verbs or nouns.

3. Compare Remaining Choices: Once you've narrowed down the choices, evaluate the remaining options in the sentence to see which one makes the most sense. Consider both meaning and connotation (whether the word has a positive, negative, or neutral tone).

4. Make an Educated Guess: After eliminating the weakest choices, select the answer that best fits the sentence context, even if you are not entirely sure of the word's exact meaning.

Example:

Original Sentence: The witness's reticent nature made it difficult to get detailed information from her.

A) talkative
B) mute
C) silent
D) nervous

Step 1: Read the sentence carefully. The sentence suggests that the witness was not providing detailed information, indicating that "reticent" likely relates to withholding or not speaking much.

Step 2: Eliminate clearly wrong choices. "Talkative" (A) is the opposite of what is described, so it can be eliminated. "Nervous" (D) may seem plausible but doesn't directly fit the context of not providing information.

Step 3: Compare remaining choices. "Silent" (C) fits well, as being quiet could explain why the witness wasn't sharing details. "Mute" (B) has the right connotation, but is generally not used to describe someone's nature or personality.

Step 4: Make an educated guess. The word "silent" (C) makes the most sense given the context of the witness's reluctance to speak.

Answer: C) silent
Explanation: "Reticent" means being reserved or unwilling to share information, which aligns with the idea of someone being quiet or silent during questioning.

When everything else fails, the elimination method is a powerful strategy that allows you to systematically rule out incorrect answers and improve your chances of making an educated guess. Even when faced with unfamiliar words, you can use context and logic to eliminate the least likely options and focus on those that fit best. In the next section, we'll discuss how to build your vocabulary over time to improve overall performance on the test.

Increase Your Vocabulary in a Hurry

Let's face the truth, a large vocabulary takes years to build. If you have only 2 to 3 months before your AFOQT test, you're running out of time to increase your vocabulary the organic way: reading widely. Don't despair though, you can still adopt efficient and focused strategies to maximize your vocabulary growth in a short period of time. Here are some realistic steps you can take:

1. Invest in a SAT or AFOQT Vocabulary Book

SAT or AFOQT vocabulary books are excellent tools for learning test-relevant words in a structured way. These books provide lists of high-frequency words, definitions, and examples to help you understand their usage. Aim to study around 20-30 words a day. Dedicate about 30-45 minutes each day to reviewing and memorizing new words. Focus on learning their meanings, usage in sentences, and common synonyms/antonyms.

2. Use Flashcards/Apps for Active Recall

Flashcards/Apps are helpful for quick vocabulary building. The easiest is to use apps like Quizlet or Anki that already have word sets for AFOQT and SAT prep. Focus on active recall by testing yourself daily. Start with 10-15 cards, and gradually increase the number as you master words. Use spaced repetition, a technique where you review words at increasing intervals, to reinforce your memory. This is crucial for retaining new vocabulary in a short timeframe.

3. Focus on Word Roots, Prefixes, and Suffixes

Learning Greek and Latin root words, along with prefixes and suffixes, helps you infer the meaning of unfamiliar words. This strategy allows you to break down words and grasp their meanings more quickly. Study 5-10 roots or affixes each day and practice applying them to unfamiliar words.

WORD KNOWLEDGE PRACTICE SET

Each question below includes an underlined word. You may be asked to determine which one of the four answer choices most closely matches the meaning of the underlined word, or which choice has the opposite meaning. If the word is used within a sentence, your task is to select the option that most accurately reflects its meaning in the context of that sentence. For practice only. This practice set does not reflect the actual number of questions in the test.

1. The word abhor most nearly means:

A) love
B) tolerate
C) detest
D) ignore

2. The word abridge most nearly means:

A) extend
B) shorten
C) complicate
D) dismiss

3. The word most opposite in meaning to cognizant is:

A) aware
B) unaware
C) mindful
D) conscious

4. The student's response was concise, giving only the necessary information.

A) long-winded
B) vague
C) brief
D) confusing

5. The artist's work was praised for its novel approach to traditional techniques.

A) outdated
B) original
C) copied
D) repetitive

6. The word fortuitous most nearly means:

A) unlucky
B) random
C) accidental
D) lucky

7. The word malleable most nearly means:

A) rigid
B) flexible
C) breakable
D) unchanging

8. The word frugal most nearly means:

A) wasteful
B) generous
C) economical
D) lavish

9. The professor's disdain for lazy students was obvious.

A) respect
B) contempt
C) affection
D) compassion

10. The committee will convene next week to discuss the proposal.

A) dismiss
B) meet
C) argue
D) ignore

11. The politician's brazen actions left the public shocked.

A) timid
B) bold
C) cautious
D) uncertain

12. The weather forecast indicated a precarious situation with the incoming storm.

A) safe
B) dangerous
C) boring
D) mild

13. The word obstinate most nearly means:

A) stubborn
B) agreeable
C) indifferent
D) flexible

14. The word infallible most nearly means:

A) imperfect
B) foolproof
C) weak
D) flawed

15. The word animosity most nearly means:

A) hatred
B) love
C) indifference
D) Support

Answers and Explanations

1. C) detest. "Abhor" refers to an intense hatred or disgust. Words like "love" and "tolerate" are opposites, making "detest" the best match.

2. B) shorten. "Abridge" means to reduce or shorten something. The opposite would be to extend, making "shorten" the correct answer.

3. B) unaware. "Cognizant" means being aware or knowledgeable, so the antonym is "unaware."

4. C) brief. "Concise" means giving a lot of information in a brief and clear manner, making "brief" the closest synonym.

5. B) original. "Novel" in this context means new and original, making "original" the best synonym.

6. D) lucky. "Fortuitous" refers to something happening by chance in a positive way, making "lucky" the best match.

7. B) flexible. "Malleable" describes something that can be shaped or changed easily, such as a flexible material or a person's opinions.

8. C) economical. "Frugal" refers to being careful with money and resources, making "economical" the closest synonym.

9. B) contempt. "Disdain" means a feeling of contempt or scorn, especially towards something regarded as unworthy.

10. B) meet. "Convene" means to come together for a meeting or gathering.

11. B) bold. "Brazen" refers to bold or shameless actions, especially in a negative sense.

12. B) dangerous. "Precarious" means unstable or dangerous, so the closest synonym is "dangerous."

13. A) stubborn. "Obstinate" means being stubborn or unyielding, unwilling to change one's opinion.

14. B) foolproof. "Infallible" means incapable of making mistakes or being wrong, making "foolproof" the best fit.

15. A) hatred. "Animosity" describes strong hostility or dislike.

CHAPTER 9. READING COMPREHENSION

INTRODUCTION

The Paragraph Comprehension subtest of the AFOQT assesses your ability to read, interpret, and extract information from written passages. This subtest evaluates how well you can understand, synthesize, and infer information from a variety of texts—a vital skill in both military and civilian careers.

In this subtest, you'll encounter four main types of questions that are designed to measure your reading comprehension skills. Each question type requires a different approach, and understanding how to tackle them effectively will help you maximize your score. The question types are as follows.

Main Idea Questions: These questions ask you to identify the primary point or central theme of a passage. You'll need to focus on what the author is trying to communicate overall, rather than on specific details.

Detail Questions: Detail questions require you to focus on specific facts or statements found within the text. These questions will test your ability to locate and recall particular pieces of information, often presented in a straightforward manner.

Inference Questions: Inference questions require you to deduce conclusions that are not explicitly stated but are implied by the text. You'll need to analyze the passage and draw logical conclusions based on the information provided.

Vocabulary in Context: These questions ask you to determine the meaning of a word based on how it is used within the passage. You'll need to analyze the surrounding words and sentences to figure out the most appropriate definition.

We now discuss these question types in greater detail.

QUESTIONS REGARDING MAIN IDEAS AND THEMES

The ability to determine central ideas and themes in a passage is pivotal for a test taker preparing for the AFOQT. This skill assesses your capacity to grasp the essence of what you read, identifying the main point or message the author intends to communicate and the underlying concepts or insights that recur throughout the text.

Understanding the central idea and themes is crucial because it forms the backbone of comprehension. It's about seeing beyond the words to grasp the 'why' and 'what' of the text. This ability is foundational, enhancing your reading comprehension and preparing you for academic success.

Strategies for Success:

- **Preview and Predict:** Start by examining titles, headings, and introductory sentences to determine the passage's main focus.
- **Note Repetitions:** Pay attention to ideas, phrases, or motifs repeated throughout the text; these often indicate key themes.
- **Summarize:** Try to encapsulate the passage's main point in a single sentence. This exercise forces you to distill the essence of what you've read.
- **Ask Questions:** While reading, continually ask yourself, "What's the main point here?" and "What themes are emerging?"
- **Context Clues:** Use context clues around unfamiliar words or concepts to understand their importance to the central idea or themes.

Application in Test Preparation:

To prepare for this question type, engage with a wide range of reading materials. After reading a piece, practice writing a brief summary of the central idea and list out potential themes. Discuss these with peers or mentors to explore different perspectives and interpretations. Utilize practice tests to familiarize yourself with how the AFOQT frames these questions, and review explanations for both correct and incorrect answers to deepen your understanding.

Common Pitfalls and How to Avoid Them:

- **Getting Lost in Details:** Don't let minor details distract you from the overall message. Always tie specifics back to the main idea or themes.
- **Overgeneralization:** Avoid too broad interpretations. Ensure your understanding of the central idea or themes is specific and supported by the text.
- **Misinterpreting the Text:** This can happen if you rush or skim too quickly. Take your time to fully engage with the passage, rereading complex or dense parts.

Example Passage 1

The development within the young of the attitudes and dispositions necessary to the continuous and progressive life of a society cannot take place by direct conveyance of beliefs, emotions, and knowledge. It takes place through the intermediary of the environment. The social environment consists of all the activities of fellow beings that are bound up in the carrying on of the activities of any one of its members. It is truly educative in its effect in the degree in which an individual shares or participates in some conjoint activity. By doing his share in the associated activity, the individual appropriates the purpose which actuates it, becomes familiar with its methods and subject matters, acquires needed skill, and is saturated with its emotional spirit.

Question: Which of the following statements best reflects the main idea of the passage?

A) The environment's role in education is mainly to ensure physical survival.
B) Young people acquire societal attitudes and skills by engaging in shared social activities.
C) Education primarily occurs through the direct transmission of knowledge and values.
D) The best way to develop societal attitudes in the young is through individual reflection.

Answer: B) Young people acquire societal attitudes and skills by engaging in shared social activities.

Explanation: The passage discusses how young individuals develop essential attitudes and dispositions for societal life through participation in social activities rather than direct transmission of beliefs or solitary study. It emphasizes that education is fostered within the social environment through shared experiences and activities. Option B accurately captures the essence of this idea. In contrast, options A, C, and D present inaccurate interpretations of the passage, suggesting either a physical, direct, or solitary approach to education, which the passage does not support.

Example Passage 2

In this Autobiography I shall keep in mind the fact that I am speaking from the grave. I am literally speaking from the grave, because I shall be dead when the book issues from the press. I speak from the grave rather than with my living tongue, for a good reason: I can speak thence freely. When a man is writing a book dealing with the privacies of his life—a book which is to be read while he is still alive—he shrinks from speaking his whole frank mind; all his attempts to do it fail, he recognizes that he is trying to do a thing which is wholly impossible to a human being. The frankest and freest and privatest product of the human mind and heart is a love letter; the writer gets his limitless freedom of statement and expression from his sense that no stranger is going to see what he is writing.

Question: The author's main purpose of this passage is to

A) Emphasize the limitations that living authors face when writing autobiographies.
B) Argue the superiority of love letters as forms of expression.
C) Illustrate the challenges of writing about one's private life for public consumption.
D) Explain the author's decision to write as if speaking from the grave in his autobiography.

Answer: D) Explain the author's decision to write as if speaking from the grave in his autobiography.

Explanation: The passage primarily focuses on the author's rationale for adopting a unique narrative perspective in his autobiography — that of speaking from the grave. This approach is chosen to achieve a level of frankness and freedom not possible when one is constrained by the considerations and repercussions of speaking as a living person. While the passage touches upon the nature of writing about private matters and compares it to the freedom found in love letters, these points serve to support the main argument rather than constitute the central idea themselves. Thus, the correct answer is D, as it directly addresses the author's intention to use posthumous narration as a means to express himself without reservation.

Questions Regarding Details

Detail-oriented questions on the AFOQT evaluate your ability to identify and understand specific information within the text. These questions test your ability to pick out facts, examples, and other particulars that support the main idea or themes of the passage.

Being adept at locating details is essential for thorough comprehension. It enables you to gather evidence, understand the structure of arguments, and appreciate the nuances of the narrative or exposition. This skill is crucial for academic research, critical analysis, and practical decision-making.

Strategies for Success:

- **Active Reading and Mental Notes:** As you read, note where the key facts, names, dates, and specific information are mentioned. In a computer-based test, you won't be able to underline the text, but mental notes will still help you quickly locate details when answering questions.
- **Practice Skimming:** Improve your skimming skills to quickly locate information within the text without having to read everything thoroughly a second time.
- **Understand Question Types:** Familiarize yourself with how detail questions are phrased. Knowing if a question asks for a fact, an example, or an explanation can guide you on what to look for.
- **Context Is Key:** Always consider the detail within the passage's context to ensure correct interpretation.

Application in Test Preparation:

Engage with texts across a variety of subjects and formats. After reading, challenge yourself to recall specific details without looking back at the text. Use practice tests to hone your ability to find and interpret details.

Common Pitfalls and How to Avoid Them:

- **Overlooking Details:** Important details can be easily missed if you read too quickly. Slow down and ensure you're fully processing the information.
- **Confusing Similar Details:** Pay attention to the nuances that differentiate similar pieces of information within the text to avoid mixing them up.
- **Relying Too Much on Memory:** Don't assume you remember all the details correctly. Double-check the passage to confirm your answers.

Example Passage 1

The *Endurance* steamed along the front of this ice-flow for about seventeen miles. The glacier showed huge crevasses and high pressure ridges, and appeared to run back to ice-covered slopes or hills 1000 or 2,000 ft. high. Some bays in its front were filled with smooth ice, dotted with seals and penguins. At 4 a.m. on the 16th we reached the edge of another huge glacial overflow from the icesheet. The ice appeared to be coming over low hills and was heavily broken. The cliff-face was 250 to 350 ft. high, and the ice surface two miles inland was probably 2,000 ft. high.

Question: According to the passage, which of the following is not true?

A) The *Endurance* traveled along the front of the ice-flow for approximately seventeen miles.
B) The glacier was devoid of any crevasses and pressure ridges.
C) Seals and penguins were spotted on some smooth ice areas in front of the glacier.
D) The ice surface two miles inland from the cliff-face was estimated to be around 2,000 ft. high.

Answer: B) The glacier was devoid of any crevasses and pressure ridges.

Explanation: The passage describes the Endurance's journey along the front of an ice-flow, detailing the presence of "huge crevasses and high-pressure ridges" in the glacier. This directly contradicts option B, which falsely claims that the glacier was devoid of crevasses and pressure ridges. Options A, C, and D are all supported by the passage: A) states the distance the *Endurance* traveled along the ice-flow, C) mentions the wildlife observed on the smooth ice, and D) provides an estimation of the ice surface's height two miles inland. Therefore, B is the correct answer as it is the only statement not corroborated by the passage, making it untrue according to the provided text.

Example Passage 2

The dock was still for a moment. Then a barrel toppled from a pile of barrels, and a figure moved like a bird's shadow across the opening between mounds of cargo set about the pier. At the same time two men approached down a narrow street filled with the day's last light. The bigger one threw a great shadow that aped his gesticulating arms behind him on the greenish faces of the buildings. Bare feet like halved hams, shins bound with thongs and pelts, he waved one hand in explanation, while he rubbed the back of the other on his short, mahogany beard.

Question: According to the passage, which is the most accurate description of the bigger man?

A) He was quietly observing from the shadows, unnoticed.
B) He carried a barrel on his shoulder as he walked.
C) He was barefoot, with his shins wrapped, and had a short mahogany beard.
D) He was wearing heavy boots and a long coat as he gestured animatedly.

Answer: C) He was barefoot, with his shins wrapped, and had a short mahogany beard.

Explanation: The passage vividly describes the bigger man's appearance and actions as he walks down a narrow street. It specifically mentions that his "bare feet like halved hams, shins bound with thongs and pelts," indicating that he is barefoot and has his shins wrapped. Additionally, it describes him as having a "short, mahogany beard," which he rubs with the back of his hand. These details collectively match option C, making it the most accurate description of the bigger man according to the passage. Options A, B, and D do not accurately reflect the details provided; A suggests he was hiding, B incorrectly mentions him carrying a barrel, and D describes attire not mentioned in the passage. Therefore, C is the correct answer.

Understanding Relationships and Making Inferences

This type of question evaluates your ability to understand the relationships between various elements within the text and to make logical inferences based on the provided information. It requires a deeper level of comprehension, moving beyond the literal content to grasp the implied meanings and connections.

The skill of making inferences and understanding relationships is crucial for real-world problem-solving and critical thinking. It allows you to read between the lines, draw conclusions from subtle cues, and connect the dots in complex situations. This skill is invaluable in academics, professional settings, and everyday life, where not everything is stated explicitly.

Strategies for Success:

- **Look for Clues:** Pay attention to the text's tone, word choice, and any hints the author might give to imply relationships or to set the groundwork for inferences.
- **Connect Ideas:** Identify how different parts of the passage relate to each other through cause and effect, contrast, or similarities.
- **Read Actively:** Ask questions as you read. Consider what is stated and what can be reasonably assumed. Question why the author included certain details and what they signify.
- **Use Background Knowledge:** Apply your own knowledge and experiences to understand unstated aspects of the passage. This can help fill in gaps and make informed inferences.
- **Practice Predicting:** Try to anticipate the author's direction or the conclusion of arguments or narratives. Such attempts prepare you to make inferences and understand relationships.

Application in Test Preparation:

Engage with a wide variety of reading materials, including those that are complex or outside your comfort zone. After reading, practice articulating the relationships you've identified and the inferences you've made. Discussing texts with others can also reveal different perspectives and deepen your understanding. Utilize practice tests to familiarize yourself with these questions' format and refine your analytical skills.

Common Pitfalls and How to Avoid Them:

- **Overreaching:** Be cautious not to extend your inferences beyond what is reasonably supported by the text. Your conclusions should always be grounded in the passage.
- **Ignoring the Context:** Every inference or relationship identified should be contextual. Avoid making assumptions based on external knowledge not supported by the passage.
- **Overlooking Subtlety:** The most crucial connections or implications are often subtle. Ensure you don't gloss over these finer points in your reading.

Example Passage 1

Once upon a time Jeremiah the prophet had asked for only one thing, that he might get away from that strange cityful of perverse men to whom it was his hard lot to be the mouthpiece of a God they were forgetting. He was tired of them. "O that I had in the wilderness a lodging place of wayfaring men that I might leave my people and go from them." Well, time passed on. The people got no wiser, and Jeremiah's burden certainly got no lighter. But the very chance he prayed for came. He had a clear and honorable opportunity to go to the lodge in the wilderness, or anywhere else he liked, away from the men who had disowned his teaching. His work was done apparently, and he had failed. Yet with the door standing invitingly open, see what Jeremiah did! He "went and dwelt among the people that were left in the land." He had his chance and he did not take it!

Question: Which of the following does this passage imply about Jeremiah?

A) Jeremiah eventually abandoned his people due to their persistent disregard for his teachings.
B) Despite his frustrations and the opportunity to leave, Jeremiah chose to stay with his people.
C) Jeremiah found solace and success in solitude, away from the city and its perverse inhabitants.
D) The people eventually embraced Jeremiah's teachings, leading to a harmonious relationship.

Answer: B) Despite his frustrations and the opportunity to leave, Jeremiah chose to stay with his people.

Explanation: The passage describes Jeremiah's deep-seated desire to escape from a difficult situation where he felt his message was not being received by a people increasingly disconnected from their faith. Despite his frustrations and the clear opportunity to leave for a quieter life in the wilderness—an opportunity he had explicitly wished for—Jeremiah decides against taking this path. When presented with a real chance to abandon the very people who had disregarded his teachings, he instead chooses to remain among them. This decision underlines a sense of duty or commitment to these people despite the personal toll it had taken on him. The passage does not suggest that he abandoned his people (A), found success in solitude (C), or that his teachings were embraced, leading to harmony (D). Instead, it clearly illustrates Jeremiah's dedication to his role and people by staying with them against his earlier desires to leave, making option B the correct and most accurate interpretation.

Example Passage 2:

In the old days Hortons Bay was a lumbering town. No one who lived in it was out of sound of the big saws in the mill by the lake. Then one year there were no more logs to make lumber. The lumber schooners came into the bay and were loaded with the cut of the mill that stood stacked in the yard. All the piles of lumber were carried away. The big mill building had all its machinery that was removable taken out and hoisted on board one of the schooners by the men who had worked in the mill. The schooner moved out of the bay toward the open lake carrying the two great saws, the traveling carriage that hurled the logs against the revolving, circular saws and all the rollers, wheels, belts and iron piled on a hull-deep load of lumber. Its open hold covered with canvas and lashed tight, the sails of the schooner filled and it moved out into the open lake, carrying with it everything that had made the mill a mill and Hortons Bay, a town.

Question: What is the likely fate of the town of Hortons Bay?

A) It will experience a revival as a tourist destination.
B) It will become a thriving fishing community.
C) It will grow into a major industrial city.
D) It will likely decline or become abandoned.

Answer: D) It will likely decline or become abandoned.

Explanation: The passage vividly depicts Hortons Bay's transition from a bustling lumbering town to a place stripped of its defining industry. With the removal of the mill's machinery, including the saws and other equipment vital for its lumber operations, and the transportation of these resources out of town, the narrative strongly implies that Hortons Bay's economic foundation has been dismantled. The specific mention that "everything that had made the mill a mill and Hortons Bay, a town" was carried away leaves little room for interpreting a future for Hortons Bay that involves economic prosperity or community sustainability in its current form. Options A, B, and C suggest potential futures that are not supported by the passage's depiction of the town's deindustrialization and the essential elements of its identity and economy being physically removed. Therefore, the most logical conclusion is option D.

VOCABULARY IN CONTEXT QUESTIONS

Vocabulary in Context questions require you to determine the meaning of a word based on how it is used within a passage. These questions test not just your knowledge of word definitions, but also your ability to infer meaning from context. The key to answering these questions is to analyze the surrounding words, phrases, and overall tone of the passage to deduce what the unfamiliar word likely means in that specific setting.

In many cases, the word's definition may vary from its usual meaning, so understanding the context is crucial. Let's explore strategies for tackling these questions and provide examples.

How to Approach Vocabulary in Context Questions:

- Read the Entire Passage: Don't just focus on the word itself. Read the sentence and the surrounding sentences to gather clues about how the word is used.

- Look for Contextual Clues: These could be definitions, synonyms, antonyms, or descriptive phrases within the passage that help define the word.
- Test Each Answer Choice: Substitute each answer choice into the sentence to see which one fits best with the meaning of the sentence.
- Eliminate Incorrect Choices: Discard any answers that don't fit logically in the context of the passage.

Example Passage 1

The towering castle ruins, weathered by centuries of storms and sun, stood majestically at the edge of the cliff, as if meant to glorify the power of time itself. Wildflowers grew in every crack of the stone walls, softening the structure's once formidable appearance. Below, the waves crashed relentlessly, echoing the relentless passage of time. Travelers who visited the site couldn't help but feel humbled, their voices hushed by the ancient beauty that seemed to radiate from every stone.

Question 1: In this passage, *glorify* most nearly means

A) Exaggerate
B) Celebrate
C) Ignore
D) Ruin

Answer: B) Celebrate

Explanation: glorify in this context suggests the castle's ruins highlight or honor the enduring passage of time, making *celebrate* the correct answer. The other options do not fit the tone or meaning.

Example Passage 2

The country near the mouth of the river is wretched in the extreme: on the south side a long line of perpendicular cliffs commences, which exposes a section of the geological nature of the country. The strata are of sandstone, and one layer was remarkable from being composed of a firmly cemented conglomerate of pumice pebbles, which must have travelled more than four hundred miles, from the Andes. The surface is everywhere covered up by a thick bed of gravel, which extends far and wide over the open plain. Water is extremely scarce, and, where found, is almost invariably brackish. The vegetation is scanty; and although there are bushes of many kinds, all are armed with formidable thorns, which seem to warn the stranger not to enter on these inhospitable regions.

Question 1: In this passage, *brackish* mostly nearly means

A) Boiling
B) Fresh
C) Salty
D) Clear

Answer: C) Salty

Explanation: *Brackish* refers to water that has more salinity than freshwater but not as much as seawater, making *salty* the correct meaning. From the context, one can also rule out A, B, or D, even if one is not familiar with the word *brackish*.

Question 2: A word that could be properly substituted for *inhospitable* most nearly means?
A) Welcoming
B) Unfriendly
C) Populated
D) Sheltered

Answer: B) Unfriendly

Explanation: *Inhospitable* refers to an environment that is harsh and difficult to live in, thus *unfriendly* is the term that most closely matches the meaning in this context.

PARAGRAPH COMPREHENSION PRACTICE SET

This section presents reading paragraphs followed by questions or incomplete statements. Your task is to read the paragraph and choose the option that best completes the statement or answers the question. For practice only. This practice set does not reflect the actual number of questions in the test.

The military force of the U.S.T. *Buford* is in command of a Colonel of the United States Army, tall and severe-looking, about fifty. In his charge are a number of officers and a very considerable body of soldiers, most of them of the regular army. Direct supervision over the deportees is given to the representative of the Federal Government, Mr. Berkshire, who is here with a number of Secret Service men. The Captain of the Buford takes his orders from the Colonel, who is the supreme authority on board.

1. Which of the following conclusions about the command structure on board of *Buford* can most reasonably be drawn from the passage?

A) The Colonel and Mr. Berkshire share equal authority on the Buford.
B) The Secret Service men are in charge of the military operations.
C) The Colonel is the supreme authority on board the Buford.
D) Mr. Berkshire commands the military personnel and the Captain.

The word "idealism" is used by different philosophers in somewhat different senses. We shall understand by it the doctrine that whatever exists, or at any rate whatever can be known to exist, must be in some sense mental. This doctrine, which is very widely held among philosophers, has several forms, and is advocated on several different grounds. The doctrine is so widely held, and so interesting in itself, that even the briefest survey of philosophy must give some account of it.

2. What's the author's attitude about the "doctrine" in this passage?

A) Dismissive
B) Critical
C) Neutral
D) Supportive

In 1857, Russell, Majors & Waddell were sending supply trains to Salt Lake for General Johnston's army, offering high wages. A respected wagon master, Lewis Simpson, invited the author to join as an "extra hand." The author found the offer appealing due to the good pay and light responsibilities, which involved covering for sick drivers while also riding his own mule and taking on minor supervisory duties.

3. What can be inferred from this passage about the author's likely next course of action?

A) The author will decline the offer due to the risks involved.
B) The author will accept the offer because of the high wages and responsibilities.
C) The author will negotiate for a higher salary.
D) The author will recommend someone else for the position.

At first glance, the horizontal strata of valleys suggest they were carved by water, but this would require the removal of massive amounts of rock. The author questions this explanation and rejects the idea of subsidence. The shape of the valleys and peaks leads to the conclusion that water action is insufficient to explain these formations. Some locals even compare the valleys to a rugged seacoast due to their structure.

4. According to the passage, how does the author view the role of water action in shaping the observed geography?

A) As the predominant force.
B) As negligible or non-contributory.
C) As causing subsidence in the valleys.
D) As influencing the drainage from the summit.

When the party came out of the Yellowstone, Adams went on alone to Seattle and Vancouver to inspect the last American railway systems yet untried. They, too, offered little new learning, and no sooner had he finished this debauch of Northwestern geography than with desperate thirst for exhausting the American field, he set out for Mexico and the Gulf, making a sweep of the Caribbean and clearing up, in these six or eight months, at least twenty thousand miles of American land and water.

5. The primary purpose of the passage is to

A) critique the American railway systems.
B) illustrate the extent of Adams' travels across America.
C) compare different geographic regions of America.
D) highlight the inadequacies of American geography.

Descartes determined that he would believe nothing which he did not see quite clearly and distinctly to be true. Whatever he could bring himself to doubt, he would doubt, until he saw reason for not doubting it. By applying this method he gradually became convinced that the only existence of which he could be quite certain was his own. He imagined a deceitful demon, who presented unreal things to his senses in a perpetual phantasmagoria; it might be very improbable that such a demon existed, but still it was possible, and therefore doubt concerning things perceived by the senses was possible.

6. Which of the following best describes what "phantasmagoria" means in the passage?

A) A systematic method of doubt
B) A state of clear and distinct truth
C) A deceptive, shifting sequence of illusions
D) A demon's improbable existence

Shirley was, I believe, sincerely glad of being relieved from so burdensome a charge as the conduct of an army must be to a man unacquainted with military business. I was at the entertainment given by the city of New York to Lord Loudoun, on his taking upon him the command. Shirley, though thereby superseded, was present also. There was a great company of officers, citizens, and strangers, and, some chairs having been borrowed in the neighborhood, there was one among them very low, which fell to the lot of Mr. Shirley. Perceiving it as I sat by him, I said, "They have given you, sir, too low a seat." "No matter," says he, "Mr. Franklin, I find a low seat the easiest."

7. The passage's author conveys Shirley's attitude by

A) highlighting his relief at no longer being in command.
B) describing his discomfort at the social event.
C) illustrating his humility in accepting a lower seat.
D) emphasizing his displeasure with Lord Loudoun.

The mansion stood isolated atop the hill, its walls painted in vibrant, mismatched colors that clashed with the surrounding landscape. A peculiar tower spiraled toward the sky in an uneven fashion. The grounds were overgrown with wild plants, yet within the disarray, there was a sense of peculiar order. Visitors were both drawn to and bewildered by its strangeness, as if the house itself reflected the personality of its mysterious and eccentric owner.

8. In this passage, *eccentric* most nearly means

A) Ordinary
B) Unusual
C) Dull
D) Stylish

I became certain that truth, sincerity, and integrity were vital to living a fulfilling life. Rather than following religious commands, my past experiences had taught me that actions were inherently good or bad based on their outcomes. These values guided me, and with some luck, I avoided significant immoral behavior during my youth, even when away from my father's guidance.

9. What convinced the author to value truth, sincerity, and integrity?

A) Reflection on life experiences.
B) Religious teachings.
C) His father's influence.
D) Observing the consequences of his actions.

Historians sometimes face great difficulty maintaining a clear narrative due to fragmented and unreliable historical records. In such cases, they must compare and conjecture from available fragments to construct a plausible account of events. For example, when studying the fall of the Roman Empire, historians might have to infer details about military campaigns from incomplete letters or inscriptions, piecing together how leaders responded to barbarian invasions.

10. What method would the author likely agree with?
A) Ignoring unreliable fragments
B) Recreating details from available records for continuity
C) Relying on archaeology
D) Collecting and comparing fragments

Answers and Explanations

1. C) The Colonel is the supreme authority on board the Buford.

Explanation: The passage explicitly states that "The Captain of the Buford takes his orders from the Colonel, who is the supreme authority on board," indicating that the Colonel holds the highest command over all operations and personnel on the Buford.

2. C) Neutral

Explanation: The author describes the doctrine of idealism as "widely held" and "interesting," and mentions the need to discuss it in a survey of philosophy, suggesting a neutral, objective attitude toward the subject.

3. B) The author will accept the offer because of the high wages and responsibilities.

Explanation: The passage mentions the high wages and the appealing nature of the "extra hand" position as significant inducements for the author, suggesting he is likely to accept the offer.

4. B) As negligible or non-contributory.

Explanation: The author dismisses the idea that water action shaped the valleys, stating that its influence is insignificant given the scale of the formations. The author argues that the structure of the valleys cannot be explained by erosion processes.

5. B) illustrate the extent of Adams' travels across America.

Explanation: The passage details Adams' extensive travels from the Yellowstone to Seattle, Vancouver, and then on to Mexico and the Caribbean, emphasizing the broad scope of his geographic exploration within America.

6. C) A deceptive, shifting sequence of illusions

Explanation: In the passage, "phantasmagoria" refers to the array of unreal things that Descartes imagined a deceitful demon presented to his senses, indicating a misleading or illusory spectacle.

7. C) illustrating his humility in accepting a lower seat.

Explanation: Shirley's response to the low seat, "No matter, Mr. Franklin, I find a low seat the easiest," suggests humility and acceptance, which the author conveys through this anecdote from the social event.

8. B) Unusual

Explanation: "Eccentric" refers to something unconventional or strange, making "unusual" the correct choice in this context. The other options do not fit the description of the house.

9. A) Reflection on life experiences.

Explanation: The author emphasizes that personal experiences led him to value these principles, rather than religion or external influences.

10. B) Recreating details from available records for continuity

Explanation: The author suggests that, in the absence of reliable historical records, historians should use conjecture to fill in gaps while ensuring that these guesses are plausible. Instead of ignoring fragments or solely relying on facts, the author supports recreating plausible details to maintain a continuous narrative, thus making B the correct choice. This method helps form a coherent story without sacrificing historical integrity entirely.

CHAPTER 10. PHYSICAL SCIENCE

The Physical Science section covers a broad spectrum of middle to high school level science concepts. These include Physics, Chemistry, Life Science, Mechanics, Electronics, and Earth Science.

The time limit is 10 minutes for 20 questions. Only 30 seconds per question may not seem like much, but the questions are typically straightforward. In many cases, you either know the answer or you do not. A process of elimination can help if you need to guess (there are no points lost for incorrect answers).

SECTION 1. METRIC SYSTEM OF MEASUREMENT

The metric system, also known as the International System of Units (SI), is the standard system of measurement used in science and most countries around the world. Developed in the late 18th century, it provides a consistent way to measure and compare physical quantities like length, mass, volume, and temperature. This system is based on multiples of ten, making it simple and efficient to use across various scientific disciplines, including physics, chemistry, and biology.

Base Units of the Metric System

In the metric system, there are seven fundamental base units that define all other measurements, which are summarized in the following table.

Seven Fundamental Base Units of the Metric System

Quantity Measured	Base Unit	Symbol
Length	Meter	m
Mass	Kilogram	kg
Time	Second	s
Electric Current	Ampere	A
Temperature	Kelvin	K
Amount of Substance	Mole	mol
Luminous Intensity	Candela	cd

Prefixes in the Metric System

The metric system uses prefixes to indicate multiples or fractions of the base units. Each prefix represents a power of ten, making it easy to convert between larger and smaller quantities. For example:

- Kilo- (k) means 1,000 times the base unit. One kilometer (km) is 1,000 meters.
- Centi- (c) means 1/100th of the base unit. One centimeter (cm) is 0.01 meters.
- Milli- (m) means 1/1,000th of the base unit. One milliliter (mL) is 0.001 liters.
- Micro- (μ) means 1/1,000,000th of the base unit. One micrometer (μm) is 0.000001 meters.

These prefixes are essential in scientific measurements because they allow scientists to work with very large or very small numbers in a manageable way. For example, measuring the distance between stars might require gigameters (Gm), while observing bacteria under a microscope would use micrometers (μm).

One of the key reasons the metric system is preferred in science is its universality. Since the metric system is used worldwide, scientists can collaborate and share their results without needing to convert between different systems of measurement. This eliminates confusion and ensures that scientific data is accurate and comparable regardless of where the research is conducted.

Additionally, the metric system is decimal-based, which simplifies calculations. Converting between units is straightforward: simply move the decimal point. For instance, converting from centimeters to meters requires moving the decimal two places to the left (since there are 100 centimeters in a meter). This simplicity contrasts with non-metric systems like the Imperial system, where conversions between units are more complex (e.g., 12 inches in a foot, 3 feet in a yard).

SECTION 2. PHYSICS FOUNDATIONS

Weight vs. mass

Weight and mass are often used interchangeably, but they refer to different concepts. Mass is the amount of matter in an object and remains constant regardless of location. It is measured in kilograms (kg) or grams. Weight, on the other hand, is the force exerted on an object due to gravity and depends on both the object's mass and the gravitational pull acting on it. It is measured in newtons (N). For example, an object's mass stays the same on Earth and the Moon, but its weight would be less on the Moon due to the weaker gravitational pull. Thus, mass is intrinsic, while weight varies with gravity.

Motion: Key Concepts

Motion is the change in position of an object over time, and it is a fundamental concept in physics. Understanding motion requires analyzing several key concepts: **velocity**, **displacement**, **momentum**, and **acceleration**.

Velocity and Displacement

Velocity is the rate at which an object changes its position. It's a vector quantity, meaning it has both magnitude (speed) and direction. Velocity differs from speed because speed only tells us how fast an object is moving, while velocity tells us both how fast and in what direction. For example, if a car is moving at 60 miles per hour north, that's its velocity. The formula for velocity is: $v = \frac{\Delta x}{\Delta t}$, Where v is velocity, Δx is displacement (change in position), and Δt is the time it takes for that displacement to occur.

Displacement refers to an object's overall change in position, considering both the distance and the direction from its starting point to its final position. Unlike distance, which only measures how much ground has been covered, displacement takes direction into account. For example, if you walk 5 meters east and then 5 meters west, your total distance is 10 meters, but your displacement is zero because you ended up in the same place you started.

Momentum

Momentum is a measure of how much motion an object has and is also a vector quantity. It depends on both the mass of an object and its velocity. The more massive an object or the faster it's moving, the more momentum it has. The formula for momentum is: $p = mv$, where p is momentum, m is mass, and v is velocity. For example, a truck moving at 20 m/s will have more momentum than a bicycle moving at the same speed because the truck has much more mass.

Momentum is important in understanding collisions and impacts. When two objects collide, the total momentum before the collision is equal to the total momentum after the collision, provided no external forces act on the system. This principle is known as the conservation of momentum.

Acceleration

Acceleration refers to the rate at which an object's velocity changes over time. It occurs when there's a change in speed, direction, or both. Like velocity, acceleration is a vector quantity, meaning it has both magnitude and direction. The formula for acceleration is: $a = \frac{\Delta v}{\Delta t}$, where a is acceleration, Δv is the change in velocity, and Δt is the time over which this change occurs. Positive acceleration means an object is speeding up, while negative acceleration (often called deceleration) means it's slowing down.

An example of acceleration can be seen when a car speeds up from rest to 60 miles per hour in 10 seconds. The car's velocity is increasing, so it is accelerating. If the car suddenly comes to a stop, it experiences negative acceleration as it slows down to zero.

Work, Force, Energy, and Power

In physics, work, force, energy, and power are fundamental concepts that explain how objects move, interact, and transfer energy in different situations. These terms are closely related, but each has its own specific meaning and application.

Force

Force is any interaction that, when unopposed, changes the motion of an object. It can cause an object to start moving, stop moving, or change its direction. Force is a vector quantity, meaning it has both magnitude and direction. It is measured in newtons (N), and the formula to calculate force is: $F = ma$, where: F is the force, m is the mass of the object, a is the acceleration.

For example, pushing a cart requires a force, and the larger the mass of the cart, the more force is needed to accelerate it.

Work

Work occurs when a force is applied to an object, causing it to move. In physics, work is only done when the force results in movement, and the movement is in the direction of the applied force.

The formula for work is: $W = Fd\cos\theta$, where: W is work (measured in joules (J)), F is the applied force, d is the distance the object moves, θ is the angle between the force and the direction of movement.

If the force is applied in the same direction as the object's movement, $\cos\theta$ equals 1, and the formula simplifies to $W = Fd$. For example, lifting a box vertically requires work because you are applying a force (upward) to move the box over a distance (upward).

If no movement occurs, no work is done—even if a large force is applied. For example, if you push against a wall with all your strength and it doesn't move, you haven't done any physical work in the scientific sense.

Energy

Energy is the capacity to do work. It comes in various forms, including kinetic energy, potential energy, and thermal energy. In physics, the most common forms of energy are: kinetic energy and potential energy.

Kinetic energy (KE): The energy an object possesses due to its motion. The faster an object moves, the more kinetic energy it has. The formula for kinetic energy is: $KE = \frac{1}{2}mv^2$, where: m is the mass of the object, v is its velocity.

Potential energy (PE): The energy stored in an object due to its position in a force field, typically a gravitational field. The formula for gravitational potential energy is: $PE = mgh$, where: m is the mass of the object, g is the acceleration due to gravity, h is the height above a reference point.

For example, a book held above a table has potential energy due to its position. If it falls, that potential energy is converted to kinetic energy.

Energy is measured in joules (J), the same unit as work. This reflects the fact that energy is the ability to perform work.

Example—Energy Calculation: A car with a mass of 1,200 kg is traveling at a velocity of 20 m/s. Calculate the car's kinetic energy.

Solution: The formula for kinetic energy (KE) is: $KE = \frac{1}{2}mv^2$

Step 1: Identify the known values

- Mass of the car, $m = 1{,}200\ \text{kg}$
- Velocity of the car, $v = 20\ \text{m/s}$

Step 2: Plug the values into the formula:

$$KE = \frac{1}{2} \times 1{,}200\ \text{kg} \times (20\ \text{m/s})^2 = \frac{1}{2} \times 1{,}200\ \text{kg} \times 400\ \text{m}^2/\text{s}^2 = 240{,}000\ \text{kg}\cdot\text{m}^2/\text{s}^2$$

Hence, the kinetic energy of the car is 240,000 joules (J).

This example demonstrates how kinetic energy is calculated based on the mass and velocity of an object. In this case, the car's motion provides a significant amount of energy due to its large mass and speed.

Power

Power is the rate at which work is done or the rate at which energy is transferred or transformed. It tells us how quickly work is completed. The formula for power is: $P = \frac{W}{t}$, where: P is power (measured in watts (W)), W is work (in joules), t is the time taken (in seconds).

Power can also be expressed as: $P = \frac{Fd}{t}$. This equation shows that the faster work is done, or the faster energy is transferred, the greater the power output. For example, if two people lift identical boxes to the same height, but one person lifts the box in half the time, that person has exerted more power.

Examples of Work, Energy, and Power in Everyday Life

Work: A person pushing a lawnmower does work because they apply force to move the lawnmower across a distance.

Energy: A moving car has kinetic energy due to its motion. When the car brakes, that kinetic energy is transformed into thermal energy (heat) through friction.

Power: A 100-watt lightbulb uses 100 joules of electrical energy per second to produce light.

To recap, force causes motion, work is done when a force moves an object, energy is the capacity to do work, and power is the rate at which work is performed or energy is transferred. These concepts are fundamental to understanding how objects move and interact with forces in the physical world.

Newton's Three Laws of Motion and the Law of Universal Gravitation

Sir Isaac Newton's three laws of motion and his law of universal gravitation are foundational principles in physics. Together, they describe how objects move and interact with forces, including the force of gravity, which governs the motion of everything from falling apples to orbiting planets.

Newton's First Law of Motion (Law of Inertia)

Newton's first law states that: "An object at rest will remain at rest, and an object in motion will remain in motion at a constant velocity unless acted upon by an external force."

This is also known as the law of inertia, which refers to an object's resistance to changes in its state of motion. If no external force acts on an object, its motion will not change—an object at rest will stay at rest, and a moving object will continue moving in a straight line at constant speed.

Example: A soccer ball lying still on the ground will not move unless someone kicks it (an external force). Once kicked, it will keep moving until forces like friction or a player's foot act to stop it.

Newton's Second Law of Motion (Force and Acceleration)

Newton's second law explains how force and acceleration are related, and is expressed by the formula: $F = ma$, where: F is the force applied to the object (in newtons, N), m is the mass of the object (in kilograms, kg), a is the acceleration of the object (in meters per second squared, m/s^2).

This law states that the acceleration of an object depends on the force applied to it and its mass. A greater force results in greater acceleration, but heavier objects require more force to accelerate at the same rate as lighter ones.

Example: Pushing an empty shopping cart requires less effort (force) than pushing a fully loaded cart. The more mass the cart has, the more force you need to apply to achieve the same acceleration.

Example: Let's assume a car has a mass of 1,200 kg, and it accelerates at a rate of $3\ \mathrm{m/s^2}$. To find the force required to produce this acceleration, we can apply Newton's Second Law.

$$F = m \cdot a = 1{,}200\ \mathrm{kg} \times 3\ \mathrm{m/s^2} = 3{,}600\ \mathrm{N}$$

So, the force needed to accelerate the car is 3,600 N (Newtons). This illustrates how the relationship between mass and acceleration dictates the force necessary to move an object, following Newton's Second Law of Motion.

Newton's Third Law of Motion (Action and Reaction)

Newton's third law states: "For every action, there is an equal and opposite reaction."

This means that whenever one object exerts a force on another, the second object exerts an equal force in the opposite direction. These forces are always paired, with both acting simultaneously.

Example: When you jump off a diving board, you push down on the board with your legs (action), and the board pushes you upward with an equal force (reaction), allowing you to leap into the air. Similarly, a rocket launches by expelling gas downward (action), which propels the rocket upward (reaction).

Newton's Law of Universal Gravitation

Newton's law of universal gravitation describes the force of gravity between two objects. The law states that: "Every particle in the universe attracts every other particle with a force directly proportional to the product of their masses and inversely proportional to the square of the distance between their centers."

The formula for this gravitational force is: $F = G\frac{m_1 m_2}{r^2}$, where: F is the gravitational force between two objects, G is the gravitational constant ($6.674 \times 10^{-11}\ \mathrm{Nm^2/kg^2}$), m_1 and m_2 are the masses of the two objects, r is the distance between the centers of the two objects.

This law explains why objects fall toward Earth and why planets orbit the Sun. Gravity is always present, pulling objects with mass toward one another, though the force weakens as distance increases.

Example: The gravitational pull of Earth keeps the Moon in orbit, and the force between Earth and an apple causes the apple to fall when it is dropped. The larger the masses of the objects, and the closer they are to one another, the stronger the gravitational force.

How These Laws Work Together

Newton's three laws of motion describe how objects behave when forces act on them, and his law of universal gravitation explains the specific force of gravity. For example, when you drop a ball, the force of gravity (as described by the law of gravitation) pulls it toward the Earth. According to Newton's first law, the ball would remain stationary unless acted on by this external force (gravity). As it falls, the ball accelerates due to gravity, following Newton's second law (F = ma). When the ball hits the ground, Newton's third law applies—the ball exerts a force on the ground, and the ground exerts an equal and opposite force on the ball, causing it to bounce.

Together, these principles give us a comprehensive understanding of motion, whether it's the motion of everyday objects or the orbits of planets in space. Newton's laws provide the foundation for much of classical physics and are crucial for solving real-world problems involving force, motion, and gravity.

Velocity

Velocity is a vector quantity that describes the speed of an object in a specific direction. It differs from speed, which only accounts for the magnitude (how fast an object is moving) and not the direction. The formula for velocity when an object is accelerating is:

$$v = u + a \cdot t$$

Where:

- v is the final velocity,
- u is the initial velocity (starting speed),
- a is the acceleration (change in velocity over time),
- t is the time over which the acceleration occurs.

Example 1: A car starts from rest (initial velocity $u = 0\,\mathrm{m/s}$) and accelerates at a rate of $2\,\mathrm{m/s^2}$ for 5 seconds. What is the car's final velocity?

Solution: Using the formula: $v = u + a \cdot t = 0\,\mathrm{m/s} + (2\,\mathrm{m/s^2} \times 5\,\text{seconds}) = 10\,\mathrm{m/s}$

Final Velocity: The car's velocity after 5 seconds is $10\,\mathrm{m/s}$.

Example 2: A cyclist is moving with an initial velocity of $10\,\mathrm{m/s}$ and accelerates at a rate of $3\,\mathrm{m/s^2}$ for 8 seconds. What is the final velocity?

Solution: Using the formula: $v = u + a \cdot t = 10\,\mathrm{m/s} + (3\,\mathrm{m/s^2} \times 8\,\text{seconds}) = 10\,\mathrm{m/s} + 24\,\mathrm{m/s} = 34\,\mathrm{m/s}$

Final Velocity: The cyclist's final velocity is $34\,\mathrm{m/s}$.

In both examples, the formula for velocity incorporates the initial velocity, the acceleration, and the time over which acceleration occurs. In the first example, the object starts from rest, making it simpler to calculate. In the second example, the object starts with an initial velocity, requiring the calculation of how much the velocity increases due to the acceleration over time.

Waves

Waves are disturbances that transfer energy from one point to another without transferring matter. They are found in many physical phenomena, including light, water, and sound. Waves can be classified into two main types: mechanical waves and electromagnetic waves. Sound waves are a specific type of mechanical wave that requires a medium, like air, water, or solids, to travel through.

Types of Waves

Mechanical Waves: These waves require a medium (like air, water, or solids) to propagate. Examples include sound waves and water waves. Mechanical waves can be further divided into:

- Transverse waves: In these waves, the motion of the medium is perpendicular to the direction of the wave. For example, in water waves, the water particles move up and down while the wave travels horizontally.
- Longitudinal waves: In these waves, the motion of the medium is parallel to the direction of the wave. Sound waves are an example of longitudinal waves, where the particles of the medium (such as air) move back and forth in the same direction as the wave.

Electromagnetic Waves: These waves do not require a medium and can travel through a vacuum. Light, radio waves, and X-rays are examples of electromagnetic waves. Unlike mechanical waves, electromagnetic waves are transverse waves.

Key Characteristics of Waves

Waves have several key characteristics that define their behavior. The **wavelength** is the distance between two consecutive points in phase on a wave, such as two crests or troughs, and is measured in meters. Another important feature is the **frequency**, which represents the number of waves that pass a given point in a specific period and is measured in hertz (Hz). A higher frequency means more waves pass through a point over time. **Amplitude** refers to the height of the wave from its rest position to its crest in transverse waves, or the maximum displacement in longitudinal waves. The amplitude determines the energy carried by the wave—the greater the amplitude, the more energy it possesses. The **speed of the wave** is how fast the wave travels through a medium and depends on the type of wave and the properties of the medium.

Parts of a Wave

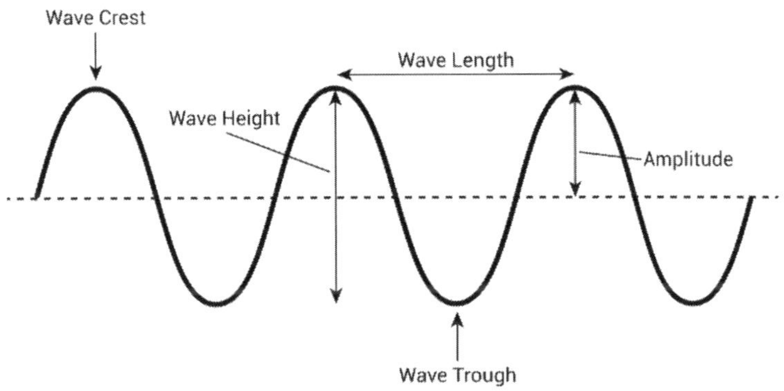

Sound Waves

Sound waves are mechanical, longitudinal waves that propagate through a medium (such as air, water, or solids) by vibrating the particles in that medium. The movement of particles creates areas of compression (where particles are close together) and rarefaction (where particles are spread apart). These compressions and rarefactions travel through the medium, allowing sound to propagate.

Sound waves differ from electromagnetic waves in that they cannot travel through a vacuum because they need a medium to carry the vibrations. The speed of sound varies depending on the medium:

- In air, sound travels at approximately 343 m/s.
- In water, sound travels faster, at about 1,480 m/s.
- In solids, sound travels even faster because the particles are more tightly packed, making it easier for them to transfer vibrations.

Pitch and Volume: The frequency of a sound wave determines its pitch. Higher frequency sound waves produce higher-pitched sounds (like a whistle), while lower frequency sound waves produce lower-pitched sounds (like a drum).

The amplitude of a sound wave is related to its volume. A wave with a greater amplitude will sound louder, while a wave with a smaller amplitude will sound quieter.

The Doppler Effect: One important phenomenon related to sound waves is the Doppler Effect. This occurs when the source of a sound is moving relative to the observer. If the source is moving toward the observer, the sound waves are compressed, resulting in a higher pitch (higher frequency). Conversely, if the source is moving away, the waves are stretched, resulting in a lower pitch (lower frequency). A classic example of the Doppler Effect is the change in pitch of a siren as an ambulance passes by.

Applications of Sound Waves: Sound waves have many practical applications. For example, ultrasound technology uses high-frequency sound waves to create images of structures inside the body, such as unborn babies in prenatal care. Another example is Sonar (Sound Navigation and Ranging), which uses sound waves to detect objects underwater, which is widely used in submarine navigation and fish detection.

In summary, waves are disturbances that transfer energy, and sound waves are a specific type of mechanical wave that propagate through a medium.

Electromagnetic spectrum

The electromagnetic spectrum encompasses all types of electromagnetic radiation, from the low-energy radio waves to the high-energy gamma rays. **Electromagnetic radiation** consists of waves of electric and magnetic fields that travel through space at the speed of light, carrying energy. These waves don't require a medium and can propagate through the vacuum of space, unlike mechanical waves like sound. The spectrum is categorized based on the wavelength and frequency of these waves, with longer wavelengths corresponding to lower frequencies and energy, and shorter wavelengths corresponding to higher frequencies and energy.

At the low-frequency end of the spectrum, **radio waves** have the longest wavelengths and are used extensively in communication technologies, such as radio, television, and cell phones. Moving up the spectrum, **microwaves** have shorter wavelengths and are used in radar technology and microwave ovens. Further along, **infrared radiation** is primarily associated with heat, as objects emit infrared light when they radiate heat. **Visible light**, the only part of the electromagnetic spectrum perceptible to the human eye, occupies a relatively small range within the spectrum, and its wavelengths correspond to the colors we see. Beyond visible light lies **ultraviolet radiation**, which has higher energy and can cause sunburn. The highest-energy waves on the spectrum are **X-rays** and **gamma rays**, both of which are used in medical imaging and cancer treatment due to their ability to penetrate matter.

Two key phenomena related to the behavior of electromagnetic waves are refraction and reflection. **Refraction** occurs when light or other electromagnetic waves pass from one medium into another (such as from air into water) and change speed, causing the wave to bend. The degree to which the wave bends depends on the wavelength of the light and the properties of the two media. This effect is why a straw in a glass of water appears bent or displaced at the surface of the water. Refraction is also responsible for phenomena like the splitting of white light into a rainbow when it passes through a prism, with each color bending by a different amount due to its wavelength.

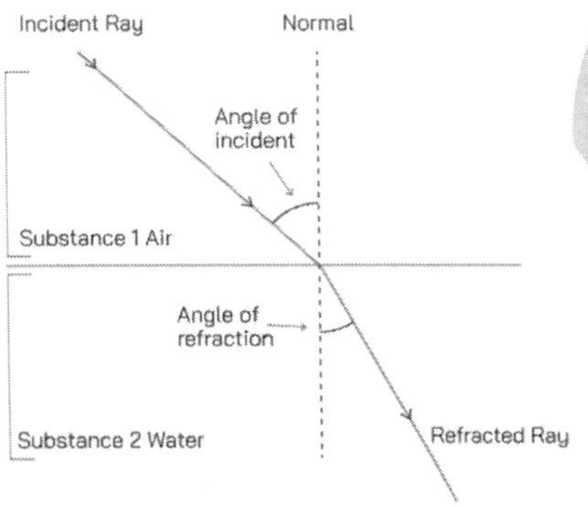

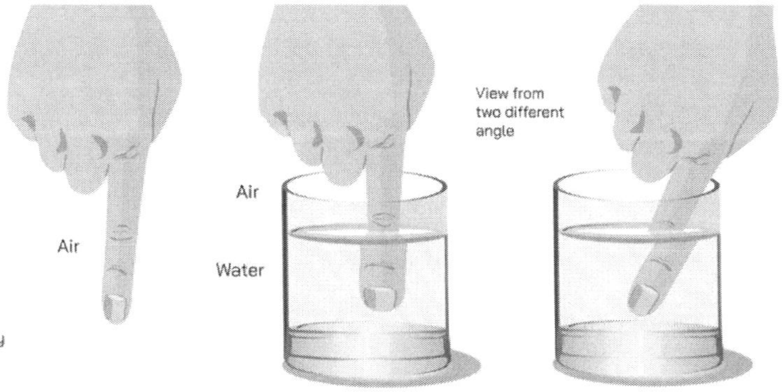

Reflection, on the other hand, occurs when electromagnetic waves encounter a surface and bounce back rather than passing through it. A common example is the reflection of light off a mirror. The angle at which the light hits the surface, known as the angle of incidence, is equal to the angle at which it reflects off the surface, called the angle of reflection. Reflection is essential in technologies such as optical instruments and communication devices, where mirrors or reflective surfaces direct waves to specific locations.

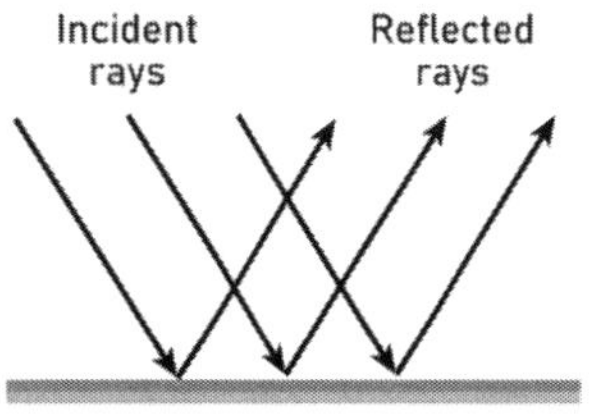

Regular Reflection

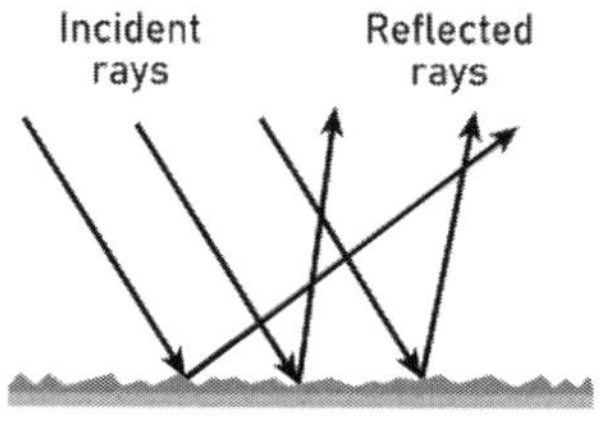

Diffuse Reflection

The interaction of electromagnetic waves with different materials and surfaces—whether through refraction, reflection, or other processes—enables the wide range of applications and technologies that depend on the electromagnetic spectrum, from wireless communication to medical imaging.

Heat and Phase Change

Heat is a form of energy that flows between objects due to a difference in their temperatures. It is crucial in understanding how matter changes between its three fundamental states: solid, liquid, and gas. The process by which matter transitions from one state to another is known as a **phase change**, and these transitions occur when heat energy is either absorbed or released by a substance.

In a **solid** state, the molecules are tightly packed together and vibrate in place, giving solids a definite shape and volume. As heat is added to a solid, the molecules begin to vibrate more rapidly, and if enough heat is supplied, the substance reaches its **melting point**, the temperature at which it transitions into a **liquid**. In the liquid state, the molecules are still close together but can move around more freely, allowing liquids to take the shape of their container while maintaining a fixed volume.

As more heat is added to a liquid, the molecules gain enough energy to break free from the forces that hold them together, leading to **evaporation** or **boiling**, where the liquid transitions into a gas. In the gaseous state, molecules are far apart and move independently, which allows gases to fill any container they are in. The temperature at which a liquid turns into a gas is called its **boiling point**.

Conversely, when heat is removed from a gas, the molecules lose energy, slow down, and move closer together, eventually condensing into a liquid. This process is called **condensation** and occurs when gas molecules release energy in the form of heat as they return to a liquid state. If even more heat is removed, the liquid can solidify through **freezing**, where the molecules arrange themselves into a rigid structure again, forming a solid.

During these phase transitions, the temperature of the substance remains constant, even though heat is being added or removed. For instance, while ice melts into water, the temperature stays at 0°C (32°F) until all the ice has melted, despite the continuous input of heat. Similarly, when water boils, its temperature stays at 100°C (212°F) until it fully transitions into steam. This constant temperature during phase changes occurs because the added heat energy is used to break molecular bonds rather than increasing the kinetic energy of the molecules.

Heat also plays a critical role in processes like **sublimation**, where a solid transitions directly into a gas without becoming a liquid, and **deposition**, where a gas turns directly into a solid, bypassing the liquid phase. Dry ice, for example, sublimates at room temperature, turning from solid carbon dioxide directly into carbon dioxide gas.

Heat Transfer: How Heat is Conducted Between Objects

Heat transfer is the process by which thermal energy moves from one object to another due to a temperature difference. Heat always flows from a hotter object to a cooler one until both objects reach thermal equilibrium. There are three primary methods of heat transfer: conduction, convection, and radiation.

Conduction is the transfer of heat through direct contact between objects. It occurs when particles in a hot object collide with particles in a cooler object, transferring energy in the form of heat. For example, when you touch a metal spoon in a hot pot, heat is conducted from the spoon to your hand. Materials like metals are good conductors of heat because their particles transfer energy easily. On the other hand, materials like wood or plastic are poor conductors, known as insulators, because they resist the transfer of heat.

In **convection**, heat is transferred through the movement of fluids (liquids or gases). As the fluid heats up, it becomes less dense and rises, while cooler, denser fluid sinks, creating a circulation pattern that transfers heat. This is why warm air rises and cooler air sinks in a room.

Radiation transfers heat through electromagnetic waves without needing a medium. For example, the Sun transfers heat to the Earth through radiation, even though space is a vacuum and there is no physical contact between the Sun and Earth.

Four Laws of Thermodynamics

The study of heat and its role in phase changes leads us naturally to the four laws of thermodynamics, which provide a deeper understanding of energy exchange in physical systems. These laws define the fundamental principles of how energy is conserved, transferred, and transformed, from the heat flowing between objects to the work done by engines or other systems.

Zeroth Law of Thermodynamics: this law establishes the concept of **thermal equilibrium**. It says that if two systems are each in thermal equilibrium with a third system, they are also in equilibrium with each other. This law explains why we can use thermometers to measure temperature—if two objects have the same temperature, no heat will flow between them.

The **First Law of Thermodynamics** states that energy cannot be created or destroyed, only transferred or transformed. The total energy in a system remains constant. For example, when you heat an object, some of the heat is used to raise its temperature, while some may be used to do work. Energy is always conserved.

The **Second Law** introduces the idea of **entropy**, which is the measure of disorder in a system. It states that natural processes tend to increase entropy, meaning energy spreads out and becomes less usable over time. This explains why heat always flows from hot objects to cold ones and why no machine can be 100% efficient.

The **Third Law** states that as a system approaches **absolute zero** (0 Kelvin), its entropy, or disorder, approaches a minimum value. In other words, at absolute zero, particles have minimal motion, and the system reaches its most ordered state. In reality, absolute zero is unreachable, but this law explains the behavior of systems at very low temperatures.

Magnetism and Polarization

Magnetism is a force generated by the movement of electric charges, causing certain materials to attract or repel each other. Magnetic fields are created by electrons moving within atoms, and these fields exert forces on other magnets or magnetic materials like iron. Magnets have two poles, north and south, where the magnetic force is strongest. Opposite poles attract, while like poles repel.

Polarization in magnetism refers to the alignment of the magnetic domains within a material. In unmagnetized materials, these domains are randomly oriented, canceling each other out. However, when exposed to a magnetic field, the domains align in the same direction, creating a magnetic effect. This process is crucial in creating magnets and in technologies like magnetic storage devices.

Magnetism and polarization are essential in many modern applications, from electric motors and generators to data storage in computers, all relying on controlled magnetic fields.

SECTION 3. CHEMISTRY FOUNDATIONS

Elements and the Periodic Table

In chemistry, **elements** are the simplest forms of matter, composed of only one type of atom. These atoms are characterized by the number of protons in their nucleus, which defines their **atomic number**. For example, hydrogen has an atomic number of 1 (one proton), while carbon has an atomic number of 6 (six protons). Every element has its unique atomic structure, which gives it specific chemical properties.

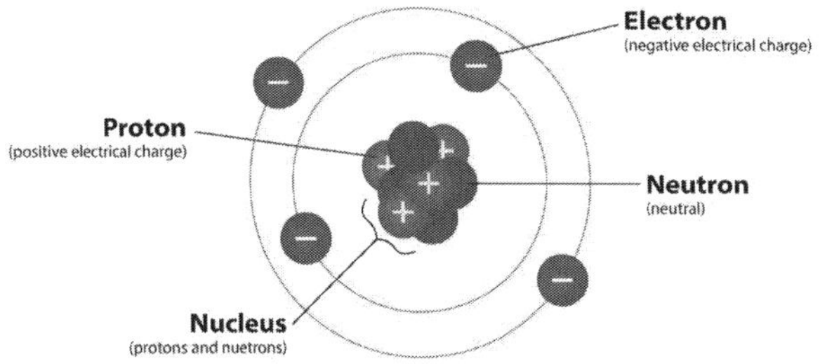

Atoms consist of three subatomic particles:

- **Protons:** These are positively charged particles found in the nucleus (center) of an atom. The number of protons in an atom determines its element and its atomic number on the periodic table.
- **Neutrons:** Neutrons are neutral (no charge) particles also located in the nucleus. They help stabilize the nucleus and, together with protons, contribute to an atom's atomic mass.
- **Electrons:** Electrons are negatively charged particles that orbit the nucleus in various energy levels or shells. They are much smaller than protons and neutrons and are responsible for the chemical reactions and bonding between atoms.

Protons and neutrons make up the dense core of the atom called the nucleus, while electrons orbit the nucleus in various energy levels or shells. The arrangement of these electrons largely determines an element's chemical reactivity.

Overview of the Periodic Table

The Periodic Table of Elements is a systematic arrangement of elements based on their atomic numbers. The table is structured in periods (horizontal rows) and groups (vertical columns). Elements within the same group have similar properties due to the same number of valence electrons, or electrons in their outer shell. The arrangement helps predict how elements behave chemically and how they will react with other elements.

PERIODIC TABLE OF THE ELEMENTS

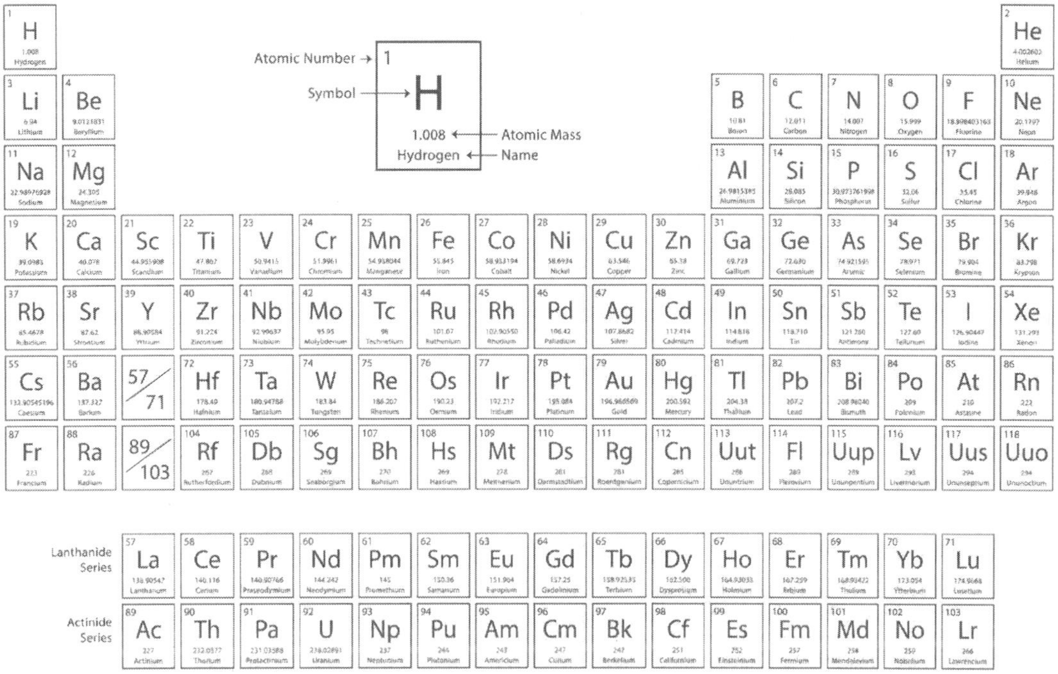

The periodic table contains a great deal of information about the atoms of each known element. Let's use the following figure as an example. This figure presents key information about an element as displayed on the periodic table. Each section of the element's box provides specific details essential for identifying the element and its properties.

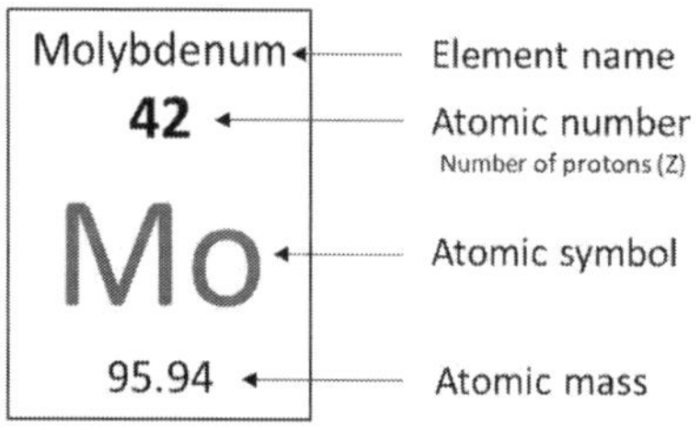

Element Name: The full name of the element, in this case, Molybdenum.

Atomic Number: This represents the number of protons in an atom's nucleus. For Molybdenum, it's 42.

Atomic Symbol: The one- or two-letter abbreviation for the element, here it's "Mo."

Atomic Mass: This is the average mass of an atom of the element, accounting for all isotopes. For Molybdenum, it is 95.94 atomic mass units (amu).

Groups and Periods

Groups are the vertical columns, numbered from 1 to 18. Elements in the same group tend to have similar chemical behaviors because of their similar electron configurations. For instance, Group 1 consists of the alkali metals, which are highly reactive due to their single valence electron. Group 18, the noble gases, is composed of elements that are generally inert because they have full valence shells.

Periods are the horizontal rows. As you move across a period from left to right, elements transition from metals to nonmetals, and their chemical properties change. For example, elements on the left side (like sodium) are metals, while those on the right (like chlorine) are nonmetals.

Element Categories

Metals: Found on the left and center of the periodic table, metals are typically malleable, conductive, and shiny. Examples include iron (Fe) and copper (Cu).

Nonmetals: Located on the right side of the table, nonmetals are usually poor conductors of heat and electricity. They are often gases or brittle solids at room temperature. Examples include oxygen (O) and carbon (C).

Metalloids: Elements that possess properties of both metals and nonmetals, making them useful in technology and industry. Silicon (Si), for example, is a metalloid often used in semiconductors.

Key Element Groups

Alkali Metals (Group 1): Extremely reactive, especially with water, alkali metals include lithium (Li) and potassium (K). They are soft metals that are highly reactive due to their single valence electron.

Alkaline Earth Metals (Group 2): These metals, such as calcium (Ca) and magnesium (Mg), are less reactive than alkali metals but still react with water and air to form oxides.

Transition Metals (Groups 3-12): Known for their ability to form colorful compounds, transition metals are often used in construction and manufacturing. Examples include iron (Fe) and gold (Au).

Halogens (Group 17): These are reactive nonmetals, such as fluorine (F) and chlorine (Cl), which readily form compounds with metals and other nonmetals.

Noble Gases (Group 18): Known for their lack of reactivity, noble gases like helium (He) and neon (Ne) have full electron shells, making them stable and chemically inert.

Periodic Trends

The periodic table reveals various trends that help predict element behavior:

- **Atomic Radius:** The size of an atom, or atomic radius, increases as you move down a group because additional electron shells are added. However, as you move across a period from left to right, the atomic radius decreases due to increasing positive charge in the nucleus pulling electrons closer.
- **Electronegativity:** This is the ability of an atom to attract electrons. Electronegativity increases across a period from left to right because atoms become more eager to fill their valence shell. Conversely, it decreases as you move down a group.
- **Ionization Energy:** This refers to the energy required to remove an electron from an atom. Ionization energy increases across a period, as atoms hold onto their electrons more tightly, and decreases down a group as the outer electrons are farther from the nucleus and easier to remove.

Importance of the Periodic Table

The periodic table is an indispensable tool in chemistry. By organizing elements based on their atomic structure and properties, it allows scientists and students to predict how elements will interact in chemical reactions. For example, understanding the periodic trends can explain why sodium reacts explosively with water or why noble gases are used in applications where chemical stability is crucial. This systematic arrangement of elements has revolutionized the way we study and understand the building blocks of matter.

Physical vs. Chemical Changes

A **molecule** is a group of two or more atoms chemically bonded together. These atoms can be of the same element, such as in oxygen (O_2), or different elements, like in water (H_2O). Molecules are the basic building blocks of chemical substances and can vary in size, complexity, and function, forming everything from simple gases to complex proteins in living organisms.

Molecules can change in physical or chemical ways. **Physical changes** affect the appearance or state of a substance without altering its chemical structure. For example, melting ice changes its state from solid to liquid, but it remains H_2O.

Chemical changes, on the other hand, involve a transformation at the molecular level. Bonds between atoms are broken and reformed, creating new substances with different properties. A classic example is the rusting of iron, where iron reacts with oxygen to form iron oxide, a completely new compound.

Compound vs. Molecule

A discussion of the difference between molecule and compound is in order here. A **compound** is a substance made up of two or more different types of atoms chemically bonded together in a fixed ratio. For example, water (H_2O) is a compound consisting of two hydrogen atoms and one oxygen atom. In contrast, Oxygen (O_2) is not a compound because it consists of only one type of element—oxygen. A compound, by definition, must contain two or more different elements chemically bonded together.

While **molecules** are groups of atoms bonded together, they can either be compounds (different types of atoms) or elements (same type of atom, like O_2). Therefore, all compounds are molecules, but not all molecules are compounds. Compounds always involve different elements, whereas molecules can consist of the same element.

Compound vs. Mixture

We already know that a compound is a substance formed when two or more different elements are chemically bonded together in fixed proportions. The chemical bonds that hold the elements together give the compound unique properties, distinct from the elements that form it. For example, water(H_2O) is a compound made of hydrogen and oxygen atoms. In a compound, the individual elements lose their original properties to create a new substance with different characteristics.

A **mixture**, on the other hand, consists of two or more substances physically combined but not chemically bonded. Each substance in a mixture retains its original properties and can be separated by physical means. For example, saltwater is a mixture of salt (NaCl) and water (H_2O). Mixtures can be **homogeneous** (uniform composition, like saltwater or wine) or **heterogeneous** (non-uniform, like a salad). Here are a couple of examples of heterogeneous mixture:

- **Oil and water**: When mixed, these liquids do not combine uniformly and instead separate into different layers, forming a heterogeneous mixture.
- **Granite**: A type of rock that is made up of visibly different minerals like quartz, feldspar, and mica, creating a non-uniform composition.

The key distinction between compound and mixture is that no chemical reaction occurs between the components in a mixture, whereas a compound involves a chemical change where bonds are formed or broken.

Acids vs. Bases

Acids are substances that release hydrogen ions (H^+) when dissolved in water, making the solution more acidic. Common acids include vinegar (acetic acid) and lemon juice (citric acid). Acids typically taste sour, can corrode metals, and have a pH less than 7. Strong acids like hydrochloric acid (HCl) completely dissociate in water, while weak acids, like acetic acid, only partially dissociate.

Bases, on the other hand, release hydroxide ions (OH^-) in water or accept hydrogen ions. Examples include baking soda (sodium bicarbonate) and bleach (sodium hypochlorite). Bases usually feel slippery, taste bitter, and have a pH greater than 7. Strong bases, such as sodium hydroxide (NaOH), fully dissociate, while weak bases only partially ionize.

Comparison: Acids and bases are opposites on the pH scale, with acids having a pH below 7 and bases above 7. While acids donate hydrogen ions, bases either accept them or release hydroxide ions. When mixed, they can neutralize each other, forming water and a salt. Both are essential in various chemical reactions, though they have contrasting properties and uses in everyday life.

Food Acidity and Alkalinity

Many foods and beverages are more acidic or basic than you might think. In fact, to human taste buds, acidic foods taste sour or tart and basic foods taste bitter. Knowing the pH of food and beverages can help you maintain a healthy diet. Research has proven that a highly acidic diet can cause long-term stomach problems such as ulcers or acid reflux, while a highly basic diet can cause digestion problems, due to an increased pH in your stomach. In addition, acidic foods and beverages can have a slow, but dangerous effect on your teeth. Your teeth are composed of a form of calcium phosphate, which can melt under acidic conditions. A list of acidic and basic foods is provided in the following table.

Food	pH
Cake, bread, rice	7.0–7.5
Vegetables (peas, cabbage, etc.)	6.0–6.5
Onions, mushrooms, eggplant	5.3–5.8
Peaches, apples, oranges	3.8–4.3
Grapes, strawberries	3.4–3.7
Soda (Coca-Cola, Sprite, Mountain Dew)	2.8–3.0

SECTION 4. MECHANICAL FOUNDATIONS

Forces and Motion

Understanding **forces and motion** is fundamental to grasping how mechanical systems function. Forces are what cause objects to move, stop, or change direction, and they play an essential role in how machines and structures operate. In this section, we will explore Newton's Laws of Motion, the forces of gravity and friction, different types of forces such as tension, compression, and shear, and the concepts of equilibrium and balance in force interactions.

Gravity and Friction

Gravity and friction are two of the most significant forces that affect the movement of objects.

Gravity

Gravity is the force that pulls objects toward the center of the Earth. It gives weight to objects and causes them to fall when dropped. In mechanical systems, gravity can be both a benefit and a hindrance. For example, gravity helps water flow in pipelines and can aid in the movement of conveyor belts going downhill. On the other hand, it can also make lifting and supporting heavy loads challenging. Engineers design cranes, elevators, and other lifting devices specifically to overcome the gravitational pull, allowing heavy materials to be moved with less effort.

Weight is the force exerted on an object due to gravity, and it can be calculated using Newton's Second Law of Motion. Newton's Second Law states that the force (F) acting on an object is the product of its mass (m) and its acceleration (a): $F = m \cdot a$

When dealing with weight, the acceleration in question is the gravitational acceleration (g). On Earth, this gravitational acceleration is approximately $9.8\,\mathrm{m/s^2}$. Therefore, we can express weight (W) as the force due to gravity by using Newton's Second Law: $W = m \cdot g$, where: W is the weight (in Newtons, N), m is the mass (in kilograms, kg), g is the gravitational acceleration, which on Earth is about $9.8\,\mathrm{m/s^2}$. This formula shows that weight is the gravitational force acting on an object's mass.

Example 1: Consider an object with a mass of 15 kg. To calculate its weight on Earth, we use the formula:

$$W = m \cdot g = 15\,\mathrm{kg} \times 9.8\,\mathrm{m/s^2} = 147\,\mathrm{N}$$

Thus, the object weighs 147 Newtons on Earth.

Example 2: Weight on Other Planets

Newton's Second Law also explains why an object weighs less on the Moon or more on a planet with stronger gravity. On the Moon, where the gravitational acceleration is only $1.62\,\mathrm{m/s^2}$, the weight of the same 15 kg object would be:

$$W = 15\,\mathrm{kg} \times 1.62\,\mathrm{m/s^2} = 24.3\,\mathrm{N}$$

This example demonstrates that weight is a force determined by the mass of the object and the local gravitational field, in line with Newton's Second Law. The object's mass remains the same, but its weight changes based on the gravitational acceleration acting upon it.

Friction

Friction is a force that resists the motion of objects or surfaces sliding past each other. It plays a vital role in many everyday activities, from walking to driving, by providing the necessary force to prevent slipping. Frictional forces can be categorized into two types: static friction and kinetic friction.

Static friction acts between surfaces that are at rest relative to each other. This type of friction must be overcome to initiate movement. It is usually stronger than kinetic friction.

Kinetic friction (also called sliding friction) acts between surfaces that are moving relative to each other. Once the object is in motion, this frictional force slows it down or prevents further acceleration unless a force continues to act on it.

Both types of friction depend on the normal force and the coefficient of friction. The **normal force** is the force exerted by a surface perpendicular to the object resting on it (typically equivalent to the object's weight when on a flat surface). The **coefficient of friction** is a dimensionless number that characterizes the interaction between two surfaces. A **dimensionless number** is a quantity that has no physical units associated with it. In other words, it is a pure number that does not depend on any system of measurement (such as meters, kilograms, or seconds). These numbers represent ratios or comparisons between quantities that cancel out any units during their calculation.

In the context of friction, the coefficient of friction (μ) is dimensionless because it is the ratio of the frictional force to the normal force. Since both forces are measured in the same unit (Newtons, for example), their units cancel out, leaving just a pure number. The coefficient of friction essentially describes how "sticky" or "slippery" two surfaces are relative to each other, without needing to refer to any specific units of measurement. For example, a coefficient of friction of $\mu = 0.5$ means that the frictional force is half the value of the normal force, regardless of the actual units involved.

The coefficient of friction is typically different for static friction (μ_s) and kinetic friction (μ_k). Different surfaces usually have different coefficients of friction. Rougher surfaces generally have higher coefficients of friction than smoother surfaces.

The force of friction F_f is calculated using the following formula: $F_f = \mu F_n$, where μ is the coefficient of friction (static or kinetic), F_n is the normal force, usually calculated as the weight of the object.

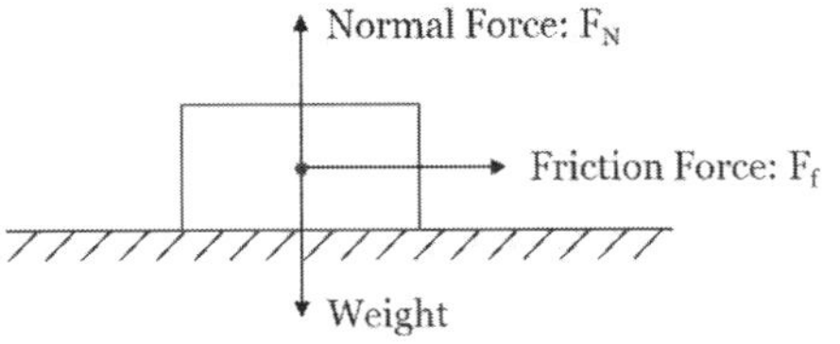

For static friction: $F_{f,static} \leq \mu_s F_n$

Static friction prevents motion until a certain threshold (equal to $\mu_s F_n$) is exceeded.

For kinetic friction: $F_{f,kinetic} = \mu_k F_n$

Once an object is in motion, kinetic friction remains constant and is typically less than static friction, i.e., $\mu_k < \mu_s$

Static friction is generally greater than kinetic friction because it takes more force to initiate the movement of an object than to keep it moving. Once an object is in motion, less force is needed to overcome the kinetic friction.

Types of Forces: Tension, Compression, and Shear

There are different types of forces that act on objects in mechanical systems, including **tension**, **compression**, and **shear**. Each type of force affects materials differently, and understanding these effects helps engineers design structures that can withstand various loads.

Tension is a force that pulls or stretches a material. It occurs when forces act in opposite directions, trying to elongate the material. For example, the force in a rope during a game of tug-of-war is a form of tension. In mechanical systems, tension is often experienced in cables, chains, or belts used to transmit force. Suspension bridges, for instance, rely on tension in their cables to support the weight of the bridge deck. Materials under tension need to have high tensile strength to resist stretching and breaking.

Compression is the force that pushes or squeezes a material. It occurs when forces act toward each other, compressing the material. A common example is the force experienced by the legs of a chair when someone sits on it. Compression is also present in columns and beams that support heavy loads, such as in buildings or bridges. Materials under compression need to be strong enough to resist buckling or crushing.

Shear is a force that causes one part of a material to slide past another. Shear forces act parallel to the surface and can cause deformation or failure when the material cannot resist the sliding action. An example of shear force is when scissors cut through paper, as the blades apply a shearing force that causes the paper to separate. In mechanical systems, shear forces can occur in bolts, rivets, and beams subjected to opposing forces.

Equilibrium and Balance of Forces

Equilibrium is a state in which all the forces acting on an object are balanced, resulting in no net force and, consequently, no acceleration. In mechanical systems, equilibrium is vital for ensuring that structures and machines remain stable and function as intended. There are two main types of equilibrium: static equilibrium and dynamic equilibrium.

Static equilibrium occurs when an object is at rest, and the forces acting on it are balanced. For example, a book lying on a table is in static equilibrium because the downward force of gravity is balanced by the upward normal force exerted by the table. In mechanical systems, static equilibrium is essential for stability. Structures like bridges and buildings must be designed to remain in static equilibrium under various loads, ensuring that they do not collapse or shift.

Dynamic equilibrium occurs when an object is moving at a constant velocity, and the forces acting on it are balanced. For instance, a car moving at a constant speed on a straight road is in dynamic equilibrium because the driving force provided by the engine is balanced by the opposing forces of friction and air resistance. In mechanical systems, maintaining dynamic equilibrium is crucial for consistent performance. For example, conveyor belts and rotating machinery need to operate in dynamic equilibrium to ensure smooth and efficient operation without sudden changes in motion.

The concept of **balance of forces** is crucial in designing mechanical systems that can handle various loads without failing. When forces are balanced, the system remains stable, and the risk of damage or failure is minimized. For instance, in a bridge, the downward force of gravity acting on the bridge deck must be balanced by the upward tension in the supporting cables and the compressive forces in the pillars. If these forces are not balanced, the bridge may collapse.

In mechanical systems, achieving equilibrium often requires the use of **counterweights**, **springs**, or **adjustable components** to ensure stability. For example, cranes use counterweights to balance the load being lifted, preventing the crane from tipping over. Springs are used in vehicle suspensions to balance the forces acting on the wheels, ensuring a smooth ride and preventing excessive bouncing or tipping.

Another critical aspect of equilibrium is the concept of **center of gravity**. The center of gravity is the point at which the entire weight of an object can be considered to act. For an object to be stable, its center of gravity must be positioned in such a way that the forces acting on it are balanced. In mechanical systems, keeping the center of gravity low and well-distributed helps maintain stability. For example, vehicles are designed with a low center of gravity to prevent them from tipping over during sharp turns or sudden maneuvers.

Energy and Power

In this section, we will explore potential and kinetic energy, conservation of energy, calculating work, power, and energy, and different types of power sources, including electrical, hydraulic, and mechanical power.

Potential and Kinetic Energy

Energy is the capacity to do work, and in mechanical systems, energy often comes in two primary forms: potential energy and kinetic energy.

Potential energy is the energy an object possesses due to its position or configuration. Essentially, it is the energy stored within an object that has the potential to be converted into another form of energy or to do work. For example, a boulder sitting at the top of a hill has gravitational potential energy because of its height above the ground. This potential energy is a result of the force of gravity acting on the boulder and the height at which it is positioned. The greater the height and mass of an object, the greater its potential energy. Springs also store energy, called **elastic potential energy**, when they are compressed or stretched.

Kinetic energy, on the other hand, is the energy of motion. Any object in motion possesses kinetic energy, which depends on its mass and velocity. For instance, a car driving down a road or a hammer hitting a nail both have kinetic energy. The faster an object moves, or the greater its mass, the more kinetic energy it will have. The relationship between kinetic energy and velocity is exponential, meaning that if you double the speed of an object, its kinetic energy will quadruple. In mechanical systems, kinetic energy is crucial for performing tasks such as driving machinery, moving parts, or transmitting force between components.

The relationship between potential energy and kinetic energy is often demonstrated by a pendulum. When the pendulum is raised to one side, it has maximum potential energy and zero kinetic energy. As it swings downward, the potential energy is converted into kinetic energy, and the pendulum gains speed. At the bottom of the swing, the potential energy is at its minimum, and kinetic energy is at its maximum. This continuous conversion between potential and kinetic energy illustrates the dynamic nature of energy in mechanical systems.

Conservation of Energy in Mechanical Systems

The **law of conservation of energy** states that energy cannot be created or destroyed, only transformed from one form to another. In mechanical systems, this principle means that the total amount of energy remains constant, even though it may change forms—such as from potential to kinetic energy, or into other forms like thermal energy.

Consider a roller coaster as an example. At the top of the track, the roller coaster car has a large amount of potential energy due to its height. As the car begins to descend, that potential energy is converted into kinetic energy, and the car gains speed. At the bottom of the hill, most of the energy is in the form of kinetic energy. As the car climbs another hill, some of that kinetic energy is converted back into potential energy. This energy conversion continues throughout the ride, with the car's speed and height changing continuously. However, the total energy remains constant, assuming we neglect losses due to **friction** and **air resistance**.

In reality, mechanical systems do experience **energy losses**, primarily due to friction and other forms of resistance. These losses convert some of the system's energy into **thermal energy** (heat). For example, in a car engine, the chemical potential energy of the fuel is converted into kinetic energy to move the car, but some of that energy is also lost as heat due to friction between the engine components. Lubrication and design optimizations are used to reduce these losses and make the system more efficient. Despite these losses, the principle of conservation of energy still holds—energy is not destroyed but rather transformed into less useful forms, such as heat.

Calculating Work, Power, and Energy

To understand the role of energy in mechanical systems, it is essential to understand how **work** and **power** are calculated.

Work is defined as the product of force and the distance over which that force is applied. In mathematical terms, work is expressed as:

$$Work = Force \times Distance$$

For work to be done, the force must be applied in the direction of the movement. If an object is not moved, no work is done, regardless of the force applied. For instance, if you push against a wall with all your strength, no work is done if the wall doesn't move, even though you are exerting a force.

Power is the rate at which work is done. It measures how quickly energy is being transferred or converted from one form to another. Power is calculated as:

$$Power = \frac{Work}{Time}$$

The unit of power is the **watt (W)**, which is equivalent to one joule per second. In practical applications, power helps us determine how efficiently a machine can perform a task. For example, an electric motor with more power can perform work faster than a less powerful motor. In mechanical systems, **horsepower (hp)** is also a common unit of power, especially in automotive applications, where it indicates the engine's ability to perform work over time. One horsepower (hp) is equivalent to 746 watts (W).

Energy can be calculated in various forms, depending on whether we are dealing with potential energy or kinetic energy. **Gravitational potential energy (PE)** is calculated as:

$$Potential\ Energy = m \times g \times h$$

where m is mass, g is the acceleration due to gravity, and h is the height above the ground.

Kinetic energy (KE) is calculated as:

$$Kinetic\ Energy = \frac{1}{2} m \times v^2$$

where m is mass and v is velocity. These equations help quantify the amount of energy in a mechanical system, allowing for more precise calculations in engineering and physics.

Example — Gravitational Potential Energy: A rock with a mass of 10 kg is lifted to a height of 5 meters. Calculate its gravitational potential energy.

Solution: $PE_g = 10\,\text{kg} \times 9.8\,\text{m/s}^2 \times 5\,\text{m} = 490\,\text{Joules}$

The gravitational potential energy of the rock is 490 Joules.

Types of Power Sources: Electrical, Hydraulic, and Mechanical

Mechanical systems often rely on different types of power sources to function. These sources provide the energy required to perform work, and each type has its unique characteristics, advantages, and applications.

Electrical Power is one of the most commonly used power sources in mechanical systems. Electrical power is generated by converting other forms of energy—such as mechanical energy from wind turbines or chemical energy from burning fossil fuels—into electricity. This electricity can then be used to power electric motors, lights, and other devices. **Electric motors** are popular in various applications, including household appliances, industrial machinery, and electric vehicles. The primary advantage of electrical power is its convenience and the ability to control power output precisely.

Hydraulic Power is used in systems that require significant force over short distances. Hydraulic systems operate based on **Pascal's Law**, which states that pressure applied to a confined fluid is transmitted equally in all directions. Hydraulic power is generated by using a **hydraulic pump** to pressurize a fluid (usually oil), which is then directed through hoses to **actuators** or **cylinders**. This pressurized fluid creates force that can move heavy loads. Hydraulic systems are widely used in construction machinery, such as excavators, cranes, and forklifts, because they provide high power output and precise control. The major advantage of hydraulic systems is their ability to generate large amounts of force efficiently, even in compact designs.

Mechanical Power refers to power generated directly by a mechanical system, often involving gears, pulleys, levers, and rotating shafts. In mechanical power transmission, energy is transferred from one component to another through direct contact, such as gears transmitting torque from a motor to a wheel. **Mechanical advantage** is achieved by using simple machines to reduce the force needed to perform work, such as a lever that multiplies the input force. Mechanical power systems are found in many machines, including bicycles, where pedaling produces mechanical power to turn the wheels, and in windmills, where the wind's kinetic energy is converted into rotational motion to generate power.

Each type of power source has specific advantages and limitations, making them suitable for different applications. Electrical power is efficient for small and precise tasks, offering easy control and versatility. Hydraulic power excels in situations requiring massive amounts of force in compact spaces, making it ideal for heavy machinery. Mechanical power provides direct and efficient energy transfer, commonly used in applications where simple and robust designs are needed.

Torque

Torque refers to the rotational equivalent of force. While force causes linear movement, torque causes rotational movement about a pivot point or axis. Torque is critical in understanding how machines and mechanical systems work, particularly those involving rotating parts such as engines, wheels, gears, and levers.

In simple terms, torque measures how much a force acting on an object causes that object to rotate. The force must be applied at a certain distance from the pivot point, and this distance is called the "lever arm." The longer the lever arm and the greater the applied force, the more torque is generated.

The mathematical expression for torque is given by:

$$\tau = r \cdot F \cdot \sin(\theta)$$

Where: τ (tau) is the torque, r is the distance (lever arm) from the pivot point to where the force is applied, F is the magnitude of the applied force, and θ is the angle between the force vector and the lever arm.

In situations where the force is applied perpendicular to the lever arm ($\theta = 90°$), the sine of $90°$ is 1, so the formula simplifies to:

$$\tau = r \cdot F$$

This is the most common scenario in basic torque problems, where the force is applied at a right angle to the lever arm.

Example: Imagine you are tightening a bolt using a wrench. The wrench is 0.25 meters long, and you apply a force of 20 Newtons at the end of the wrench, perpendicular to the handle. To calculate the torque generated, you use the simplified torque formula:

$$\tau = r \cdot F = 0.25\,\text{m} \times 20\,\text{N} = 5\,\text{Nm}$$

The torque generated is 5 Newton-meters (Nm). This means that you are applying a rotational force of 5 Nm to the bolt, which will cause it to rotate and tighten.

If you were to increase the length of the wrench (lever arm) to 0.5 meter, keeping the force the same, the torque would increase: $\tau = 0.5\,\text{m} \times 20\,\text{N} = 10\,\text{Nm}$

In this case, doubling the lever arm doubles the torque, making it easier to tighten the bolt with less effort.

Torque is a crucial concept in various fields, including automotive engineering (e.g., tightening bolts, engine mechanics), construction (e.g., using levers and pulleys), and even sports (e.g., the torque generated when swinging a bat or golf club). In summary, torque is a measure of rotational force and is calculated by multiplying the force by the lever arm distance. The longer the lever arm or the greater the force, the more torque is generated, leading to more efficient rotational movement.

SECTION 5. MECHANICAL COMPREHENSION

Simple Machines

Simple machines are fundamental mechanical devices that help make work easier by providing a mechanical advantage—allowing users to apply less force over greater distances to move or lift loads. These simple tools form the basis of many complex machines and are key concepts in mechanical engineering and physics. Below, we explore the different types of simple machines, their properties, and their applications.

Lever

A lever is a simple machine consisting of a rigid bar that rotates around a fixed point called the **fulcrum**. Levers help move or lift loads by converting a small applied force into a larger output force, making work easier.

There are three types of levers, each distinguished by the position of the **fulcrum**, **load**, and **effort**:

1. **First-Class Lever**: In a first-class lever, the fulcrum is positioned between the load and the effort. Examples include seesaws and crowbars. By adjusting the distance between the fulcrum and the points of force application, first-class levers can either increase force or speed, depending on the requirements of the task.
2. **Second-Class Lever**: In a second-class lever, the load is positioned between the fulcrum and the effort. Examples include wheelbarrows and nutcrackers. These levers always increase force but decrease speed. By applying effort at a greater distance from the fulcrum than the load, the user can move heavier loads with less effort.
3. **Third-Class Lever**: In a third-class lever, the effort is applied between the fulcrum and the load. Examples include tweezers and baseball bats. These levers are designed to increase the speed of the output rather than force. They require more input effort but allow for greater output distance and speed, making them useful in situations like swinging a bat.

Three Lever Classes

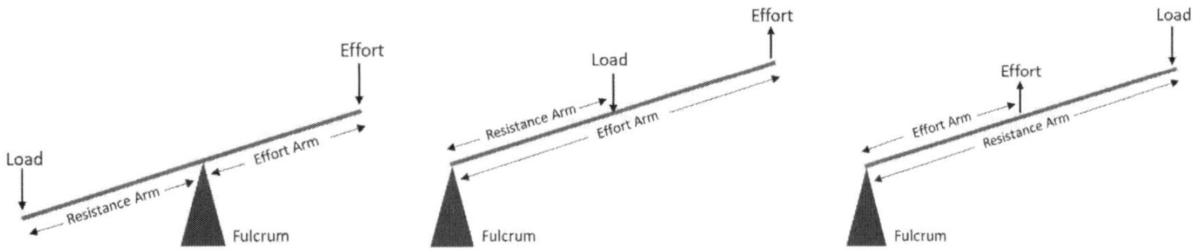

First Class Lever | Second Class Lever | Third Class Lever

Levers are extensively used in everyday applications, such as lifting objects, prying open boxes, and even in muscle movements in the human body. Understanding the types of levers and their applications helps to optimize the amount of force required in different situations.

Pulley

A pulley is a simple machine consisting of a wheel with a groove along its edge through which a rope or cable can pass. Pulleys change the direction of the applied force and can also provide mechanical advantage, making it easier to lift heavy objects.

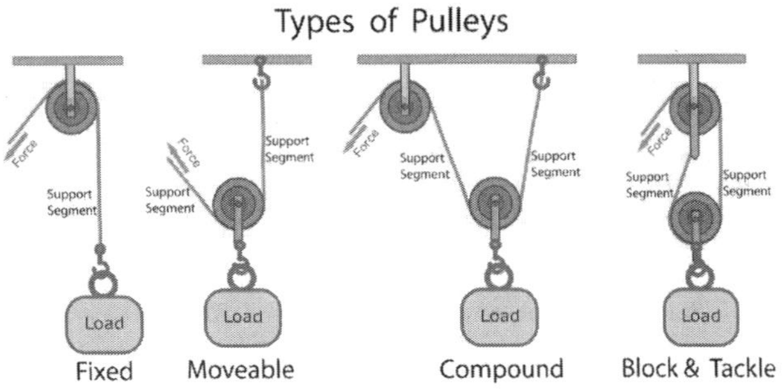

The foregoing figure shows four types of pulley systems: Fixed Pulley, Moveable Pulley, Compound Pulley, and Block & Tackle. Each type of pulley helps reduce the effort required to lift a load by redistributing or multiplying the force, making it easier to lift or move heavy objects. Let's discuss how each type of pulley reduces force needed.

1. Fixed Pulley: A fixed pulley has a wheel attached to a fixed support, such as a ceiling or beam. The rope passes through the wheel, and the load is on one end, while force is applied at the other end.

A fixed pulley does not reduce the amount of force needed to lift the load, but it changes the direction of the force. Instead of lifting the load directly upward, you pull the rope downward, which is often more convenient. As illustrated, there is only one rope segment supporting the load. The road segment to the left-hand side of the pulley does not directly support the load, and it is NOT counted as a supporting rope segment. The mechanical advantage of a fixed pulley is 1, meaning the force required to lift the load is the same as the load itself.

2. Moveable Pulley: A moveable pulley is attached directly to the load itself, and the pulley moves with the load as it is lifted.

A moveable pulley reduces the amount of force needed to lift the load by distributing the weight between two supporting segments of the rope. This effectively halves the effort required. For example, if the load weighs 100 N, the force needed to lift it would be 50 N. The mechanical advantage of a moveable pulley is 2.

Note that the mechanical advantage is equal to the number of supporting rope segments.

3. Compound Pulley: A compound pulley system is a combination of both fixed and moveable pulleys. It allows for a significant reduction in the effort required to lift a load.

By combining fixed and moveable pulleys, a compound pulley provides a higher mechanical advantage. The force needed is reduced based on the number of pulleys in the system, allowing the load to be lifted with much less effort. For instance, with multiple pulleys, the effort needed may be only a fraction of the load's weight. The mechanical advantage can be calculated by counting the number rope segments supporting the load, which, in the figure above is 2. Hence, if the load is 100 N, one will only need to apply a downward force of 50 N to pull it up using the compound pulley as shown.

4. Block & Tackle: A block and tackle system uses multiple pulleys in both fixed and moveable blocks, increasing the mechanical advantage. The **block** refers to the set of pulleys that are housed together in a casing. A block can contain one or more pulleys, and there are usually two blocks in a block and tackle system—one **fixed block** (attached to a support) and one **movable block** (attached to the load). The **tackle** refers to the ropes or cables that are threaded through the pulleys. The tackle is what transfers the force applied by the user to the load.

In a block and tackle system, the load's weight is distributed across multiple sections of the rope, significantly reducing the effort required. The mechanical advantage is equal to the number of rope segments supporting the load. In the figure given before, there are 2 rope segments in this particular block and tackle system. So, the force required is reduced to 1/2 of the load's weight. With a number of pulleys in the block and tackle system, the system allows a user to lift very heavy loads with minimal effort, making it ideal for heavy-duty lifting applications.

Example: The image on the right shows a block and tackle pulley system. If the weight being lifted is 120 Newtons, how much force is needed to pull the load up?

Solution: This pulley system has three supporting ropes, which means the mechanical advantage is 3. The mechanical advantage tells us that the force required to lift the load is divided by the number of ropes supporting it.

To calculate the force needed: $Force = \frac{Load}{Mechanical\ Advantage}$ Given that the load is 120 N and the mechanical advantage is 3: $Force = \frac{120\,N}{3} = 40\,N.$

Wheel and Axle

The wheel and axle is another simple machine that helps reduce friction and makes movement more efficient. It consists of a larger wheel attached to a smaller cylindrical axle, allowing both to rotate together.

The mechanical advantage of a wheel and axle comes from the difference in radius or diameter between the wheel and the axle. When a force is applied to the wheel, it is transferred to the axle, allowing it to exert a larger force over a shorter distance. This principle allows the wheel and axle to convert a small input force into a greater output force, which is useful for tasks such as moving heavy loads or generating rotational movement.

For example, in a car, the steering wheel acts as a wheel and axle system. Turning the large steering wheel requires minimal effort but generates enough force to move the smaller axle, effectively turning the car's wheels. Similarly, a doorknob is a wheel and axle, where the larger knob allows the user to exert less force while applying enough torque to move the latch.

Wheel and axle systems are commonly used in vehicles, pulleys, gears, and tools like screwdrivers. By reducing the effort needed and minimizing friction, this simple machine helps increase efficiency in various applications.

Inclined Plane: Calculating Effort and Load Relationships

An inclined plane is a flat surface set at an angle, used to facilitate the movement of heavy objects from a lower to a higher elevation or vice versa. The mechanical advantage of an inclined plane comes from spreading the effort needed to move the load over a longer distance, reducing the force required.

The relationship between effort and load on an inclined plane depends on the length of the plane and the height it rises. The longer the length of the inclined plane relative to its height, the less force is needed to move the load. For example, if you need to load a heavy box onto a truck, using a long ramp requires less force than lifting the box directly. However, while less force is needed, the distance over which the box must be moved is increased.

The mechanical advantage of an inclined plane can be calculated by dividing the length of the plane by its height:

$$Mechanical\ Advantage = \frac{Length\ of\ Inclined\ Plane}{Height}$$

Inclined planes are found in many everyday situations, such as wheelchair ramps, loading docks, and roadways on hills. By reducing the required effort to move an object, inclined planes make it possible to accomplish tasks that would otherwise be challenging.

Wedge: Application of Force and Use in Everyday Life

A **wedge** is a type of inclined plane that moves. It consists of two inclined planes joined back-to-back, forming a sharp edge that can be used to split or cut objects. The wedge converts a force applied to its blunt end into forces perpendicular to its inclined surfaces, effectively amplifying the input force to perform tasks like cutting, splitting, or holding materials in place.

The mechanical advantage of a wedge depends on its angle. The thinner and sharper the wedge, the greater the mechanical advantage. For example, an **axe** used to chop wood is a wedge. When force is applied to the axe head, the sharp edge splits the wood with greater force than was applied by the user. Similarly, **chisels** and **knives** are examples of wedges that amplify the force applied to cut or shape materials.

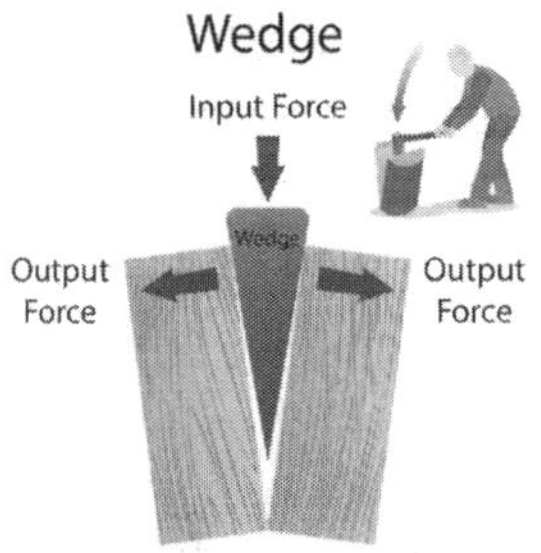

Wedges are used in a variety of everyday tools, from kitchen knives to doorstops. They are indispensable for many tasks, such as cutting, carving, splitting, and even holding objects in place, highlighting their versatility and efficiency in transferring and amplifying force.

Screw

A screw is a simple machine that consists of an inclined plane wrapped around a cylinder, forming a spiral. The **threads** of the screw help convert **rotational force (torque)** into linear motion, allowing it to move through or hold objects in place.

The **pitch** of a screw refers to the distance between the threads. The smaller the pitch, the more force can be generated with each rotation, as the inclined plane wraps more tightly, effectively reducing the required input force. This is why screws with finer threads are easier to turn but take more rotations to fully insert compared to screws with coarser threads.

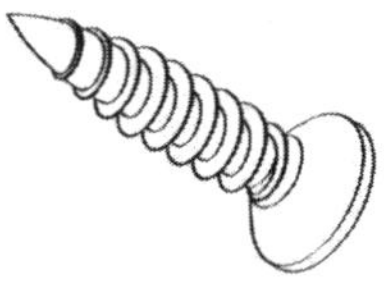

In practical applications, screws are used to fasten objects together or lift heavy loads. For example, a **car jack** uses a screw mechanism to lift the vehicle with relatively little effort. By turning the screw, the rotational force is converted into upward linear movement, lifting the car. Similarly, **wood screws** are used to securely join materials, as the threads create friction that resists removal, ensuring a tight and lasting connection.

Screws are one of the most widely used simple machines in the modern world, offering versatility in fastening, lifting, and securing applications, all while providing a mechanical advantage through the conversion of rotational force into linear movement.

Gears

Gears are mechanical components used to transfer power and change the speed or direction of motion between machine parts. There are different types of gears for various applications. **Spur gears** have straight teeth and are the most common type, used in applications requiring simple, parallel movement like clocks or conveyor systems. **Bevel gears** have angled teeth, allowing them to transfer power between shafts at different angles, making them useful in car differentials. **Worm gears** consist of a screw-like worm and a matching gear, which provide high reduction ratios and are used in applications requiring large torque, like elevators.

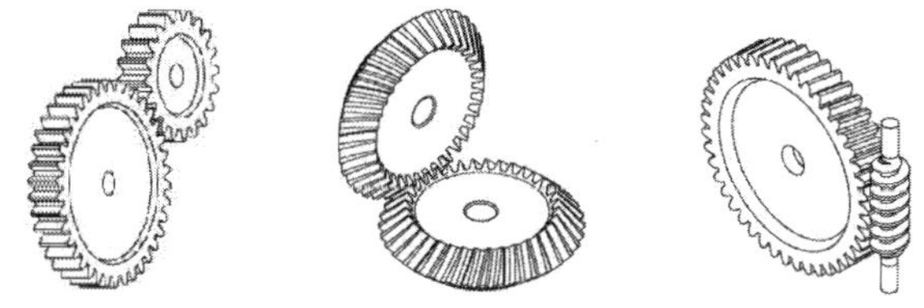

Spur Gear Bevel Gear Worm Gear

The **gear ratio** determines how gears affect the output speed and torque in a mechanical system. The gear ratio is determined by the number of teeth on the gears involved, not by their diameter. Specifically, it is calculated by dividing the number of teeth on the **driven gear** by the number of teeth on the **driving gear**.

If the gear ratio is greater than 1:1, it means that the driven gear has more teeth than the driving gear. As a result, the driven gear rotates more slowly compared to the driving gear, which leads to an increase in torque and a decrease in speed. For instance, if the driving gear has 20 teeth and the driven gear has 80 teeth, the gear ratio is: $Gear\ Ratio = \frac{80}{20} = 4:1$. This means that for every four rotations of the driving gear, the driven gear rotates only once. Such a configuration is ideal when more force is needed, such as in the first gear of a car to provide more power for starting from a stop.

Conversely, if the gear ratio is less than 1:1, the driven gear has fewer teeth than the driving gear. In this scenario, the driven gear rotates faster, resulting in increased speed but reduced torque. For example, if the driving gear has 60 teeth and the driven gear has 20 teeth, the gear ratio is: $Gear\ Ratio = \frac{20}{60} = 1:3$. This means that for every rotation of the driving gear, the driven gear will rotate three times. This setup is useful when speed is more important than force, such as when a cyclist shifts to a higher gear on flat terrain to move faster with less pedaling effort.

It is important to note that while diameter may affect the physical design and placement of gears, it does not determine the gear ratio. The number of teeth alone determines how speed and torque are modified. When the gear ratio is greater than 1:1, the system produces more torque but moves more slowly. When the gear ratio is less than 1:1, the system moves faster but generates less torque.

Fluid Power/Hydraulics

Fluid power systems use fluids—either liquids or gases—to transmit power. **Hydraulic systems** utilize liquids, typically oil, while **pneumatic systems** use compressed air. Hydraulics are ideal for applications that require high force with precision, such as lifting heavy objects, whereas pneumatics are used for tasks that require fast, lightweight, and repetitive power.

Pressure is a measure of the amount of force exerted per unit area. It tells us how much force is applied on a surface and is usually measured in units such as Pascals (Pa), pounds per square inch (psi), or atmospheres (atm), depending on the context. The formula for calculating pressure is:

$$Pressure = \frac{Force}{Area}, \text{ or } P = \frac{F}{A}$$

where Pressure (P) is measured in Pascals (Pa) or other units like psi, Force (F) is measured in Newtons (N) or pounds (lbs), Area (A) is measured in square meters (m^2) or square inches (in^2).

Example 1: Let's say you are applying a force of 100 Newtons on a surface that has an area of 2 square meters. To find the pressure exerted on the surface, you would use the formula: $P = \frac{F}{A} = \frac{100\ \text{N}}{2\ \text{m}^2} = 50\ \text{Pa}$.

Example 2: Imagine you need to apply a pressure of 200 Pascals (Pa) on a surface that has an area of 4 square meters. What amount of force is required to generate this pressure?

> To calculate the force, we rearrange the pressure formula to solve for force: $Force = Pressure \times Area$
>
> Now, plug in the known values: $Force = 200\ Pa \times 4\ m^2 = 800\ N$

A key concept in hydraulic systems is **Pascal's Law**, which states that when pressure is applied to a confined fluid, it is transmitted equally in all directions. This means that if you apply force to one part of the system, that pressure is transferred throughout the entire fluid, which can then be used to create movement elsewhere. For example, in a **hydraulic jack**, a small force applied to a pump piston creates pressure that is transmitted through the fluid, allowing a larger piston to lift a heavy load, such as a car. The hydraulic jack effectively multiplies the input force, making it possible to lift large weights with minimal effort.

The figure below illustrates a simple hydraulic system. The relationship between the two pistons is governed by the following formula:

$$\frac{F_1}{A_1} = \frac{F_2}{A_2}$$

Where:

- F_1 is the force applied on the smaller piston.
- A_1 is the area of the smaller piston.
- F_2 is the force exerted on the larger piston.
- A_2 is the area of the larger piston.

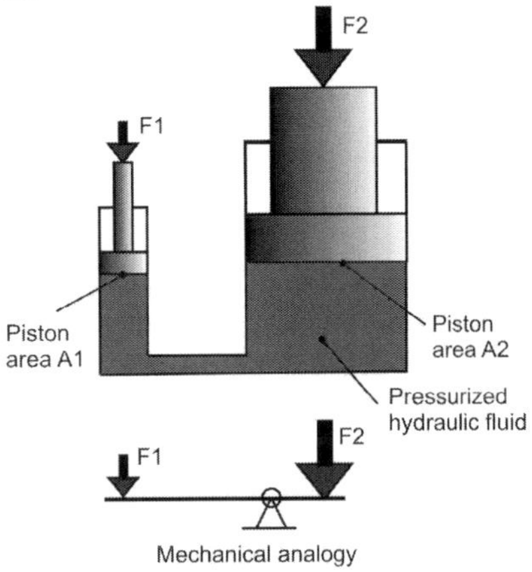

Example: Let's go through an example using Pascal's Law.

Suppose James applies 200 pounds of force to the smaller piston, and the area of the larger piston is 5 times that of the smaller piston. What's the force exerted on the larger piston?

Solution: We can calculate the force exerted on the larger piston using the formula: $F_2 = F_1 \times \frac{A_2}{A_1}$

Given: $F_1 = 200$ pounds, $A_2 = 5 \times A_1$, substituting into the formula: $F_2 = 200 \times \frac{5A_1}{A_1}$

The A_1 terms cancel out: $F_2 = 200 \times 5 = 1000$ pounds. So, the force exerted on the larger piston is 1000 pounds.

This example demonstrates how a hydraulic system can multiply force. In this case, applying 200 pounds of force on the smaller piston results in 1000 pounds of force being exerted on the larger piston.

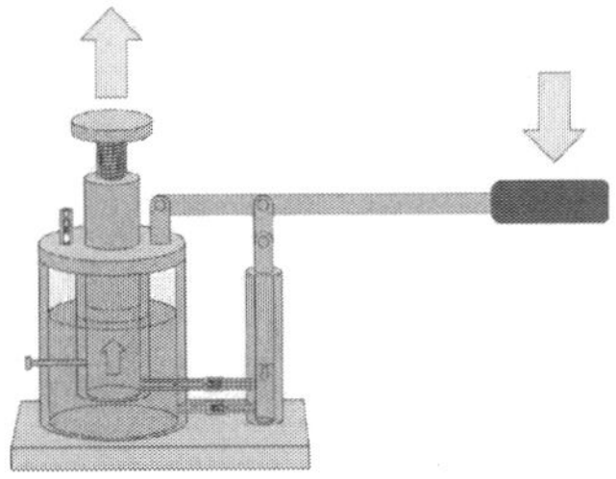

Structure of a Typical Hydraulic Jack

Hydraulic and pneumatic systems are used in many types of machinery. Hydraulic systems are seen in applications like car brakes and heavy machinery such as cranes, where high force is required. On the other hand, pneumatic systems are often used in assembly lines for tasks that need quick and repetitive actions, such as pressing or packaging. For example, a **pneumatic jackhammer** uses compressed air to generate powerful blows for breaking concrete.

Section 6. Electronics Foundations

Electricity is a form of energy that results from the movement of electrons, which are subatomic particles found in atoms. At its most fundamental level, electricity is the flow of electrical charge, typically through a conductor, which allows electrons to move easily.

To understand how electricity exists, we must first look at the structure of an atom. Atoms are the basic building blocks of all matter and consist of three primary particles: protons, neutrons, and electrons. **Protons** have a positive charge, **neutrons** have no charge (they are neutral), and **electrons** carry a negative charge. Protons and neutrons are found in the **nucleus**, or center, of the atom, while electrons orbit the nucleus in shells or energy levels.

Electricity forms when electrons are freed from their orbits around the nucleus and flow from one atom to another. This flow of electrons is what we refer to as an electric current. In materials like metals, certain electrons, called **free electrons**, are loosely bound to atoms, making them easier to move. When a voltage (electrical force) is applied, these free electrons move in a directed way, creating an electrical current.

The ability of a material to conduct electricity depends on how easily its electrons can move. Materials fall into three main categories based on their electrical conductivity: conductors, insulators, and semiconductors.

Conductors are materials that allow electrons to flow freely. Metals, such as copper, aluminum, and silver, are good conductors because they have many free electrons that can easily move from atom to atom. This is why metals are commonly used in electrical wiring and circuits.

Insulators are materials that do not allow electrons to move freely, making them poor conductors of electricity. Examples of insulators include rubber, glass, and plastic. In these materials, electrons are tightly bound to their atoms, making it difficult for an electric current to flow. Insulators are important in preventing electrical currents from flowing where they are not wanted, such as in the protective coatings around electrical wires.

Semiconductors fall between conductors and insulators. They do not conduct electricity as well as metals, but under certain conditions (such as the introduction of heat or the addition of certain impurities), they can become conductive. Silicon and germanium are common semiconductor materials, and they are crucial in the manufacture of electronic devices like transistors and integrated circuits. Semiconductors are the foundation of modern electronics, allowing devices like computers and smartphones to function.

Electric Current

Electric current refers to the flow of electric charge through a conductor, typically measured in the movement of electrons. When a voltage is applied across a conductor, it creates an electric field that drives the flow of electrons from a region of higher potential to lower potential. This movement of electrons constitutes what we call an electric current. Essentially, electric current is how electric energy travels through circuits to power devices, from small electronics like smartphones to large machines.

There are two primary types of electric current:

- **Direct Current (DC):** In this type of current, electrons flow in a single, constant direction. DC is commonly found in batteries, where current flows from the positive terminal to the negative terminal through a device.
- **Alternating Current (AC):** In AC, the direction of electron flow periodically reverses. This type of current is used in homes and businesses, as it is more efficient for transmitting electricity over long distances.

An **elementary charge** is the smallest unit of electric charge that is carried by a single proton or electron. It is a fundamental constant in physics, representing the basic quantum of electric charge. Electric current is measured in amperes (A), commonly referred to as "amps." One ampere represents the flow of one **coulomb** of charge per second through a point in a circuit. A coulomb (C) is the standard unit of electric charge in the International System of Units (SI), and one coulomb is equivalent to approximately 6.242×10^{18} **elementary charges**. To measure electrical current in a circuit, one needs an **ammeter** or a multimeter set to the current (ampere) mode.

Voltage

Electric voltage, often referred to simply as voltage, is a measure of the **electric potential difference** between two points in an electric circuit. Voltage can be thought of as the "pressure" that pushes electric charges through a conductor, such as a wire. It is a critical concept in understanding how electrical energy moves through a circuit, powering devices and allowing them to operate.

At its core, voltage represents the **electric potential energy** per unit charge at a point in a circuit. This electric potential can be thought of as the ability of a charge to do work, like moving through a circuit to power a device. When there is a difference in electric potential between two points, this creates an electrical potential difference, which is what drives the movement of electrons, or electric current.

In some contexts, the term **electromotive force (EMF)** is used to describe the voltage generated by a source, such as a battery or a generator. Although it includes the word "force," EMF is actually a type of voltage and not a physical force. It represents the energy provided to each charge as it moves through the source of the voltage.

To measure voltage, an instrument called a **voltmeter** is used. This device is connected across two points in a circuit and provides a reading of the electrical potential difference between them. Voltage is measured in units called **volts** (V), and it is one of the most important parameters in any electrical system, dictating how much energy is available to move charges through a circuit.

Resistance

Electric resistance is a measure of how much a material or component opposes the flow of electric current. It determines how difficult it is for the current to pass through a conductor. In simpler terms, resistance can be thought of as the "friction" that electric charges experience as they move through a circuit.

Different materials have different levels of resistance. **Conductors**, like copper or aluminum, have very low resistance, allowing current to flow easily through them. On the other hand, **insulators**, such as rubber or glass, have high resistance, which prevents the flow of current. Resistance is a key factor in controlling the amount of current that flows through a circuit.

Several factors affect the resistance of a material. The type of material plays a significant role, as metals generally have lower resistance than non-metals. The length of the conductor also matters; the longer the conductor, the higher the resistance. Similarly, the cross-sectional area of the material affects its resistance; a thinner wire has more resistance than a thicker one. Temperature also influences resistance, as most materials increase in resistance as their temperature rises.

Devices called **resistors** are specifically designed to introduce a known amount of resistance into a circuit. Resistors are used to control current, protect components, and adjust signal levels.

Electrical Effects

Electrical energy, when applied to different components or materials, can lead to various electrical effects, primarily chemical, heat, and magnetic effects. These effects form the basis of many everyday applications and devices. Here is a discussion of each type:

1. Chemical Effect

When an electric current is passed through certain liquids, it can cause chemical changes. This process is known as electrolysis. The chemical effect of electricity is used in various applications, such as:

- Electroplating: A process in which a layer of metal is deposited onto a surface using an electric current. For example, gold-plating jewelry.
- Battery Operation: In rechargeable batteries, chemical energy is converted to electrical energy when discharging and vice versa when charging.

2. Heating Effect

When electric current flows through a conductor with resistance, heat is generated due to the resistance opposing the current flow. This is known as the Joule heating effect or resistive heating. Applications of the heating effect include:

- Electric Heaters: Devices like electric stoves and room heaters convert electrical energy into heat to warm a space.
- Incandescent Bulbs: Current passes through a thin tungsten filament, heating it up until it glows and produces light.
- Fuses: The heating effect is used in fuses to break the circuit when excessive current flows, protecting electrical appliances.

3. Magnetic Effect

When electric current flows through a conductor, it creates a magnetic field around it. This is the basis of the magnetic effect of electricity, and it is used in several important applications:

- Electromagnets: Electric current through a coil produces a magnetic field, and this effect is used in lifting heavy metallic objects in junkyards.

- Electric Motors: Current passing through coils in motors generates magnetic fields that create motion, allowing devices like fans, washing machines, and drills to operate.
- Transformers: Magnetic fields produced by alternating current in coils allow transformers to step up or step down voltages in power systems.

These electrical effects are fundamental to various technologies and devices and understanding them allows for their practical and safe use in electrical and electronic systems.

SECTION 7. CIRCUITS

An electric circuit is a path through which electric current flows, enabling the transfer of energy from a source to a device or load. At its core, a circuit consists of a few fundamental elements, including a voltage source, conductors (wires), loads (such as resistors or light bulbs), and switches.

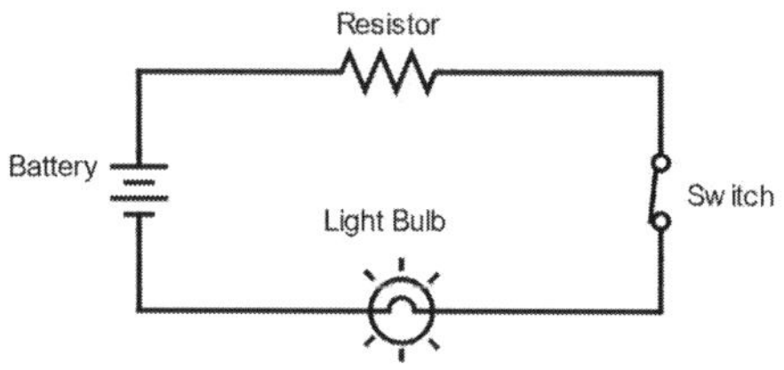

In the image provided, the circuit features a battery as the **voltage source**, supplying the electrical energy needed to push electrons through the system. A voltage source creates an electric potential difference, which drives current through the circuit.

The **load** in this case includes a resistor and a light bulb. The resistor limits the current, protecting the components and ensuring that they operate within safe parameters. The light bulb serves as an example of an electrical load, converting electrical energy into light and heat. Loads in circuits consume the energy provided by the source to perform useful work.

A **switch** is also present in the circuit. When the switch is open, it creates an open circuit, meaning the path for the current is interrupted, and no current flows. In contrast, when the switch is closed, it completes the circuit, creating a closed circuit where current can flow uninterrupted from the voltage source through the components.

Electric circuits can vary in complexity, but the basic principles remain the same: energy flows from the source, passes through the circuit's components, and returns to the source. The flow of electricity can be indicated using two different notations: electron flow notation and conventional current notation. **Electron flow notation** depicts the actual physical movement of electrons, which flow from the negative terminal to the positive terminal. This notation reflects the fact that electrons, which are negatively charged, are responsible for carrying electric current through conductive materials.

Conventional current notation represents the flow of current as moving from the positive terminal to the negative terminal of a power source. This direction was assumed before the discovery of electrons and is based on the movement of positive charges. Conventional current notation is typically used in diagrams today.

Ohm's Law

Ohm's Law is a fundamental principle that describes the relationship between voltage, current, and resistance in an electrical circuit. It states that the current flowing through a conductor is directly proportional to the voltage applied across it and inversely proportional to the resistance.

The formula for Ohm's Law is: $V = I \times R$, where: V represents the voltage (measured in volts), I is the current (measured in amperes), R is the resistance (measured in ohms).

This equation allows you to calculate any one of the three variables if the other two are known. For instance, if you know the voltage and resistance in a circuit, you can calculate the current using: $I = \frac{V}{R}$.

Example: Imagine you have a simple circuit with a resistor, and the voltage across it is 12 volts, while the resistance is 4 ohms. To find the current flowing through the circuit, you can apply Ohm's Law: $I = \frac{12\text{ V}}{4\,\Omega} = 3\text{ A}$. So, the current in the circuit is 3 amperes.

Series Circuits

A series circuit is one of the simplest types of electrical circuits where components are connected end-to-end, providing only one path for the current to flow. In a series circuit, all the components, such as resistors, are arranged in a single loop. This means that the same electric current flows through each component without branching. The image provided shows a battery connected to three resistors — Resistor 1, Resistor 2, and Resistor 3 — in a series configuration.

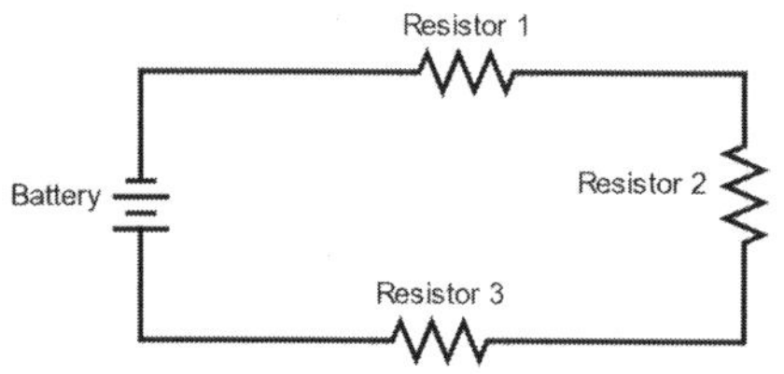

Characteristics of Series Circuits

Current: One of the defining characteristics of a series circuit is that the current (I) is the same at all points in the circuit. Since the current has only one path to take, it must pass through each resistor in turn. This means that if you were to measure the current at any point along the circuit — before Resistor 1, after Resistor 3, or anywhere in between — the current would be the same.

Resistance: In a series circuit, the total resistance (R) is the sum of the individual resistances of the components. As you add more resistors in series, the total resistance of the circuit increases, making it harder for the current to flow. In the example image, the total resistance R_{total} is the sum of the three resistors:

$$R_{\text{total}} = R_1 + R_2 + R_3$$

Because the resistances add up, increasing the number of resistors will reduce the overall current, given a constant voltage. This is important in designing circuits where controlling the current is necessary.

Voltage: Unlike the current, the voltage (V) in a series circuit is not the same across all components. The voltage from the battery is divided across each resistor, depending on its resistance. The total voltage across all the resistors must add up to the total voltage provided by the battery. This is known as **voltage drop**.

For example, if the battery provides 12 volts, and the resistors have different values, the voltage drop across each resistor will be different but will sum up to 12 volts. The voltage drop across each resistor depends on its resistance, with larger resistances experiencing a larger voltage drop.

Example: Given a series circuit as the one in the above image, a battery with a voltage of 18 V is connected to three resistors in series. Resistor 1 is 2 ohms, Resistor 2 is 3 ohms, and Resistor 3 is 5 ohms. What is the total resistance, the current flowing through the circuit, and the voltage drop across Resistor 2?

Solution:

Step 1: Calculate the Total Resistance.

To find the total resistance in the circuit, simply add the individual resistances:

$$R_{\text{total}} = R_1 + R_2 + R_3 = 2\,\Omega + 3\,\Omega + 5\,\Omega = 10\Omega$$

Step 2: Calculate the Total Current Using Ohm's Law

With the total resistance known, we can calculate the current flowing through the circuit using Ohm's Law:

$$I = \frac{V}{R_{\text{total}}} = \frac{18\,\text{V}}{10\,\Omega} = 1.8\,\text{A}$$

So, the current in the circuit is 1.8A and this current is the same through each resistor.

Step 3: Calculate the Voltage Drop Across Resistor 2

In a series circuit, the voltage drop across each resistor can be calculated using Ohm's Law: $V_{\text{drop}} = I \times R$

For Resistor 2, with a resistance of 3 ohms and the current $I = 1.8$A, the voltage drop is:

$$V_{\text{drop, R2}} = I \times R_2 = 1.8\,\text{A} \times 3\,\Omega = 5.4\,\text{V}$$

Step 4: Conclusion:

- Total Resistance: 10 ohms
- Current in the Circuit: 1.8 amperes
- Voltage Drop Across Resistor 2: 5.4 volts

This means that out of the total 18 volts provided by the battery, 5.4 volts are dropped across Resistor 2. You could similarly calculate the voltage drop across Resistor 1 and Resistor 3 using the same method.

In summary, a series circuit provides a single path for current to flow. The current is the same at all points, but the total resistance is the sum of all individual resistances. Voltage is divided across each component, with each resistor experiencing a voltage drop proportional to its resistance. Understanding these concepts is key for solving typical series circuit problems that might appear on the AFOQT or other exams.

Parallel Circuits

A parallel circuit is a type of electrical circuit where components are connected across the same voltage source in such a way that there are multiple paths for the current to flow. Unlike in series circuits where there is just one path, each component in a parallel circuit lies on a separate branch, and the total current from the source is divided among these branches.

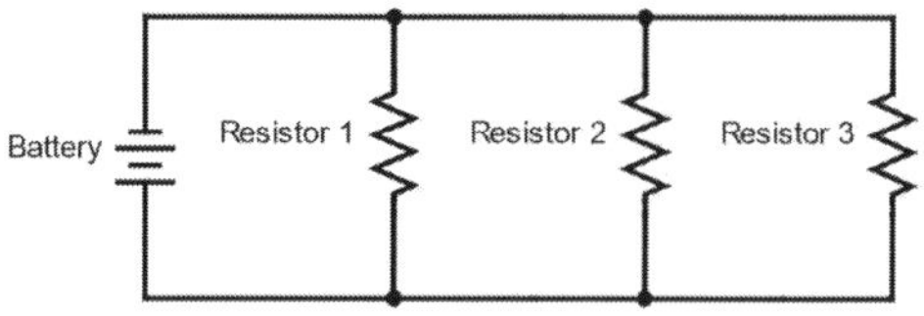

Characteristics of Parallel Circuits

Voltage: In a parallel circuit, the voltage across each component (resistor, light bulb, etc.) is the same. This means that each resistor in the attached diagram experiences the same voltage drop as provided by the battery, regardless of its resistance or the current flowing through it.

Current: The total current flowing from the source (battery) is split among the branches. The amount of current flowing through each branch depends on the resistance of the branch. A lower resistance draws more current, and vice versa.

Resistance: The total resistance in a parallel circuit is less than the resistance of the least resistive branch. This is because the presence of multiple paths allows more current to flow through the circuit than would be possible through a single resistor.

Calculating Parameters in Parallel Circuits

In parallel circuits, the formula for total resistance (R_{total}) is not simply the sum of the resistances. Instead, the total resistance is determined by the reciprocal of the sum of the reciprocals of each individual resistance:

$$\frac{1}{R_{\text{total}}} = \frac{1}{R_1} + \frac{1}{R_2} + \frac{1}{R_3}$$

This formula shows that adding more resistors in parallel decreases the total resistance.

Example: A battery provides a voltage of 12 V to a parallel circuit consisting of three resistors. Resistor 1 has a resistance of 6 ohms, Resistor 2 has a resistance of 3 ohms, and Resistor 3 has a resistance of 2 ohms. Calculate the total resistance of the circuit and the current through each resistor.

Solution:

Step 1: Calculate Total Resistance: The total resistance R_{total} can be calculated by the reciprocal formula for parallel circuits. Using the provided resistances:

$$\frac{1}{R_{\text{total}}} = \frac{1}{6} + \frac{1}{3} + \frac{1}{2} = \frac{1}{6} + \frac{2}{6} + \frac{3}{6} = \frac{6}{6} = 1$$

Step 2: Calculate Current through Each Resistor: Using Ohm's Law $I = \frac{V}{R}$ and knowing the voltage is 12 V for each resistor:

$$I_1 = \frac{12\,V}{6\,\Omega} = 2\,A$$
$$I_2 = \frac{12\,V}{3\,\Omega} = 4\,A$$
$$I_3 = \frac{12\,V}{2\,\Omega} = 6\,A$$

The total current from the battery is $I_{\text{total}} = I_1 + I_2 + I_3 = 2\,A + 4\,A + 6\,A = 12\,A$

This example shows how voltage remains constant across each resistor in a parallel circuit while the current varies depending on the resistance.

Series-Parallel Circuits

A series-parallel circuit is a type of electrical circuit that combines elements of both series and parallel configurations. This allows different components in the circuit to exhibit characteristics of both series and parallel circuits, leading to a more complex behavior of current, voltage, and resistance. The figure here shows an example of a series-parallel circuit where Resistor 1 is in series with the parallel combination of Resistors 2 and 3. We'll use this figure throughout this section to discuss Series-Parallel Circuits.

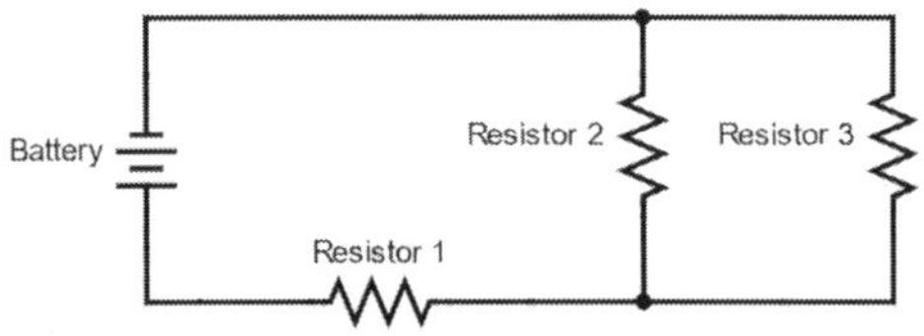

In a series-parallel circuit, some components are connected in series, while others are connected in parallel:

Series components: Components that share the same current, as there is only one path for current to flow through them. In the image, Resistor 1 is in series with the parallel section of the circuit (Resistors 2 and 3).

Parallel components: Components that are connected across the same voltage, but the current splits between them. In this case, Resistor 2 and Resistor 3 are connected in parallel.

Characteristics of Series-Parallel Circuits

Current: In a series-parallel circuit, the current behaves differently in the series and parallel portions. In the series portion of the circuit (the part that includes Resistor 1), the current is the same through all components because there is only one path for the current to follow.

In the parallel portion of the circuit (the part that includes Resistor 2 and Resistor 3), the current splits between the branches. The amount of current in each branch depends on the resistance of the individual components in that branch. Components with lower resistance will carry more current.

Resistance: Calculating the total resistance in a series-parallel circuit involves combining the rules for both series and parallel resistances:

1. First, calculate the equivalent resistance of the components in parallel (Resistors 2 and 3). The formula for parallel resistances is: $\frac{1}{R_{\text{parallel}}} = \frac{1}{R_2} + \frac{1}{R_3}$

2. Once the equivalent resistance of the parallel section is known, add it to the resistance of the series component (Resistor 1) to find the total resistance of the circuit: $R_{\text{total}} = R_1 + R_{\text{parallel}}$

Voltage: Voltage behaves differently in the series and parallel portions of the circuit. In the series portion, the voltage drop is proportional to the resistance of each component. The voltage across Resistor 1 will depend on its resistance and the total current flowing through the circuit.

In the parallel portion, the voltage across each parallel branch (Resistors 2 and 3) is the same and equal to the voltage applied to that portion of the circuit.

Example: A battery supplies 24 volts to a series-parallel circuit. Resistor 1 has a resistance of 4 ohms, Resistor 2 has a resistance of 6 ohms, and Resistor 3 has a resistance of 12 ohms. What is the total resistance of the circuit, the current through Resistor 1, and the voltage across Resistor 2?

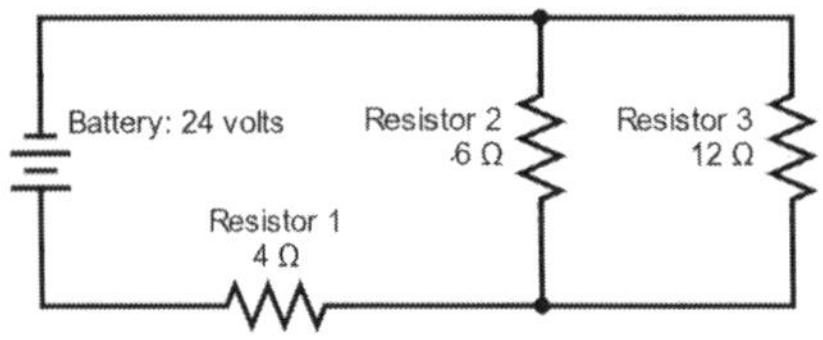

Solution:

Step 1: Calculate the Equivalent Resistance of the Parallel Section. First, find the combined resistance of Resistor 2 and Resistor 3, which are in parallel:

$$\frac{1}{R_{\text{parallel}}} = \frac{1}{6\,\Omega} + \frac{1}{12\,\Omega} = \frac{2}{12} + \frac{1}{12} = \frac{3}{12}$$

$$R_{\text{parallel}} = \frac{12}{3} = 4\Omega$$

Step 2: Find the Total Resistance. The total resistance of the circuit is the resistance of Resistor 1 (which is in series with the parallel section) plus the equivalent resistance of the parallel section:

$$R_{\text{total}} = R_1 + R_{\text{parallel}} = 4\,\Omega + 4\,\Omega = 8\Omega$$

Step 3: Calculate the Total Current. Using Ohm's Law $I = \frac{V}{R}$ the total current in the circuit is: $I = \frac{24\text{ V}}{8\,\Omega} = 3\text{ A}$

Step 4: Find the Voltage Across Resistor 2. The voltage across the parallel section (Resistors 2 and 3) is the same as the voltage drop across each resistor in that section. The voltage drop across Resistor 1 is:

$$V_1 = I \times R_1 = 3\,\text{A} \times 4\,\Omega = 12\,\text{V}$$

The remaining voltage is across the parallel section: $V_{\text{parallel}} = 24\,\text{V} - 12\,\text{V} = 12\,\text{V}$

So, the voltage across Resistor 2 is 12V.

As illustrated in the above example, series-parallel circuits combine the characteristics of both series and parallel circuits, making them versatile and more complex than purely series or parallel circuits.

Measuring Electric Power

Electric power is the rate at which electrical energy is transferred in a circuit. It is measured in **watts** (W), where one watt equals one joule of energy transferred per second. The basic formula to calculate electric power is: $P = V \times I$, where: P is the power in watts, V is the voltage in volts, I is the current in amperes.

Power increases as either the voltage or current increases. This formula applies to all electrical devices, from small household items to large industrial machines. For larger power measurements, we often use **kilowatts** (kW) and **megawatts** (MW). 1 kilowatt (kW) is equal to 1,000 watts. This unit is commonly used to measure the power output of household appliances and small machines. 1 megawatt (MW) is equal to 1,000 kilowatts or 1,000,000 watts. Megawatts are used to measure large-scale power generation, such as power plants or large industrial equipment.

Example: A light bulb operates with a current of 2 A and a voltage of 120 V. What is the power consumed by the light bulb?

Solution: Using the formula $P = V \times I$: $P = 120\,\text{V} \times 2\,\text{A} = 240\,\text{W}$. The power consumed is 240 watts or 0.24 kW.

SECTION 8. ELECTRICAL AND ELECTRONIC SYSTEMS

Electrical and electronic systems are fundamental to modern life, powering everything from household appliances to sophisticated digital devices. **Electrical systems** are designed to generate, transmit, and use electrical energy, focusing on large-scale power distribution and consumption. In contrast, **electronic systems** deal with controlling the flow of electrons through smaller circuits, used in devices such as computers, mobile phones, and many automated control systems. Together, these systems form the backbone of modern infrastructure and technology.

AC vs. DC

Alternating Current (AC) and Direct Current (DC) are the two main types of electric current used in electrical and electronic systems.

Alternating Current (AC): In AC, the direction of the flow of electrons alternates periodically. This means that the current changes direction at regular intervals, typically many times per second. The frequency of AC is measured in hertz (Hz), which indicates the number of cycles per second. AC is primarily used for power transmission and distribution because it can be easily transformed to higher or lower voltages, making it efficient for transmitting electricity over long distances. Most of the electricity supplied to homes and businesses is AC.

Direct Current (DC): In DC, the flow of electrons is unidirectional, meaning it moves in a constant direction from the negative terminal to the positive terminal. DC is commonly used in batteries, electronic devices, and automotive systems. It is preferred for applications requiring stable and consistent voltage, such as charging mobile phones or powering electronic circuits.

The key difference between AC and DC is how they deliver electrical energy. AC is ideal for efficiently transmitting power over long distances, while DC is more suited for low-voltage, portable applications.

Grounding

Grounding is an essential safety practice in electrical systems that involves creating a direct physical connection between electrical devices or circuits and the earth. The purpose of grounding is to prevent electrical shock, protect equipment, and ensure the safe operation of electrical systems by providing a controlled path for excess or stray electrical current.

In any electrical system, unintended faults or malfunctions can cause electrical current to flow in unintended ways, potentially leading to dangerous situations. Grounding directs these currents safely into the earth, reducing the risk of electric shock or fire. For example, if an exposed wire accidentally touches a metal part of an appliance, grounding ensures that the excess current flows to the ground instead of energizing the appliance's casing and posing a shock hazard to anyone touching it.

There are two primary types of grounding in electrical systems:

System Grounding: This involves grounding the neutral point of electrical power systems, such as transformers or generators. It stabilizes voltage during normal operation and ensures that the power system has a reference point, which helps in maintaining consistent voltage throughout the circuit.

Equipment Grounding: This involves connecting the metal parts of appliances and devices to the ground to protect users. The grounding wire carries away excess current in case of a fault, preventing damage and shock.

Grounding systems typically involve ground rods or other conductors driven into the earth, providing a low-resistance path for electricity. **Ground fault circuit interrupters (GFCIs)** are often used alongside grounding to provide an additional layer of safety, especially in wet environments like kitchens or bathrooms, where the risk of electric shock is higher.

Proper grounding is a crucial part of any electrical installation, making it safer for both people and equipment by reducing the risks associated with electrical faults.

Impedance

Impedance is the measure of opposition that an alternating current (AC) experiences in a circuit, similar to resistance but with additional complexities that come from components like capacitors and inductors. Unlike resistance, which only affects direct current (DC) by reducing current flow, impedance considers both resistance and the effects of changing current and voltage in AC circuits.

Capacitive Reactance: Capacitive reactance refers to the opposition created by a capacitor in an AC circuit. Capacitors are components that store energy in the form of an electric field and tend to resist changes in voltage. The opposition offered by capacitors decreases as the frequency of the AC signal increases. This means that at higher frequencies, capacitors allow more current to pass through, making them effective for filtering and managing low-frequency signals. Essentially, the faster the current alternates, the less the capacitor opposes it.

Inductive Reactance: Inductive reactance is the opposition provided by an inductor in an AC circuit. Inductors store energy in the form of a magnetic field, and they resist changes in current. In contrast to capacitors, the opposition presented by an inductor increases as the frequency of the AC signal rises. This means that inductors are more effective at blocking high-frequency signals while allowing lower-frequency signals to pass. As the current alternates more rapidly, inductors create greater opposition to the flow of current.

Impedance in AC Circuits: In an AC circuit, impedance combines the effects of resistance, capacitive reactance, and inductive reactance. Resistance affects both AC and DC in the same way, while capacitive and inductive reactance change based on the frequency of the AC signal. Together, these factors determine how much opposition the current faces in the circuit. Capacitors and inductors cause a phase difference between the current and voltage, meaning the two are not perfectly in sync—sometimes the current leads or lags behind the voltage depending on the component. This phase relationship is a crucial aspect of how AC circuits behave.

Electric and Electronic Components

Electric and electronic components are the building blocks of all electrical and electronic circuits, enabling the functionality of devices ranging from simple household appliances to complex communication systems. These components help control, direct, and transform electrical energy to perform tasks like switching, amplifying signals, or storing energy. Below is a list of common components.

Power Supply

A power supply is a crucial component that provides the necessary electrical energy to power electronic devices and circuits. It converts electrical energy from a source, such as a wall outlet, into the appropriate voltage, current, and type (AC or DC) required by the system. Power supplies can be linear or switching types, with linear power supplies

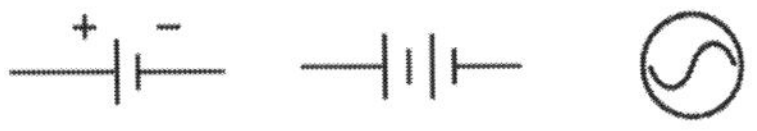

providing a stable output voltage and switching power supplies being more energy efficient.

Wire

A wire is a conductor that provides a pathway for electrical current to flow through a circuit. Typically made of copper or aluminum, wires are chosen for their excellent conductivity and ability to carry current with minimal resistance. Wires are insulated with materials like plastic or rubber to prevent accidental contact with other conductive surfaces and to protect users from electric shock.

Wires come in various **gauges**, which refer to the thickness or diameter of an electrical wire, which is measured using standards like the American Wire Gauge (AWG) system. In AWG, a smaller number indicates a thicker wire, while a higher number indicates a thinner wire.

Wire gauge affects both the current-carrying capacity and the resistance of the wire. Thicker wires can carry more current with less resistance, making them suitable for high-power applications, while thinner wires are used for lower current needs. Choosing the correct gauge is essential for safety and efficiency, as using a wire with too small a gauge could lead to overheating and potential fire hazards.

Electrical wires are **color-coded** to indicate their specific functions, ensuring safety and proper connections during installation and maintenance. The color-coding of wires varies slightly between regions, but certain standards are widely recognized.

- Black, Red, or Blue wires are typically used for live (hot) wires, carrying current from the power source to devices or appliances.
- White or Gray wires are used as neutral wires, providing the return path for current and completing the circuit.
- Green, Green with Yellow Stripes, or Bare Copper wires are used as ground wires, which connect the system to the earth to prevent electrical shock in case of faults.

Color codes may differ slightly in different countries or regions, so it is important to follow local electrical standards.

In an electrical diagram, different notations are used to indicate whether wires are connected or not connected.

- Connected Wires: A solid dot is placed at the intersection of wires to indicate they are electrically connected.
- Not Connected Wires: Wires that cross without connecting are shown without a dot, often with one wire having a small arc over the other to indicate they are not joined.

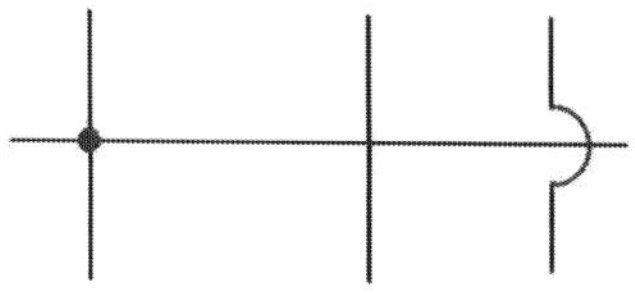
Connected   Not Connected

These symbols ensure clear interpretation of electrical pathways and prevent errors in circuit construction.

Switch

A switch is an electromechanical component that controls the flow of electrical current in a circuit. It works by opening or closing the circuit, effectively turning devices on or off. When the switch is open, the circuit is incomplete, and no current flows; when it is closed, the circuit is complete, allowing current to pass through. Switches come in various types, including toggle switches, push-button switches, and rotary switches, each designed for different applications and levels of current.

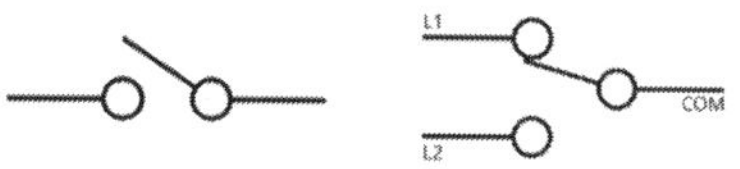

Switch, single-pole single-throw   Switch, single-pole double-throw

In addition to controlling devices manually, switches are also used in automation systems to respond to certain conditions, such as temperature or pressure changes, making them versatile in both simple and complex electrical systems.

Fuse and Circuit Breaker

Fuses and circuit breakers are safety devices designed to protect electrical circuits from excessive current that could cause damage or create a fire hazard.

Fuse   Fuse

A fuse is a component containing a metal wire that melts when the current exceeds a specified level, breaking the circuit and stopping the current flow. Fuses are inexpensive and simple but need to be replaced once blown.

A circuit breaker is a switch that automatically opens when it detects an overload or short circuit, interrupting the current. Unlike fuses, circuit breakers can be reset and used again. Circuit breakers are often used in residential and industrial electrical panels to protect wiring and equipment.

Both fuses and circuit breakers are crucial for preventing damage to electrical systems and ensuring safety by stopping the flow of excess current before it causes harm. They play a critical role in protecting both people and equipment from the risks associated with electrical faults.

Amplifier

An amplifier is an electronic device that increases the power, voltage, or current of an input signal. Its main purpose is to make a weak signal stronger, allowing it to drive a larger load, such as a speaker, or to be processed by other components in an electronic system. Amplifiers are found in various applications, from audio systems, where they amplify sound signals, to radio transmitters, where they boost the power of signals for long-distance communication.

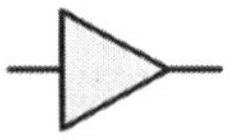

Amplifier

There are different types of amplifiers depending on their application and the signals they process. Audio amplifiers are used in sound systems to enhance audio signals, allowing music and voice to be heard loudly through speakers. Radio frequency (RF) amplifiers are used in communication systems to boost high-frequency signals for transmission. There are also operational amplifiers (op-amps), which are versatile components used in signal processing, filtering, and analog computation.

Amplifiers are characterized by their gain, which is the ratio of the output signal strength to the input signal strength. A higher gain means a more substantial amplification effect. However, amplifiers must maintain signal quality, so distortion and noise are crucial factors in determining amplifier performance.

Semiconductor

A semiconductor is a material that has electrical conductivity between that of a conductor (like copper) and an insulator (like glass). The conductivity of a semiconductor can be manipulated by adding impurities, a process known as **doping**, which creates either an excess of electrons (n-type) or a shortage of electrons (p-type). This makes semiconductors highly versatile for controlling current.

Semiconductors are the foundation of most electronic components, including diodes, transistors, and integrated circuits (ICs). Common materials used in semiconductors are silicon and germanium, with silicon being the most widely used due to its abundance and favorable properties.

The key advantage of semiconductors is their ability to control electrical behavior with precision. By combining n-type and p-type materials, components can be made to amplify signals, switch currents, or regulate voltage. The p-n junction formed in diodes and transistors is a core aspect of their functionality.

Semiconductors are the backbone of the electronics industry, enabling the development of computers, smartphones, and virtually all modern electronic devices.

Resistor

A resistor opposes the flow of electric current, allowing control over the amount of current passing through a circuit. Resistors are used to limit current, divide voltage, and protect components from excess current. They are available in variable and non-variable (fixed) types.

Resistor

Rheostat

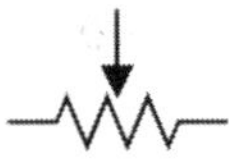

Potentiometer

Fixed resistors have a specific resistance value that does not change. They are used where a constant level of resistance is required, such as setting current in a circuit. **Variable resistors**, also known as potentiometers or rheostats, allow the user to adjust the resistance value manually. Potentiometers are often used as volume controls in audio devices, while rheostats are used for adjusting the brightness of lights or motor speeds.

The resistance value of a resistor is measured in ohms (Ω), and its power-handling capacity determines how much energy it can dissipate as heat. Resistors are typically color-coded to indicate their resistance value, making them easy to identify and use in circuit design.

Whether fixed or variable, resistors are crucial in controlling and stabilizing current and voltage, ensuring that circuits operate safely and effectively.

Diodes

A diode is an electronic component that allows current to flow in only one direction, acting as a one-way valve for electrical current. Diodes are used to protect circuits by blocking reverse currents that could damage components, and they are also fundamental in **rectification**, which converts alternating current (AC) into direct current (DC).

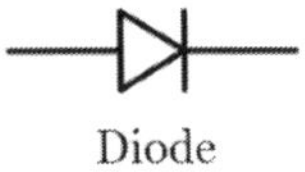

Diode

Diodes are made from semiconductor materials, typically silicon, and have two terminals: the **anode** (positive side) and the **cathode** (negative side). Current flows from the anode to the cathode when a forward voltage is applied, but it is blocked if a reverse voltage is applied. This directional behavior makes diodes useful in power supplies and signal processing.

There are different types of diodes designed for specific purposes. **Light Emitting Diodes (LEDs)** emit light when current flows through them and are used for displays and indicators. **Zener diodes** are used for voltage regulation, allowing current to flow in the reverse direction when a specific breakdown voltage is reached.

Rectifier and Inverter

A rectifier is an electrical device that converts alternating current (AC) into direct current (DC). This process, called **rectification**, is fundamental in power supply systems, where many electronic devices require DC power to operate, even though the electricity available from the power grid is AC. Rectifiers are used in devices like phone chargers, power adapters, and other power supplies that convert household AC into usable DC.

The key component used in a rectifier is a diode. In a basic rectification process, the diode is used to block the negative portion of the AC signal, allowing only the positive portion to pass, effectively creating a pulsating DC signal. There are different types of rectifiers:

- **Half-wave rectifiers** use a single diode to convert half of the AC cycle, resulting in DC output that pulses.
- **Full-wave rectifiers** use multiple diodes (often four in a bridge rectifier) to convert both the positive and negative halves of the AC cycle, resulting in a smoother DC output.

In short, a diode is the core component that makes rectification possible, and a rectifier is a circuit that uses one or more diodes to achieve the conversion from AC to DC, crucial for powering electronic devices.

An **inverter** is an electrical device that converts direct current (DC) into alternating current (AC). It is used in various applications where AC power is needed but only a DC source, such as a battery or solar panel, is available.

The primary function of an inverter is to provide AC power for devices like home appliances, tools, or grid systems from DC sources. Inverters are commonly used in renewable energy systems, uninterruptible power supplies (UPS), and electric vehicles, ensuring that the generated DC can power standard AC devices efficiently.

Transistor

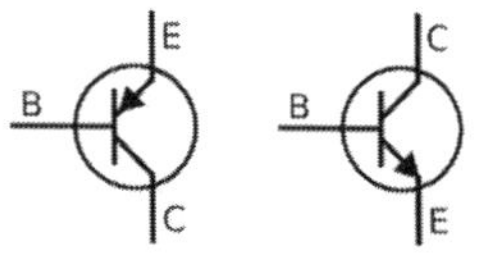

PNP BJT Transistor NPN BJT Transistor

A transistor is a semiconductor device used to amplify or switch electronic signals. It has three terminals: the emitter, base, and collector. By applying a small input current to the base, a transistor can control a much larger current flowing between the collector and emitter. This ability makes transistors fundamental in both analog and digital circuits.

Transistors come in two main types: **Bipolar Junction Transistors (BJTs)** and **Field-Effect Transistors (FETs)**. BJTs use both electron and hole charge carriers, while FETs control current flow using an electric field. FETs are widely used in integrated circuits due to their efficiency and small size.

Transistors are used as amplifiers to increase signal strength and as switches in digital circuits, such as those found in microprocessors. Their switching capability allows them to represent binary states (on and off), making them essential for building logic gates and digital devices. The invention of the transistor revolutionized electronics, enabling the miniaturization and integration of complex circuits.

Transformer

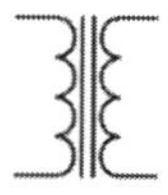
Transformer

A transformer is an electrical device used to transfer electrical energy between two or more circuits through the principle of electromagnetic induction. Its primary function is to change the voltage levels, either stepping up (increasing) or stepping down (decreasing) the voltage, depending on the requirement. Transformers consist of two coils, known as the primary and secondary windings, which are wound around a magnetic core. When alternating current (AC) flows through the primary winding, it creates a changing magnetic field, which induces a voltage in the secondary winding.

The ratio of the number of turns in the primary coil to the secondary coil determines whether the transformer steps up or steps down the voltage. For example, if the secondary winding has more turns than the primary, the transformer will increase the voltage, making it a step-up transformer. Conversely, if the primary has more turns, the transformer will reduce the voltage, making it a step-down transformer.

Transformers are widely used in power distribution systems to efficiently transmit electricity over long distances. High-voltage transmission reduces energy loss, and step-down transformers are then used to bring the voltage to safer, usable levels for homes and businesses. They are also used in electronic devices, such as power adapters and chargers, to convert the high-voltage supply to a lower voltage suitable for the device.

Transistor vs Amplifier: Compared with a transistor, an **amplifier** is a circuit or device that increases the power, current, or voltage of an input signal. It usually consists of multiple components, including transistors, resistors, and capacitors, which work together to boost a signal. For example, in an audio system, an amplifier increases the weak input signal from a microphone or music player so that it can drive a speaker. In essence, a transistor is a component that can be used within an amplifier circuit, but an amplifier itself is a broader assembly designed for the specific purpose of boosting signals. Transistors are thus building blocks, while amplifiers are complete systems that perform a defined task.

Capacitor

A capacitor is an electrical component that stores energy in an electric field and releases it when needed. It consists of two conductive plates separated by an insulating material called the **dielectric**. When voltage is applied across the plates, an electric charge builds up, storing energy that can be quickly released.

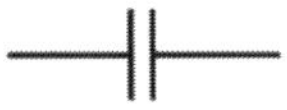

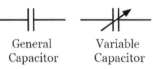

General Capacitor

Variable Capacitor

Capacitors are used in a variety of applications, such as filtering (removing unwanted frequencies from signals), energy storage (providing power in short bursts), and timing circuits (creating delays). A capacitor handles AC (alternating current) and DC (direct current) differently due to its ability to store and release energy in an electric field.

Capacitor with DC: When DC is applied, a capacitor initially allows current to flow as it charges up to the applied voltage. Once fully charged, the capacitor blocks further current because there is no changing voltage across its terminals. Essentially, after charging, a capacitor acts like an open circuit for DC, meaning no current flows after the capacitor is fully charged.

Capacitor with AC: When AC is applied, the voltage across the capacitor is constantly changing direction. This causes the capacitor to continuously charge and discharge, which allows current to flow through the circuit. As the voltage changes in each cycle, the capacitor opposes these changes, creating a phase difference between the voltage and current. It effectively allows AC to pass while blocking DC.

In summary, a capacitor blocks DC after charging, acting like an open circuit, while it allows AC to pass by continuously charging and discharging in response to the alternating voltage.

There are different types of capacitors, such as electrolytic capacitors, which have high capacitance values and are often used in power supplies, and ceramic capacitors, which are smaller and used in signal processing. A capacitor is measured by its **capacitance**, which indicates the capacity to store electrical charge. The unit of capacitance is the farad (F), named after the English scientist Michael Faraday.

Inductor

Inductor

An inductor is a passive electrical component that stores energy in the form of a magnetic field when electric current flows through it. It typically consists of a coil of wire, often wound around a core made of magnetic material, such as iron. When current passes through the coil, it creates a magnetic field, and the energy is stored in that magnetic field. The property that defines an inductor is called inductance, measured in henrys (H), which indicates how effectively the inductor can store energy.

Inductors resist changes in current, meaning they oppose a rapid increase or decrease in the current flowing through them. This property makes inductors useful in applications such as filters, tuning circuits, and energy storage. They are often used in conjunction with capacitors to create oscillators or in power supplies to smooth out voltage changes.

Inductors play a critical role in AC circuits, where their inductive reactance depends on the frequency of the alternating current. They are commonly found in transformers, motors, and electrical chokes.

A key fact to remember is that a capacitor stores energy in an electric field, while an inductor stores energy in a magnetic field:

- A capacitor stores energy in an electric field between its plates when a voltage is applied across it.
- An inductor stores energy in a magnetic field that is generated around it when current flows through the coil.

Transducer

A transducer is a device that converts one form of energy into another. In electronics, transducers are used to convert physical quantities, like pressure, temperature, or light, into electrical signals or vice versa. Examples include microphones (which convert sound waves into electrical signals) and speakers (which convert electrical signals into sound).

Transducers are classified based on the type of energy they convert. **Sensors** are transducers that convert physical conditions into electrical signals, such as thermistors for temperature measurement. **Actuators**, on the other hand, convert electrical signals into mechanical motion, such as in motors.

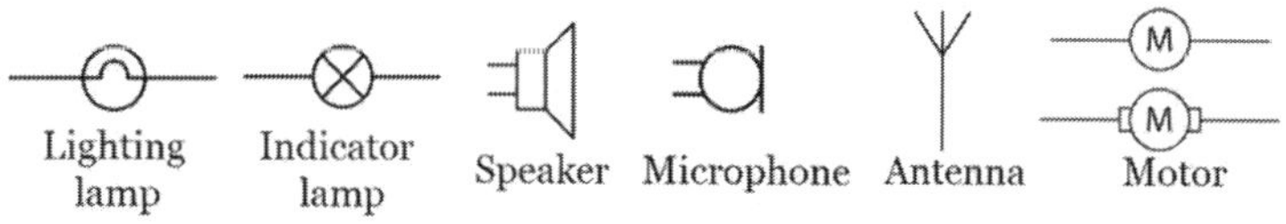

Transducers are essential in a wide range of applications, including automation, measurement, and control systems. They enable machines to interact with the physical world, making them crucial for applications ranging from medical instruments to industrial robots.

SECTION 9. ELECTRICITY AND MAGNETISM

Electricity Generation and Distribution

Electricity generation is the process of converting various forms of energy into electrical energy, which powers homes, industries, and cities. The most common method involves using turbines, which are rotated by steam, wind, or water. For instance, in a thermal power plant, fossil fuels like coal, natural gas, or oil are burned to produce steam, which drives a turbine connected to an electric generator.

In **hydroelectric power plants**, falling or flowing water turns turbines, while in wind farms, wind spins the blades of wind turbines. **Nuclear power plants** use nuclear reactions to produce heat that creates steam to turn turbines. **Renewable energy sources** like solar panels convert sunlight directly into electricity using photovoltaic cells, while geothermal power taps the heat from beneath the Earth's surface.

The generator in each of these processes uses **electromagnetic induction** to convert mechanical energy into electrical energy. Inside the generator, a coil of wire rotates within a magnetic field, inducing an electric current. Once generated, electricity is transmitted over the power grid to reach consumers. Different generation methods vary in efficiency, environmental impact, and availability, providing a diverse mix of energy sources for reliable electricity production.

Electricity is distributed through a complex network called the **power grid**, which ensures that power generated at power plants reaches homes, businesses, and industries. The distribution process begins with electricity generation at power stations, which produce high-voltage electricity. To minimize energy loss during transmission, the voltage is significantly increased using step-up transformers.

The high-voltage electricity then travels through **transmission lines**, which carry power over long distances. Transmission lines connect to substations, where step-down transformers reduce the voltage to safer levels for local distribution.

From the **substations**, electricity flows into a network of distribution lines that deliver power to neighborhoods and individual buildings. These local lines further lower the voltage to the levels suitable for residential and commercial use. Finally, **service transformers** near homes and businesses step down the voltage to a level that can be safely used by electrical appliances and equipment.

Throughout the grid, switches and circuit breakers are used to manage and protect the flow of electricity. The power grid is monitored and controlled to ensure a consistent and reliable supply, responding to changes in demand and rerouting electricity in case of faults or maintenance needs. This system makes modern, reliable electricity supply possible.

Electric Generator and Motor

Electric generators and motors are essential devices that operate on the principle of electromagnetic induction. Although their purposes are opposite, they share similar components and working principles.

An **electric generator** converts mechanical energy into electrical energy. It works based on Faraday's **law of electromagnetic induction**, which states that a changing magnetic field within a coil of wire induces an electric current. Generators typically consist of a rotor (the rotating part) and a stator (the stationary part). When a turbine or another mechanical force rotates the rotor inside a magnetic field, it creates a flow of electricity in the coils. Generators are widely used in power plants, including thermal, hydro, wind, and nuclear, to produce the electricity that powers homes and industries.

On the other hand, an electric motor performs the reverse function—converting electrical energy into mechanical energy. Motors rely on the interaction between a magnetic field and electric current to generate torque, which then produces rotational motion. Motors are composed of similar components to generators, with a rotor and stator, but instead of producing electricity, they use it to create movement. Electric motors are found in various applications, from household appliances to industrial machinery and electric vehicles.

In essence, electric generators and motors are closely related, with generators supplying the electricity needed for motors to perform useful mechanical tasks. Both are integral to energy production, consumption, and numerous everyday applications, contributing significantly to modern technology and convenience.

SECTION 10. LIFE SCIENCES

Biology Basics and Key Concepts

Biology is the study of life and living organisms, encompassing everything from microscopic cells to vast ecosystems. It seeks to understand the structure, function, growth, evolution, and interactions of all forms of life. By grasping the essential principles of biology, we can better comprehend how life operates and interacts within the natural world. Below are the key concepts that provide the foundation for biological understanding.

The Cell as the Fundamental Unit of Life

All living things, regardless of their complexity, are composed of **cells**, the smallest unit of life. Cells perform essential functions that sustain life, including respiration, energy production, and growth. There are two major types of cells: **prokaryotic** cells (which lack a true nucleus and are found in organisms like bacteria) and **eukaryotic** cells (which have a nucleus and are found in plants, animals, and fungi). While some organisms consist of just one cell (unicellular), others, like humans, are made up of trillions of specialized cells (multicellular), each playing a specific role.

DNA: The Genetic Blueprint

At the heart of all living organisms is **DNA (Deoxyribonucleic Acid)**, a long molecule that contains the genetic instructions necessary for growth, development, and reproduction. DNA is organized into structures called **chromosomes**, which are located in the nucleus of eukaryotic cells. Each segment of DNA, known as a **gene**, codes for a specific protein, and these proteins perform essential tasks within the organism. The unique combination of genes inherited from both parents shapes the organism's traits, from eye color to susceptibility to diseases.

Metabolism: The Energy System of Life

Living organisms need a continuous supply of **energy** to carry out their life processes. This energy is obtained from food and is used to fuel **metabolism**, the series of chemical reactions within cells that break down or build up molecules. Metabolism can be divided into **anabolic** processes, which build larger molecules from smaller ones (such as muscle growth), and **catabolic** processes, which break down molecules to release energy (such as digestion). **Cellular respiration** is a key metabolic process that occurs in the **mitochondria**, where glucose is broken down to release energy in the form of **ATP** (adenosine triphosphate), the energy currency of the cell.

Homeostasis: Maintaining Internal Balance

To survive, organisms must regulate their internal environment to remain stable despite changes in the external environment. This process is called **homeostasis**. For example, humans maintain a constant body temperature of about 98.6°F (37°C) by sweating when it's too hot or shivering when it's too cold. Similarly, organisms maintain other crucial conditions like pH, water balance, and glucose levels. This ability to regulate internal processes is vital for maintaining health and function across varying environmental conditions.

Growth, Development, and Reproduction

All living organisms undergo **growth** and **development** as they progress through different stages of life. Growth involves an increase in size and mass, while development refers to changes in an organism's form and function as it matures. For instance, a seed grows into a tree, and a fertilized egg develops into a mature adult. In addition to growth, organisms have the ability to **reproduce**, ensuring the continuation of their species. **Asexual reproduction** involves a single parent producing offspring identical to itself, while **sexual reproduction** requires the fusion of male and female gametes, leading to genetic variation in offspring.

Evolution: Change Over Time

A core principle of biology is **evolution**, the process by which populations of organisms change over generations through inherited genetic variations. This process, driven largely by **natural selection**, explains the vast diversity of life on Earth. Organisms with traits better suited to their environments are more likely to survive and reproduce, passing on those advantageous traits to future generations. Over long periods, these changes can result in the formation of new species. **Evolutionary biology** provides insight into the shared ancestry of all living organisms and the adaptations that have allowed them to thrive in different ecosystems.

Ecological Interdependence

Living organisms do not exist in isolation; they are part of larger systems called **ecosystems**, where they interact with both living (biotic) and non-living (abiotic) elements. **Ecology**, the study of these interactions, examines how organisms affect and are affected by their environment. For example, plants produce oxygen through **photosynthesis**, which animals need to breathe, while animals release carbon dioxide, which plants use to grow. These interrelationships form a web of life, where each species plays a crucial role in maintaining the balance of the ecosystem. Disruptions to this balance, such as habitat destruction or climate change, can have significant consequences for biodiversity and ecosystem health.

Biology, with its focus on understanding the principles of life, forms the basis for exploring the complexities of living organisms. From the cellular level, where DNA and metabolism govern life's processes, to the larger ecological systems that demonstrate the interdependence of organisms, these foundational concepts are critical for understanding biology. As we delve deeper into subjects like cell structure, genetics, and human physiology in the subsequent sections, these core principles will continue to guide our exploration of life.

Plant and Animal Cell Structure

Cells are the basic units of life, and while plant and animal cells share many common structures, they also have key differences that reflect their unique functions within living organisms. Understanding the structure of plant and animal cells is fundamental to the study of biology and helps explain how these cells perform the vital functions necessary for life. In this section, we will explore the similarities and differences between plant and animal cell structures, focusing on the various organelles that contribute to their functions.

Common Features of Plant and Animal Cells

Both plant and animal cells are **eukaryotic**, meaning they have a true nucleus and other membrane-bound organelles. This is in contrast to **prokaryotic** cells, such as bacteria, which lack a nucleus and other complex structures. Below are the key organelles found in both plant and animal cells:

Cell Membrane: The cell membrane, also known as the plasma membrane, is a thin, flexible barrier that surrounds the cell. It is primarily composed of a **phospholipid bilayer** with embedded proteins. The cell membrane plays a critical role in controlling the movement of substances in and out of the cell, maintaining the cell's internal environment (homeostasis), and facilitating communication with other cells. The **selectively permeable** nature of the cell membrane allows essential nutrients to enter while keeping harmful substances out.

Nucleus: The nucleus is the control center of the cell, housing the cell's genetic material in the form of **DNA**. Within the nucleus, DNA is organized into structures called **chromosomes**, which carry the instructions for all cellular activities. The nucleus is surrounded by a **nuclear envelope**, a double membrane that contains **nuclear pores** to regulate the exchange of materials between the nucleus and the cytoplasm. Inside the nucleus, the **nucleolus** is responsible for producing **ribosomes**, which are essential for protein synthesis.

Cytoplasm: The cytoplasm is the jelly-like substance that fills the interior of the cell, where all the organelles are suspended. It consists mainly of water, salts, and proteins and provides a medium for chemical reactions to occur. The cytoplasm helps in the movement of materials within the cell and supports the organelles.

Mitochondria: Often referred to as the "powerhouses" of the cell, mitochondria are responsible for generating the energy that cells need to function. Mitochondria convert **glucose** and **oxygen** into **ATP (adenosine triphosphate)** through a process called **cellular respiration**. This energy is used for various cellular activities, including growth, repair, and movement. Mitochondria have their own DNA, which is inherited maternally, and they can replicate independently within the cell.

Ribosomes: Ribosomes are small, round structures either floating freely in the cytoplasm or attached to the **endoplasmic reticulum (ER)**. They are responsible for assembling proteins by linking together **amino acids** according to instructions provided by messenger RNA (mRNA). These proteins play a crucial role in nearly every cellular function, including enzyme production, structural support, and cell signaling.

Endoplasmic Reticulum (ER): The endoplasmic reticulum (ER) is a network of membrane-bound tubules that extends from the nuclear envelope into the cytoplasm. There are two types of ER: **rough ER** and **smooth ER**. The rough ER is studded with ribosomes, making it the site of protein synthesis and modification. The smooth ER, which lacks ribosomes, is involved in lipid synthesis, detoxification of harmful substances, and storage of ions.

Golgi Apparatus: The Golgi apparatus functions as the cell's packaging and distribution center. After proteins and lipids are synthesized in the ER, they are transported to the Golgi apparatus, where they are further modified, sorted, and packaged into **vesicles**. These vesicles are then sent to their destinations, either within the cell or to the cell membrane for secretion.

Lysosomes: Found predominantly in animal cells, lysosomes are membrane-bound organelles that contain digestive enzymes. They are responsible for breaking down waste materials, old cell parts, and foreign invaders like bacteria. Lysosomes play a key role in **autophagy**, the process by which cells recycle their own components, helping to maintain cellular health.

Cytoskeleton: The cytoskeleton is a network of protein fibers that provides structural support to the cell. It helps maintain the cell's shape, anchors organelles in place, and assists with cell movement and division. The cytoskeleton is composed of three main types of fibers: **microfilaments**, **intermediate filaments**, and **microtubules**, each playing distinct roles in cellular processes.

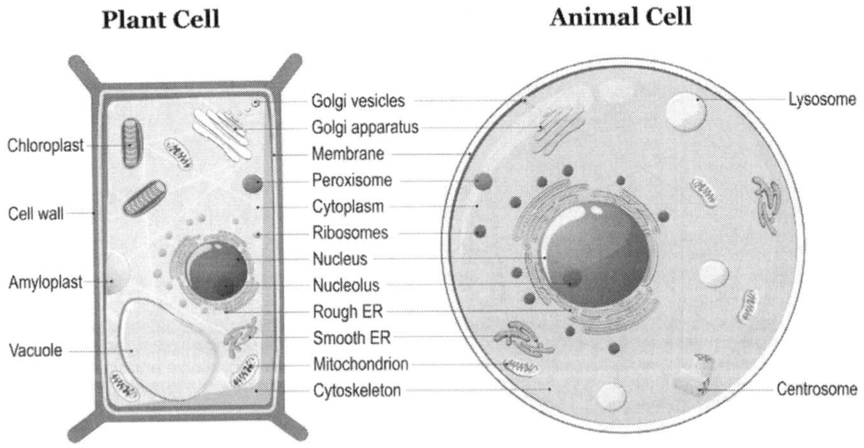

Unique Features of Plant Cells

In addition to the organelles shared with animal cells, plant cells possess several unique structures that allow them to carry out specific functions, such as photosynthesis and maintaining structural integrity.

Cell Wall: Unlike animal cells, plant cells have a cell wall that surrounds the cell membrane. The cell wall is primarily made of **cellulose**, a strong carbohydrate that provides rigidity and support. The cell wall allows plants to maintain their shape and stand upright, which is essential for maximizing exposure to sunlight. Additionally, the cell wall offers protection against mechanical stress and pathogens.

Chloroplasts: One of the most distinctive features of plant cells is the presence of chloroplasts, which are the organelles responsible for **photosynthesis**. Chloroplasts contain the pigment **chlorophyll**, which captures light energy from the sun and converts it into chemical energy in the form of glucose. This process not only provides energy for the plant but also produces oxygen, which is released into the atmosphere. Chloroplasts have their own DNA and replicate independently, similar to mitochondria.

Large Central Vacuole: Plant cells contain a large central vacuole, which is a membrane-bound sac that stores water, nutrients, and waste products. The vacuole plays a crucial role in maintaining **turgor pressure**, the internal pressure exerted by the vacuole against the cell wall. This pressure helps keep the plant upright and prevents wilting. The vacuole also stores important molecules and contributes to the breakdown of waste materials, functioning somewhat like lysosomes in animal cells.

Unique Features of Animal Cells

While plant cells have chloroplasts, cell walls, and large vacuoles, animal cells contain structures that are either absent or not as prominent in plant cells. Below are some key differences:

Centrioles: Animal cells contain structures known as centrioles, which are involved in **cell division**. Centrioles are cylindrical structures made of microtubules that help organize the assembly of the **mitotic spindle** during cell division (mitosis and meiosis). Plant cells typically do not have centrioles, although they still undergo cell division through a slightly different mechanism.

Smaller Vacuoles: While plant cells have a large central vacuole, animal cells contain smaller vacuoles or vesicles. These smaller vacuoles are involved in storage, transport, and waste removal, but they do not play as significant a role in maintaining cell structure as the large vacuole in plant cells.

Comparison of Plant and Animal Cells

Although plant and animal cells share many of the same organelles, their unique structures reflect their differing roles in the natural world. For example, the **cell wall** and **chloroplasts** in plant cells enable them to capture sunlight and perform photosynthesis, a process that animal cells cannot do. On the other hand, **centrioles** and **lysosomes** in animal cells are specialized for tasks related to digestion and cell division, functions that are less emphasized in plant cells.

Both cell types are essential to life on Earth, with plant cells forming the basis of ecosystems by producing energy through photosynthesis, while animal cells play roles in consuming and distributing that energy within the food chain.

Understanding the structures of plant and animal cells is fundamental to the study of biology. While they share many common features, such as the cell membrane, nucleus, mitochondria, and endoplasmic reticulum, they also have unique components that enable them to perform specialized functions. Plant cells, with their cell walls, chloroplasts, and large central vacuole, are optimized for photosynthesis and structural support, whereas animal cells rely on lysosomes, centrioles, and a flexible membrane to support their more dynamic and diverse functions. Together, these cells represent the building blocks of life in plants and animals, each adapted to their environment and role within biological systems.

Plant Physiology

Plant physiology studies how plants function, including how they grow, reproduce, and respond to their environment. Plants are complex organisms that rely on a variety of processes to survive, such as absorbing water and nutrients, producing food through photosynthesis, and reproducing to create new plants. In this section, we'll explore the main physiological processes that keep plants alive and help them thrive.

Photosynthesis: How Plants Make Their Food

One of the most important processes for plants is **photosynthesis**, which is how they produce their own food. Photosynthesis takes place in the **chloroplasts**, which contain the pigment **chlorophyll**. Chlorophyll allows plants to capture energy from sunlight and use it to convert water and carbon dioxide into glucose (a type of sugar) and oxygen. The glucose provides energy for the plant to grow, while the oxygen is released into the air. Photosynthesis is vital not only for the plant itself but also for the ecosystem, as plants provide oxygen and serve as a food source for many animals.

Respiration: Using Energy for Growth and Survival

While photosynthesis is how plants make food, **respiration** is the process they use to break down that food and turn it into energy. This energy is used for all the plant's activities, including growing, repairing tissues, and transporting nutrients. Respiration takes place in the plant's cells, specifically in the **mitochondria**, and it happens day and night, unlike photosynthesis, which only occurs when there's sunlight.

Water and Nutrient Transport: The Role of Xylem and Phloem

Plants need water and nutrients to grow, and they have special tissues to transport these throughout the plant. These tissues are called **xylem** and **phloem**.

Xylem carries water and minerals from the roots to the rest of the plant. Water enters the roots from the soil and travels upward through the xylem to the stems and leaves, where it's used in photosynthesis. **Phloem** transports sugars made during photosynthesis from the leaves to other parts of the plant, such as the roots, stems, and developing flowers or fruits. This ensures that all parts of the plant have the energy they need to grow and function.

Stomata: Tiny Openings for Gas Exchange

Plants take in carbon dioxide from the air and release oxygen through tiny openings on their leaves called **stomata**. These openings are essential for photosynthesis and respiration. The **guard cells** surrounding the stomata can open or close them depending on the plant's needs. For example, the stomata may close on a hot day to prevent too much water loss through evaporation, helping the plant conserve water.

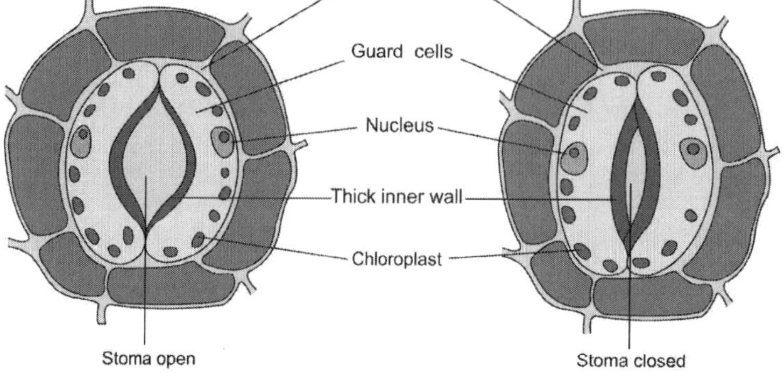

Plant Hormones: How Plants Grow and Respond to Their Environment

Plants produce special chemicals called **hormones** that help regulate their growth, development, and responses to environmental changes. For example, some hormones promote stem and root growth, while others help the plant conserve water during droughts by closing the stomata. Hormones also guide plants to grow toward light or help fruits ripen. These substances ensure plants can adapt to changes in their surroundings and thrive under different conditions.

Plant Reproduction: Sexual and Asexual Methods

Plants have two main ways of reproducing: **sexual reproduction** and **asexual reproduction**.

Sexual reproduction occurs when pollen from the male part of the plant (the anther) is transferred to the female part of the plant (the stigma). This can happen through wind, water, or pollinators like bees. Once the pollen reaches the stigma, fertilization occurs, and seeds are formed. These seeds can then grow into new plants.

Asexual reproduction, or **vegetative reproduction**, happens without seeds. Instead, new plants grow from parts of the parent plant, like stems, roots, or leaves. For example, a potato plant can grow from a single potato tuber, and strawberry plants send out runners that develop into new plants. Asexual reproduction produces plants that are genetically identical to the parent plant.

Plant Responses to the Environment

Plants are sensitive to their surroundings and can change their behavior or growth patterns based on environmental factors. Some of the main ways plants respond to their environment include:

Phototropism: Plants tend to grow toward a light source. This is because they need light for photosynthesis, and growing toward light ensures they can capture as much as possible.

Gravitropism: Plants also respond to gravity. Roots typically grow downward (toward gravity), while stems grow upward (against gravity), helping the plant stay anchored and upright.

Thigmotropism: Some plants, like vines, respond to touch. This is called thigmotropism, and it allows climbing plants to wrap around objects for support as they grow.

Plants can also show **nastic movements**, which are not related to the direction of the stimulus. For example, the Mimosa pudica plant closes its leaves when touched, and some flowers open and close at certain times of the day based on temperature or light.

Basics of Genetics

Genetics studies how traits are passed from one generation to the next in both plants and animals. It explains why offspring resemble their parents and how certain characteristics, like eye color in humans or flower color in plants, are inherited. The basic principles of genetics apply to all living organisms, whether they are plants, animals, or even microorganisms.

DNA: The Genetic Code for Life

At the center of genetics is **DNA (deoxyribonucleic acid)**, which acts as the blueprint for all living organisms. DNA is made up of a double-helix structure composed of four chemical bases: adenine (A), thymine (T), cytosine (C), and guanine (G). These bases pair up in specific ways (A with T, C with G) to form the rungs of the DNA "ladder."

Each segment of DNA that codes for a specific trait is called a **gene**. For example, a gene might determine an animal's fur color or a plant's height. These genes are organized on **chromosomes**, which are long strands of DNA found in the nucleus of every cell.

DNA Structure

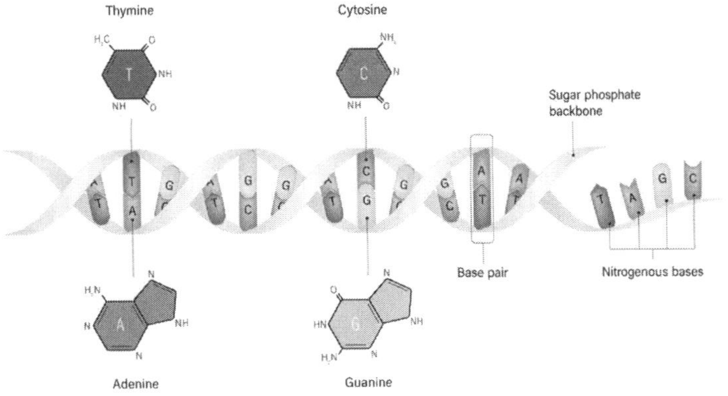

Genes and Inheritance

Genes come in pairs, with one copy inherited from each parent. In both plants and animals, this combination of genes from the mother and father determines the traits of the offspring. The different versions of a gene are known as **alleles**. For example, a gene responsible for flower color in a plant might have a "red" allele and a "white" allele.

If an organism inherits two of the same alleles for a trait (one from each parent), it is said to be **homozygous** for that trait. If it inherits two different alleles, it is **heterozygous**. Dominant alleles can mask the effects of recessive alleles, so even if an organism has one dominant and one recessive allele, it will exhibit the trait of the dominant allele. For example, in humans, brown eyes are dominant over blue eyes, so a person with one brown-eye allele and one blue-eye allele will have brown eyes.

DNA Replication: Copying Genetic Information

For both plants and animals, cells must copy their DNA before they divide to ensure that each new cell has the same genetic information. This process is called **DNA replication**.

During replication, the DNA molecule unwinds, and the two strands separate. Each strand serves as a template for the formation of a new complementary strand, ensuring that each new cell receives an exact copy of the original DNA. This process is essential for growth, repair, and reproduction in living organisms.

Mutations: Changes in Genetic Information

Sometimes, mistakes happen during DNA replication, leading to changes in the genetic code. These changes are called **mutations**. Mutations can be caused by environmental factors like radiation or chemicals, but they can also happen randomly.

Most mutations are harmless, but some can affect the organism's traits. In plants, a mutation might cause a flower to develop a different color, while in animals, it might lead to changes in fur patterns or even health conditions. Occasionally, mutations can be beneficial, providing an advantage that helps an organism survive better in its environment. Over time, beneficial mutations can lead to evolutionary changes in species.

Selective Breeding and Genetics

Humans have long used genetics to improve both plants and animals through **selective breeding**. By choosing individuals with desired traits and breeding them together, we can enhance certain characteristics over generations. For example, farmers might breed plants that produce more fruit, or dog breeders might select for specific coat colors or temperaments.

This practice has been crucial in agriculture, helping to develop crop varieties that are more resistant to pests, drought, or disease. In animals, selective breeding has given rise to the wide variety of dog breeds and livestock that meet specific needs for food or work.

Genetic Variation and Diversity

In both plants and animals, genetic diversity is essential for the survival and adaptation of species. Sexual reproduction increases genetic variation by combining genes from two parents, leading to offspring with unique genetic makeups. This diversity allows species to adapt to changing environments and helps protect against diseases that might otherwise wipe out genetically identical populations.

In summary, genetics provides the blueprint for life, guiding the development, reproduction, and diversity of both plants and animals. From the copying of DNA to the inheritance of traits, the study of genetics helps explain the continuity and variation seen in all living things.

Ecology: Fundamental Concepts and Principles

In this section, we'll discuss how living organisms interact with each other and their environment, which is the focus of ecology. Ecology helps us gain insight into how life on Earth is sustained, how ecosystems function, and how humans impact natural systems.

Levels of Ecological Organization

Ecology is often studied at various levels of organization, each representing a different scale of interaction between organisms and their environment. These levels include:

Organism: The individual living being. This level focuses on how a single organism interacts with its environment, including how it obtains food, reproduces, and responds to external conditions.

Population: A group of individuals of the same species living in a specific area. Ecologists study population dynamics, such as growth rates, birth and death rates, and how populations are affected by factors like competition for resources.

Community: A collection of different populations that live in the same area and interact with one another. This level of organization examines how species coexist and the effects of predation, competition, and symbiosis on community structure.

Ecosystem: A community of living organisms and the non-living components of their environment (such as water, soil, and air) that interact as a system. Ecosystem ecology looks at how energy flows through the system and how nutrients cycle between organisms and their environment.

Biosphere: The largest level of organization in ecology, encompassing all ecosystems on Earth. It includes every living organism and the environments they inhabit. The biosphere is essentially the global ecological system.

Energy Flow in Ecosystems

One of the most fundamental principles in ecology is how energy flows through ecosystems. Energy originates from the sun and is captured by producers (usually plants) through photosynthesis. These producers convert sunlight into chemical energy, which is stored in their tissues as food.

Energy then flows through the ecosystem in the following way:

Producers (autotrophs): Plants, algae, and some bacteria that produce their own food by converting sunlight into energy.

Consumers (heterotrophs): Organisms that cannot produce their own food and must consume other organisms to obtain energy. Consumers are further classified into primary consumers (herbivores), secondary consumers (carnivores that eat herbivores), and tertiary consumers (carnivores that eat other carnivores).

Decomposers: Fungi, bacteria, and other organisms that break down dead organic material, returning nutrients back to the soil. This recycling of nutrients is critical for the continuation of life in an ecosystem.

Energy transfer between these levels is not 100% efficient. Typically, only about 10% of the energy from one level is passed on to the next, while the rest is lost as heat. This is known as the **10% rule** and explains why ecosystems typically have fewer large carnivores than herbivores.

Nutrient Cycling

Another key principle in ecology is the cycling of nutrients, such as carbon, nitrogen, and phosphorus, which are essential for life. Unlike energy, which flows in one direction through an ecosystem, nutrients are recycled. The **carbon cycle**, for example, describes how carbon moves between the atmosphere, organisms, and the Earth. Plants take in carbon dioxide during photosynthesis, which is then passed through the food chain when consumers eat plants. Carbon is returned to the atmosphere through respiration, decomposition, and combustion (burning fossil fuels).

Similarly, the **nitrogen cycle** involves nitrogen being fixed from the atmosphere by certain bacteria, made available to plants, passed on to consumers, and eventually returned to the atmosphere or soil through waste and decay. These cycles ensure that essential elements are available for organisms to survive.

Habitat and Niche

A **habitat** is the physical environment where an organism lives, while a **niche** refers to the role an organism plays within its ecosystem. A species' niche includes how it obtains food, its behavior, and how it interacts with other species.

For example, a bird's habitat might be a forest, but its niche could include eating insects from the trees, nesting in tree branches, and contributing to seed dispersal. Understanding niches helps ecologists determine how species coexist and how they affect one another. When two species have overlapping niches, they may compete for the same resources, which can lead to **competitive exclusion**, where one species outcompetes the other.

Population Dynamics

The study of **population dynamics** explores how populations change in size and composition over time. Factors such as birth rates, death rates, immigration, and emigration influence population growth. Populations tend to grow rapidly when resources are abundant, but as resources become limited, growth slows and may eventually reach a stable state, known as **carrying capacity**. Carrying capacity is the maximum population size that an environment can sustainably support.

Populations can experience different growth patterns. **Exponential growth** is rapid population increase under ideal conditions, where resources are unlimited. This type of growth is typically seen in populations introduced to a new environment. **Logistic growth** is the type of growth that slows as resources become limited, eventually stabilizing at the carrying capacity. Natural population fluctuations are common due to environmental changes, predation, disease, and competition for resources.

Species Interactions

Species within an ecosystem interact in various ways, including:

Predation: One organism (the predator) hunts and kills another (the prey). Predation helps regulate population sizes and maintain balance within ecosystems.

Competition: Species compete for the same resources, such as food, water, or space. This competition can be either **interspecific** (between different species) or **intraspecific** (within the same species).

Symbiosis: A close and long-term interaction between different species. There are several types of symbiotic relationships. **Mutualism** refers to the interaction where both species benefit from the interaction (e.g., bees pollinating flowers). **Commensalism** is the interaction when one species benefits, and the other is neither harmed nor helped (e.g., birds nesting in trees). Finally, **parasitism** is when one species benefits at the expense of the other (e.g., ticks feeding on mammals).

Human Impact on Ecosystems

Humans have a profound effect on ecosystems through activities such as deforestation, pollution, and climate change. These actions can disrupt the balance of ecosystems, leading to habitat loss, species extinction, and changes in nutrient cycling and energy flow. Understanding the principles of ecology helps us identify ways to mitigate our impact and promote the sustainability of natural systems.

Classification of Living Things

With so many plants and animals in such a diverse ecological system, how do we put them into different categories to better understand the natural world? The answer lies in taxonomy, the science of classifying living things. Taxonomy organizes organisms into groups based on shared characteristics, helping us understand their relationships and how they have evolved over time. This system not only helps scientists study biodiversity but also aids in identifying and categorizing new species as we continue to explore the vast variety of life on Earth.

The Taxonomic Hierarchy

The taxonomic hierarchy categorizes living organisms into increasingly specific groups. The primary levels, from the most general to the most specific, are:

1. **Domain**: The highest rank, which divides all life into three broad categories based on cell structure: Bacteria, Archaea, and Eukarya.
2. **Kingdom**: A broad classification that groups organisms based on their basic characteristics, such as whether they are plants, animals, fungi, or microorganisms.
3. **Phylum**: Organisms in a kingdom are further divided into phyla based on more specific traits.
4. **Class**: Each phylum is divided into classes. For example, mammals and birds are different classes within the animal kingdom.
5. **Order**: Classes are divided into orders based on similarities in body structure and other characteristics.
6. **Family**: Orders are divided into families, grouping organisms with even more closely related characteristics.
7. **Genus**: A genus groups species that are very closely related and often resemble each other.
8. **Species**: The most specific level of classification, a species refers to a group of individuals that can breed and produce fertile offspring.

The Three Domains of Life

All living organisms can be classified into one of the three domains:

1. **Bacteria**: This domain includes prokaryotic, single-celled organisms that lack a nucleus. Bacteria are incredibly diverse and can be found in nearly every environment on Earth.
2. **Archaea**: Like bacteria, archaea are prokaryotic and single-celled, but they differ in their genetic makeup and can often survive in extreme environments, such as hot springs and salt lakes.
3. **Eukarya**: This domain includes all organisms with eukaryotic cells, which have a nucleus and other membrane-bound organelles. Eukarya includes animals, plants, fungi, and protists.

The Five Kingdoms

Within the Eukarya domain, organisms are classified into one of five kingdoms:

1. **Animalia**: Multicellular organisms that obtain food by consuming other organisms (heterotrophs). This kingdom includes mammals, birds, fish, insects, and more.
2. **Plantae**: Multicellular organisms that produce their own food through photosynthesis (autotrophs). This kingdom includes flowering plants, ferns, and mosses.

3. **Fungi**: Organisms that absorb nutrients from decaying organic matter. Fungi include mushrooms, molds, and yeasts.
4. **Protista**: A diverse group of mostly single-celled organisms, such as algae and protozoa, that do not fit into the other kingdoms.
5. **Monera**: This kingdom, often used to classify bacteria, includes single-celled prokaryotes.

As an example, the table below summarizes the classification hierarchy of a human being (Homo sapiens) and a sunflower (Helianthus annuus):

Taxonomic Rank	Human (Homo sapiens)	Sunflower (Helianthus annuus)
Domain	Eukarya	Eukarya
Kingdom	Animalia	Plantae
Phylum	Chordata	Angiosperms
Class	Mammalia	Dicotyledons
Order	Primates	Asterales
Family	Hominidae	Asteraceae
Genus	Homo	Helianthus
Species	Homo sapiens	Helianthus annuus

Binomial Nomenclature

The system of **binomial nomenclature** is used to name species, giving each organism a two-part name based on its genus and species. For example, the scientific name of a human is **Homo sapiens**, where *Homo* is the genus and *sapiens* is the species. This naming system helps avoid confusion, as many organisms may have multiple common names but only one scientific name.

SECTION 11. GEOLOGY

Geology is the study of the Earth, including its materials, processes, and history.

Structure of Earth

One key aspect of geology is understanding the structure of the Earth, which consists of four distinct layers: the crust, mantle, outer core, and inner core, each varying in composition and depth.

The **crust** is the outermost layer, extending from 5 to 70 kilometers (3 to 43 miles) deep. It is thinner beneath oceans (oceanic crust) and thicker under continents (continental crust). The crust is where all life exists, and it's composed of solid rocks like granite and basalt.

Beneath the crust lies the **mantle**, which extends from 70 kilometers to about 2,900 kilometers (43 to 1,800 miles) deep. The mantle is semi-solid, composed mainly of silicate minerals, and slowly moves due to convection currents, driving plate tectonics.

Below the mantle is the **outer core**, a liquid layer of iron and nickel, extending from 2,900 to 5,100 kilometers (1,800 to 3,200 miles) deep. The motion in this layer generates Earth's magnetic field.

At the very center is the **inner core**, a solid sphere of iron and nickel, reaching depths of 6,371 kilometers (3,959 miles). Despite extreme temperatures (up to 5,400°C or 9,800°F), the inner core remains solid due to the immense pressure exerted upon it.

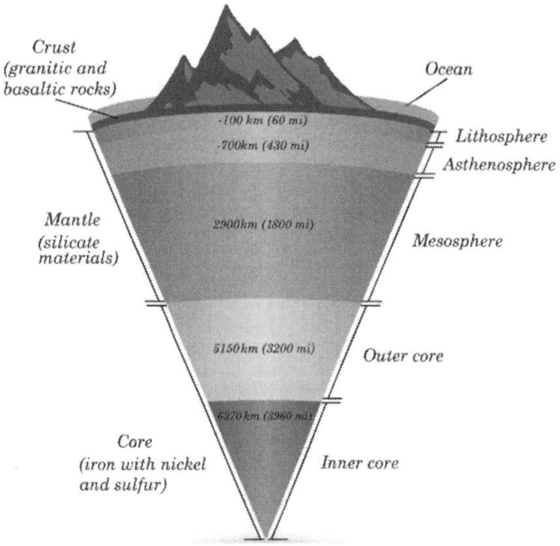

Plate Tectonics

Plate tectonics describes the movement of large, rigid plates that make up Earth's outer layer, or **lithosphere**, which is divided into several major and minor plates. These plates float atop the semi-fluid **asthenosphere**, a part of the upper mantle. The convection currents in the semi-fluid asthenosphere of the mantle allows the plates to move slowly over time.

There are three main types of plate boundaries: **divergent**, **convergent**, and **transform**. At **divergent boundaries**, plates move away from each other, allowing magma from the mantle to rise and form new crust, such as at the mid-Atlantic Ridge. At **convergent boundaries**, plates collide, and one plate may be forced beneath the other in a process called **subduction**, which can form mountain ranges or cause volcanic activity, as seen in the Andes Mountains. **Transform boundaries** occur where plates slide past one another, leading to earthquakes, such as those along California's San Andreas Fault.

Plate tectonics is responsible for shaping Earth's surface over millions of years, leading to the formation of continents, oceans, mountains, and the occurrence of natural phenomena like earthquakes and volcanic eruptions.

Earthquake

Earthquakes occur when stress builds up along **fault lines** or at plate boundaries, causing the Earth's crust to suddenly release energy. This energy travels through the Earth as seismic waves, which can cause the ground to shake. Most earthquakes happen along tectonic plate boundaries, where plates either collide, move apart, or slide past each other. The release of energy happens because the plates get stuck due to friction and then break free, releasing the stored stress in the form of an earthquake.

Earthquakes are measured by their intensity and magnitude. The **Richter scale** is one common method for measuring an earthquake's **magnitude**, which quantifies the amount of energy released during the event. Each number on the Richter scale represents a tenfold increase in amplitude of the seismic waves and roughly 32 times more energy released. A **seismograph** is the instrument used to detect and record seismic waves, helping scientists pinpoint the earthquake's epicenter and strength.

In addition to the Richter scale, the **Modified Mercalli Intensity (MMI) scale** measures the intensity of an earthquake based on its observed effects, such as structural damage and human perception. Earthquakes can cause significant destruction, making understanding their mechanics crucial for preparedness and safety.

Types of Rock

Rocks are classified into three main types based on how they form: igneous, sedimentary, and metamorphic.

Igneous rocks form from the cooling and solidification of molten rock, either magma (beneath the Earth's surface) or lava (on the surface). These rocks are further divided into intrusive and extrusive types. **Intrusive igneous rocks**, such as granite, form when magma cools slowly beneath the Earth's surface, resulting in large crystals. **Extrusive igneous rocks**, like basalt, form when lava cools quickly on the surface, leading to smaller crystals or a glassy texture.

Sedimentary rocks form from the accumulation and compaction of sediments, which can include fragments of other rocks, minerals, and organic materials. Over time, these sediments are deposited in layers and harden into rock. Examples of sedimentary rocks include **limestone** and **sandstone**. These rocks often contain fossils, providing valuable information about Earth's past environments.

Metamorphic rocks are formed when existing rocks—whether igneous, sedimentary, or other metamorphic rocks—are subjected to intense heat, pressure, or chemically active fluids, causing them to transform. This process, known as **metamorphism**, alters the rock's structure and mineral composition. Marble, for instance, is a metamorphic rock formed from limestone, while gneiss is a metamorphic rock that can form from granite.

Each rock type plays a significant role in the rock cycle, contributing to the dynamic processes that shape Earth's surface over time.

Geologic Time Scale

The geologic time scale is a system that geologists use to organize the vast history of Earth into manageable intervals. It divides Earth's 4.6-billion-year history into a series of eons, eras, periods, and epochs based on significant geological and biological events, such as the appearance and extinction of species, continental shifts, and climate changes. This system helps scientists understand the timing and relationships between events in Earth's past.

The **largest division** of time is the **eon**, and Earth's history is split into four eons: the **Hadean**, **Archean**, **Proterozoic**, and **Phanerozoic**. The Phanerozoic Eon is the most recent and is divided into three major **eras**: the

Paleozoic, **Mesozoic**, and **Cenozoic**. Each era is characterized by major shifts in life forms and Earth's environment.

For example, the **Paleozoic Era** is known for the development of early life forms, including fish and amphibians, while the **Mesozoic Era** is often called the "Age of Reptiles" due to the dominance of dinosaurs. The **Cenozoic Era**, which continues today, is known as the "Age of Mammals" and has seen the rise of humans.

Each era is divided further into **periods**, such as the **Jurassic Period** during the Mesozoic Era, and **epochs** within periods, particularly in the more recent Cenozoic Era. The geologic time scale allows scientists to piece together the events that shaped the Earth and the evolution of life.

Water Cycle

The water cycle, also known as the hydrologic cycle, is a continuous process that moves water throughout Earth's systems, connecting the atmosphere, land, and oceans. This cycle plays a crucial role in regulating climate, supporting ecosystems, and maintaining the availability of fresh water.

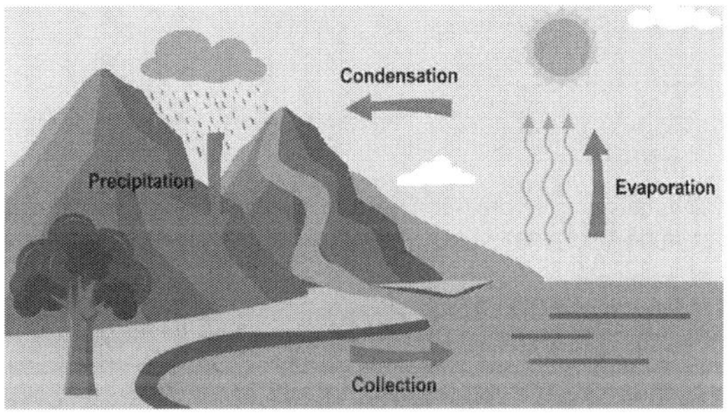

The water cycle consists of several key processes: evaporation, condensation, precipitation, and collection. The cycle begins with **evaporation**, where heat from the Sun causes water from oceans, lakes, and rivers to change from a liquid to a gas (water vapor). Plants also contribute to evaporation through a process called **transpiration**, where water is released from their leaves into the atmosphere.

As water vapor rises and cools in the atmosphere, it undergoes **condensation**, forming clouds. These clouds, when saturated with moisture, eventually lead to **precipitation**, which falls back to the Earth's surface in the form of rain, snow, sleet, or hail.

Once water reaches the ground, it follows different paths. It may flow as **runoff** into rivers, lakes, or oceans, or it may infiltrate into the ground, replenishing groundwater supplies. Groundwater can be stored in underground aquifers, and over time, this water may return to the surface through springs or be drawn up by plants.

The cycle repeats continuously, driven primarily by solar energy and gravity. The water cycle is critical in maintaining Earth's weather patterns, supporting life, and shaping landscapes. It also plays a role in transporting nutrients and minerals throughout ecosystems, making it essential for sustaining life on Earth.

Carbon Cycle

The carbon cycle describes the movement of carbon among the atmosphere, oceans, land, and living organisms. Carbon is a fundamental element in life and plays a significant role in regulating Earth's climate by controlling the concentration of carbon dioxide (CO_2) in the atmosphere.

The cycle begins when carbon dioxide from the atmosphere is absorbed by plants through photosynthesis. During this process, plants convert CO_2 into glucose, a form of organic carbon, which they use for growth. As animals consume plants, carbon is transferred to them and becomes part of their bodies. When plants and animals respire, they release CO_2 back into the atmosphere.

When organisms die, **decomposition** breaks down their bodies, releasing carbon back into the soil or atmosphere. In some cases, organic matter is buried and subjected to intense heat and pressure over millions of years, forming **fossil fuels** like coal, oil, and natural gas. When these fossil fuels are burned, they release stored carbon back into the atmosphere in the form of CO_2, contributing to the **greenhouse effect** and **global warming**.

The oceans also play a key role in the carbon cycle. They absorb large amounts of CO_2 from the atmosphere, where it dissolves and forms carbonic acid. Marine organisms, such as plankton, use dissolved carbon to build their shells. When these organisms die, their shells settle on the ocean floor, where the carbon can be stored for long periods in marine sediments.

Human activities, such as deforestation and the burning of fossil fuels, have significantly altered the natural balance of the carbon cycle, leading to an increase in atmospheric CO_2 levels and contributing to climate change.

SECTION 12. PALEONTOLOGY

Paleontology is the scientific study of life that existed in the distant past, primarily through the examination of fossils. **Fossils** are the preserved remains or traces of ancient organisms, including bones, shells, imprints, and even evidence of their behavior, such as footprints or burrows. Paleontology bridges the gap between biology and geology, helping us understand the history of life on Earth and the evolutionary processes that have shaped it.

By studying fossils, paleontologists can reconstruct past ecosystems, track the evolution of species, and investigate how organisms interacted with each other and their environments over millions of years. Fossil evidence is crucial for understanding events like **mass extinctions**, such as the one that wiped out the dinosaurs, as well as the emergence of new species.

Paleontology is not limited to studying animals; it also includes plant fossils, microorganisms, and traces of ancient life forms. Through various dating techniques, including **carbon dating** for younger fossils and **radiometric dating** for older fossils, paleontologists can estimate the age of fossils and their surrounding rocks. This information is vital for constructing a timeline of Earth's biological and geological history, offering invaluable insights into the evolution of life and environmental changes over time

Paleontology is closely linked to **geology** because the remains of ancient organisms are preserved in layers of rock known as **strata**. Geologists study the composition, structure, and processes that have shaped the Earth's crust, while paleontologists analyze the fossils within these rock layers to interpret the history of life. The process of **stratigraphy**, the study of rock layers, is crucial for paleontology. Fossils are often found in sedimentary rocks, which form through the deposition of material over time. By examining different layers of rock, paleontologists can determine the relative ages of fossils and understand the timeline of life on Earth.

Paleontology also helps geologists understand the conditions that shaped ancient environments. For example, the discovery of marine fossils in desert regions suggests that these areas were once covered by oceans. Fossils also reveal information about climate change, mass extinctions, and the migration of species. In this way, paleontology and geology complement each other, with geology providing the context in which paleontologists interpret the fossil record. Together, they offer a detailed history of life and Earth's geological changes over time.

SECTION 13. METEOROLOGY

Meteorology is the scientific study of the atmosphere and the processes that produce weather and climate. It involves understanding various elements of the Earth's atmosphere, the impact of fronts, cloud formation, and interpreting temperature changes. Meteorology plays a vital role in predicting weather patterns, studying climate, and understanding how atmospheric conditions influence life on Earth.

Earth's Atmosphere

The Earth's atmosphere is a layer of gases surrounding the planet, held by Earth's gravity. It extends about 10,000 kilometers (6,200 miles) above Earth and is composed of several layers, each with distinct characteristics. The atmosphere is mainly made up of nitrogen (78%), oxygen (21%), and trace gases such as carbon dioxide and argon. These layers include:

- **Troposphere**: This is the lowest layer where all weather occurs. It extends from Earth's surface to about 8-15 kilometers (5-9 miles). The temperature decreases with altitude in this layer, and it contains most of the atmosphere's water vapor, which forms clouds and precipitation.
- **Stratosphere**: Extending from the top of the troposphere to about 50 kilometers (31 miles) above Earth, this layer contains the ozone layer, which absorbs and scatters ultraviolet solar radiation.
- **Mesosphere**: Located above the stratosphere, the mesosphere extends from 50 kilometers to 85 kilometers (31 to 53 miles) and is where most meteors burn up upon entering the atmosphere.
- **Thermosphere**: This layer extends up to 600 kilometers (373 miles) and is where auroras occur. Temperatures increase significantly with altitude due to the absorption of high-energy solar radiation.
- **Exosphere**: The outermost layer, extending from the thermosphere up to 10,000 kilometers (6,200 miles). It gradually fades into space, and its particles are sparse, often escaping into space.

Magnetosphere

The magnetosphere is the region surrounding Earth where its magnetic field dominates. Earth's magnetic field is generated by the movement of molten iron in the outer core and extends far into space. The magnetosphere protects the planet from harmful solar wind and cosmic radiation by deflecting charged particles from the Sun. When some particles are trapped in the magnetosphere, they collide with gases in the upper atmosphere, creating the natural phenomenon known as the **auroras** (Northern and Southern Lights).

The magnetosphere is crucial for life on Earth because it shields the planet from solar radiation that could otherwise strip away the atmosphere and make Earth uninhabitable. Understanding the magnetosphere helps scientists track space weather and predict disruptions caused by solar storms, which can affect satellites, GPS systems, and power grids on Earth.

Fronts

In meteorology, a front is the boundary between two air masses of different temperatures and densities. Fronts are responsible for a variety of weather conditions, including precipitation, storms, and temperature changes. There are four main types of fronts:

Cold front: This occurs when a cold air mass moves into a region occupied by warmer air. Cold fronts often lead to rapid temperature drops, strong winds, and thunderstorms. After a cold front passes, the air is usually cooler and drier.

Warm front: A warm front forms when a warm air mass slides over a cooler air mass. Warm fronts move more slowly than cold fronts and typically bring steady rain or snow over a large area. The temperature increases gradually as the warm front passes.

Stationary front: When neither a cold air mass nor a warm air mass has the strength to replace the other, a stationary front forms. This front can result in prolonged cloudy, rainy weather that may last for several days.

Occluded front: This occurs when a cold front overtakes a warm front, lifting the warm air mass off the ground. Occluded fronts can bring a mix of weather patterns, including precipitation and changing temperatures.

Fronts are essential in meteorology because they often trigger significant weather events, including storms, changes in wind direction, and varying levels of precipitation.

Clouds

Clouds are a visible collection of tiny water droplets or ice crystals suspended in the atmosphere. Clouds form when moist air rises and cools, causing water vapor to condense into liquid droplets. Clouds are classified into different types based on their appearance and altitude:

- **Cumulus clouds**: Puffy, white clouds with flat bases, often associated with fair weather. However, when they grow taller, they can develop into cumulonimbus clouds, which bring thunderstorms.
- **Stratus clouds**: These are uniform, gray clouds that often cover the entire sky and resemble fog. They are typically associated with overcast conditions and light rain or drizzle.
- **Cirrus clouds**: High-altitude clouds made of ice crystals, appearing wispy or feather-like. Cirrus clouds are usually a sign of fair weather, but they can indicate that a change in the weather is coming.
- **Nimbus clouds**: Clouds that bring precipitation. When combined with other cloud types (like cumulonimbus or nimbostratus), these clouds produce rain, snow, or thunderstorms.

Cloud formation plays a critical role in Earth's weather systems, as clouds regulate the distribution of sunlight and heat and are the source of precipitation.

Temperature Conversions

In meteorology, temperature is an important variable that influences weather conditions. Temperature is typically measured in degrees Celsius (°C) or Fahrenheit (°F), depending on the region. To convert between these two units, use the following formulas:

Celsius to Fahrenheit:
$T(°F) = T(°C) \times \frac{9}{5} + 32$. For example, to convert 25°C to Fahrenheit: $T(°F) = 25 \times \frac{9}{5} + 32 = 77°$.

Fahrenheit to Celsius:
$T(°C) = (T(°F) - 32) \times \frac{5}{9}$. For example, to convert 77°F to Celsius: $T(°C) = (77 - 32) \times \frac{5}{9} = 25°$.

SECTION 14. ASTRONOMY

Astronomy is the study of the universe beyond Earth, exploring celestial objects such as stars, planets, galaxies, and the vast expanse of space. It is one of the oldest sciences, as humans have long looked to the skies to understand their place in the cosmos. Astronomy seeks to explain the origins, evolution, and structure of the universe and the various phenomena observed in space.

The universe itself is an incredibly vast and expanding space filled with billions of **galaxies**, each containing countless stars, planets, and other celestial objects. It began approximately 13.8 billion years ago with the **Big Bang**, a massive explosion that marked the origin of space and time. Since then, the universe has been expanding, creating the complex structures and celestial bodies we observe today.

The Milky Way

Among the billions of galaxies in the universe, we live in the Milky Way Galaxy, a large, spiral-shaped galaxy that contains our solar system. The Milky Way is estimated to be about 100,000 light-years in diameter and has four main spiral arms. At the center of the Milky Way is a dense region thought to contain a supermassive black hole.

Our solar system is located in one of the outer spiral arms, called the Orion Arm, about 27,000 light-years from the galactic center. The Milky Way is part of a local group of galaxies, which includes neighboring galaxies like Andromeda. From Earth, we can see the Milky Way as a faint, milky band of light across the night sky, which is made up of the light from countless stars too distant to distinguish individually.

The Sun

The Sun is the central star of our solar system and the primary source of energy for life on Earth. It is classified as a G-type main-sequence star or a yellow dwarf, though it appears white from space. The Sun is an enormous ball of hydrogen and helium undergoing **nuclear fusion**, where hydrogen atoms fuse to form helium, releasing vast amounts of energy in the process. This energy radiates outward in the form of light, heat, and solar wind, driving weather systems on Earth and supporting photosynthesis in plants.

The Sun has a diameter of about 1.39 million kilometers (865,000 miles) and accounts for about 99.86% of the total mass of the solar system. The surface temperature of the Sun is approximately 5,500°C (9,932°F), while its core reaches temperatures of about 15 million°C (27 million°F), where nuclear fusion occurs.

The Sun is about 4.6 billion years old and is expected to remain stable for another 5 billion years before evolving into a **red giant** and eventually shedding its outer layers to form a **white dwarf**. The Sun influences everything in the solar system, from planetary orbits to space weather, and supports all forms of life on Earth.

The Solar System

The solar system consists of the Sun, eight planets, and a variety of other celestial objects that orbit around it. The Sun's gravitational pull keeps everything in the solar system in orbit. Surrounding the Sun are the planets, dwarf planets, moons, asteroids, comets, and other small bodies.

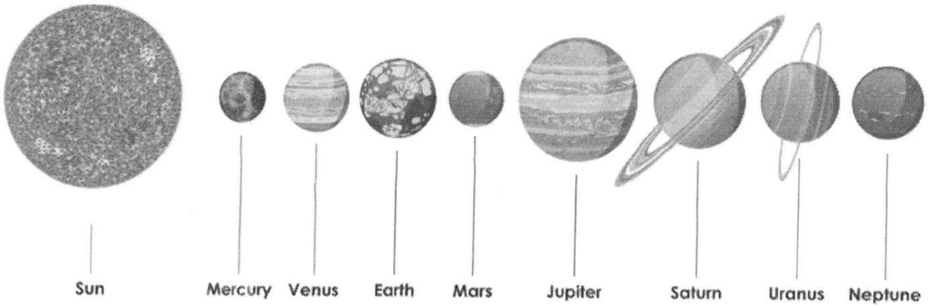

The eight planets of the solar system are divided into two groups: terrestrial planets and gas giants. The four **terrestrial planets**—Mercury, Venus, Earth, and Mars—are rocky and have solid surfaces. Mercury is the closest planet to the Sun and has extreme temperatures. Venus has a thick, toxic atmosphere and is the hottest planet. Earth is the only planet known to support life, and Mars is often called the Red Planet due to its iron oxide surface.

Beyond Mars lies the **asteroid belt**, a region filled with rocky debris. Past the asteroid belt are the gas giants—Jupiter and Saturn—and the ice giants—Uranus and Neptune. Jupiter is the largest planet, with a strong magnetic field and dozens of moons. Saturn is famous for its stunning ring system. Uranus and Neptune are colder and have atmospheres rich in hydrogen, helium, and methane, giving them a bluish tint.

Beyond Neptune is the **Kuiper Belt**, home to dwarf planets like Pluto and other icy objects. Farther still lies the **Oort Cloud**, a distant region filled with comets that marks the outermost boundary of the solar system.

The solar system also contains numerous moons orbiting the planets, asteroids, and comets. These objects, along with the planets and Sun, make up a dynamic and interconnected system that has existed for billions of years, continually evolving and influencing each other through gravitational forces.

Meteors, Comets, and Asteroids

Meteors, comets, and asteroids are small celestial bodies with distinct characteristics. A **meteor** is the flash of light seen when a **meteoroid** (a small space rock) enters Earth's atmosphere and burns up due to friction. If a meteoroid survives the atmosphere and reaches Earth's surface, it's called a **meteorite**.

Comets are icy bodies from the outer solar system that develop a glowing coma and a tail as they approach the Sun and their ice vaporizes. They follow long, elliptical orbits and are visible from Earth periodically. The tail always points away from the Sun due to solar wind.

Asteroids are rocky remnants mostly found in the asteroid belt between Mars and Jupiter. Unlike comets, they don't develop tails because they are made of rock and metal. While most asteroids remain in the belt, some can be nudged into Earth's path, with rare impacts throughout history.

These celestial bodies provide valuable insight into the early solar system and its formation.

The Seasons on Earth

The seasons on Earth are a result of the planet's 23.5-degree axial tilt and its elliptical orbit around the Sun. This tilt causes different areas of Earth to receive varying amounts of sunlight at different times of the year, which leads to the familiar cycle of seasons: spring, summer, fall (autumn), and winter.

As Earth orbits the Sun, the tilt of its axis means that one hemisphere is tilted toward the Sun while the other is tilted away. This is what causes the seasons to change. For instance, during the **summer solstice** (around June 21), the Northern Hemisphere is tilted toward the Sun, experiencing more direct sunlight and longer days, which marks the beginning of summer. At the same time, the Southern Hemisphere is tilted away from the Sun, receiving less sunlight and shorter days, resulting in winter. Six months later, during the **winter solstice** (around December 21), the roles reverse: the Northern Hemisphere tilts away from the Sun, entering winter, while the Southern Hemisphere enjoys summer.

In between the solstices are the **equinoxes** (around March 21 and September 21), when Earth's tilt is such that both hemispheres receive nearly equal amounts of sunlight. This balance between day and night length marks the transition into spring or fall, depending on the hemisphere.

The degree of seasonal change varies with latitude. Regions near the equator experience little seasonal variation because they consistently receive strong sunlight throughout the year. In contrast, areas near the poles endure extreme seasonal differences, with long, cold winters and short, cool summers. These seasonal cycles influence weather patterns, ecosystems, and human behavior worldwide.

Phases of the Moon

The phases of the Moon occur as the Moon orbits Earth, changing the portion illuminated by the Sun. This cycle takes approximately 29.5 days and includes eight main phases:

1. New Moon: The Moon is between the Earth and the Sun, making it invisible from Earth.
2. Waxing Crescent: A small crescent of the Moon becomes visible.
3. First Quarter: Half of the Moon's right side is illuminated.
4. Waxing Gibbous: More than half of the Moon is visible but not yet full.
5. Full Moon: The entire face of the Moon is illuminated.
6. Waning Gibbous: The visible portion begins to shrink after the full moon.
7. Last Quarter: Half of the Moon's left side is illuminated.
8. Waning Crescent: A small crescent is visible before transitioning back to a new moon.

These phases repeat in a regular cycle, influencing Earth's tides and human culture.

Solar and Lunar Eclipses

Solar and lunar eclipses occur when the Sun, Earth, and Moon align in specific ways, temporarily blocking sunlight.

A **solar eclipse** happens when the Moon passes between the Earth and the Sun, blocking some or all of the Sun's light. The following is an illustration of how a solar eclipse occurs.

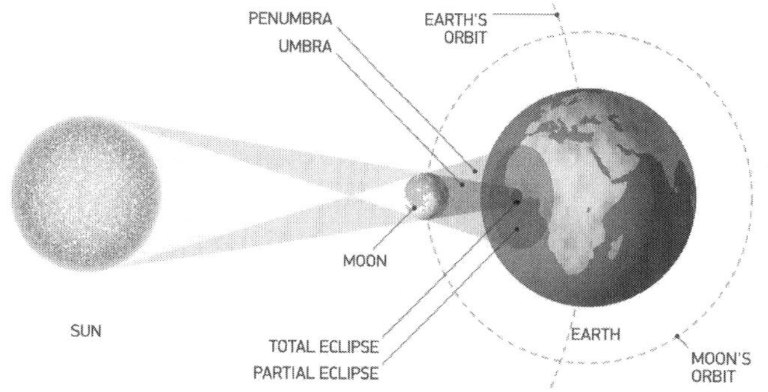

In the above figure, umbra is the darkest part of the shadow where the Sun is completely obscured. In a solar eclipse, observers within the umbra experience a total eclipse. Penumbra is the partially shaded outer region of the shadow. In a solar eclipse, observers in the penumbra experience a partial eclipse.

There are three types of solar eclipses:

- Total solar eclipse: The Moon completely covers the Sun, casting a shadow on Earth, and the sky darkens as if it were night. This occurs in a narrow path where the Moon's shadow falls.
- Partial solar eclipse: The Moon only partially covers the Sun, creating a visible crescent of sunlight.
- Annular solar eclipse: The Moon is farther from Earth and doesn't completely cover the Sun, leaving a ring of sunlight visible around the Moon.

Solar eclipses are rare and only occur during the new moon phase.

A **lunar eclipse** occurs when the Earth passes between the Sun and the Moon, casting Earth's shadow on the Moon. There are two types:

- Total lunar eclipse: The Earth completely blocks sunlight from reaching the Moon, and the Moon takes on a reddish color, often called a "blood moon".
- Partial lunar eclipse: Only part of the Moon enters Earth's shadow, and the rest remains illuminated.

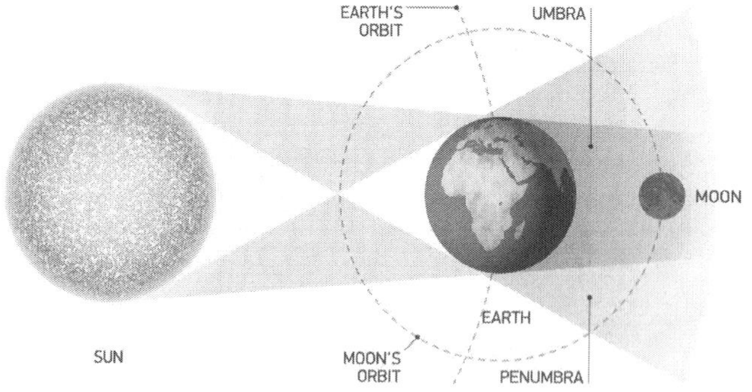

In a lunar eclipse, observers within the umbra experience a total eclipse, while observers in the penumbra experience a partial or penumbral lunar eclipse.

Lunar eclipses can occur only during a full moon and are visible from anywhere on Earth where the Moon is above the horizon. Unlike solar eclipses, lunar eclipses are safe to watch without eye protection.

PHYSICAL SCIENCE PRACTICE SET 1

This set includes questions related to Physics, Chemistry, Life Science, Geology, Meteorology, Astronomy, and Earth Science. Select the correct answer from the choices given. For practice only. This practice set does not reflect the actual number of questions in the test.

1. What is the base unit of length in the metric system?

A) Gram
B) Meter
C) Kelvin
D) Ampere

2. What does the prefix 'kilo-' represent in the metric system?

A) 100
B) 1,000
C) 10
D) 10,000

3. What is the Second Law of Thermodynamics?

A) Energy can be created from nothing
B) Energy transfers increase the disorder (entropy) of a system
C) Heat naturally flows from colder objects to hotter objects
D) Energy is always conserved in a closed system

4. What is the basic principle of refraction?

A) Light reflects off a surface at an equal angle
B) Light bends as it passes from one medium to another
C) Light travels faster in denser materials
D) Light cannot travel through a vacuum

5. Which of the following is a longitudinal wave?

A) Light wave
B) Sound wave
C) Microwave
D) Radio wave

6. Which phenomenon occurs when sound waves are reflected?

A) Refraction
B) Absorption
C) Echo
D) Polarization

7. What property of a sound wave determines its pitch?

A) Wavelength
B) Amplitude
C) Frequency
D) Speed

8. Which of the following describes convection?

A) Heat transfer through direct contact
B) Heat transfer through fluids moving due to temperature differences
C) Heat transfer through electromagnetic waves
D) Heat transfer through solid materials only

9. Which type of radiation does not require a medium to travel through?

A) Conduction
B) Convection
C) Electromagnetic
D) Thermal

10. What is potential energy?

A) The energy of motion
B) The energy stored due to position
C) The energy transferred by heat
D) The energy transferred by work

11. What determines an element's atomic number?

A) Number of electrons
B) Number of protons
C) Number of neutrons
D) Total mass

12. Which subatomic particle is negatively charged?

A) Proton
B) Neutron
C) Electron
D) Nucleus

13. What are neutrons responsible for in an atom?

A) Chemical bonding
B) Stabilizing the nucleus
C) Conducting electricity
D) Determining reactivity

14. What is the gravitational potential energy of a 10 kg object raised to a height of 5 meters? (Assume $g = 9.8\ \mathrm{m/s^2}$)

A) 98 J
B) 49 J
C) 490 J
D) 980 J

15. How is power defined in physics?

A) The rate at which work is done
B) The amount of energy stored in a system
C) The force applied over a distance
D) The total energy in a system

16. What is the base unit for mass in the metric system?

A) Pound
B) Kilogram
C) Gram
D) Ton

17. Which of the following describes a metalloid?

A) Highly reactive
B) Only conducts electricity at high temperatures
C) Exhibits properties of both metals and nonmetals
D) Completely nonreactive

18. What happens to atomic radius as you move across a period from left to right?

A) It increases
B) It decreases
C) It stays the same
D) It fluctuates

19. Which type of rock forms from pre-existing rocks undergoing heat and pressure?

A) Igneous
B) Sedimentary
C) Metamorphic
D) Organic

20. What is the cause of Earth's magnetic field?

A) Motion in the outer core
B) Rotation of the Earth
C) Movement of the tectonic plates
D) Solar radiation

21. What is the primary function of DNA in living organisms?

A) To provide structural support to cells
B) To transport oxygen through the body
C) To code for proteins and genetic traits
D) To break down food for energy

22. Which of the following is an example of an anabolic process in metabolism?

A) Digestion of food into smaller molecules
B) Cellular respiration to release energy
C) Building muscle tissue from amino acids
D) The release of carbon dioxide from cells

23. In photosynthesis, which substance do plants use to capture sunlight?

A) Glucose
B) Oxygen
C) Chlorophyll
D) Carbon dioxide

24. Which organelle is responsible for producing energy in both plant and animal cells?

A) Nucleus
B) Mitochondria
C) Ribosome
D) Golgi apparatus

25. Which process is primarily responsible for plants producing their own food?

A) Cellular respiration
B) Photosynthesis
C) Fermentation
D) Protein synthesis

Answers and Explanations

1. B) Meter. Explanation: The meter is the base unit of length in the metric system, used universally in science to measure distance and displacement.

2. B) 1,000. Explanation: 'Kilo-' means 1,000 times the base unit. For example, one kilometer (km) equals 1,000 meters, and one kilogram (kg) equals 1,000 grams.

3.B) Energy transfers increase the disorder (entropy) of a system. Explanation: The Second Law of Thermodynamics states that natural processes tend to increase the entropy, or disorder, in a system.

4.B) Light bends as it passes from one medium to another. Explanation: Refraction occurs when light changes speed and direction as it moves between different mediums, such as air and water.

5. B) Sound wave. Explanation: Sound waves are longitudinal, meaning the oscillations of particles are parallel to the direction of the wave's travel. Electromagnetic waves, such as light, radio waves, and microwave are transverse waves.

6. C) Echo. Explanation: An echo occurs when sound waves bounce off a surface and return to the listener, creating a reflected sound.

7. C) Frequency. Explanation: The pitch of a sound is determined by its frequency. Higher frequencies produce higher-pitched sounds, while lower frequencies produce lower-pitched sounds.

8. B) Heat transfer through fluids moving due to temperature differences. Explanation: Convection is the transfer of heat through the movement of fluids (liquids or gases) due to temperature-driven density differences.

9. C) Electromagnetic. Explanation: Electromagnetic radiation, such as light and radio waves, does not require a medium and can travel through a vacuum, unlike conduction and convection.

10. B) The energy stored due to position. Explanation: Potential energy is the stored energy an object has due to its position, such as an object elevated above the ground, which has gravitational potential energy.

11. B) Number of protons. Explanation: The atomic number is based on the number of protons in an atom's nucleus, which uniquely identifies the element and defines its chemical behavior.

12. C) Electron. Explanation: Electrons carry a negative charge and orbit the nucleus in various energy levels. Their arrangement affects how atoms interact and bond with others.

13. B) Stabilizing the nucleus. Explanation: Neutrons help stabilize the nucleus by balancing the repulsive forces between positively charged protons, contributing to an atom's stability.

14. C) 490 J. Explanation: Gravitational potential energy is calculated as $PE = mgh = 10\,\text{kg} \times 9.8\,\text{m/s}^2 \times 5\,\text{m} = 490\,\text{J}$.

15. A) The rate at which work is done. Explanation: Power is the rate at which work is performed or energy is transferred, and is measured in watts (W). It is calculated as $P = \frac{W}{t}$.

16. B) Kilogram. Explanation: The kilogram is the SI base unit for mass. It is commonly used in scientific and everyday measurements, with 1 kilogram equal to 1,000 grams.

17. C) Exhibits properties of both metals and nonmetals. Explanation: Metalloids have mixed properties, making them useful in electronics. For example, silicon is a semiconductor, conducting electricity under specific conditions.

18. B) It decreases. Explanation: As you move across a period, the increasing positive charge in the nucleus pulls the electrons closer, resulting in a smaller atomic radius.

19. C) Metamorphic. Explanation: Metamorphic rocks are created when igneous or sedimentary rocks are subjected to intense heat and pressure, altering their structure and mineral composition.

20. A) Motion in the outer core. Explanation: Earth's magnetic field is generated by the movement of liquid iron and nickel in the outer core.

21. C) To code for proteins and genetic traits. Explanation: DNA contains genetic information in the form of genes, which are segments of DNA that code for specific proteins. These proteins perform essential functions in the body and help determine an organism's traits, such as eye color or susceptibility to certain diseases.

22. C) Building muscle tissue from amino acids. Explanation: Anabolic processes involve the building up of larger molecules from smaller ones. In this case, the body builds muscle tissue by assembling amino acids into proteins, which is a typical anabolic activity. In contrast, catabolic processes break down molecules, such as in digestion.

23. C) Chlorophyll. Explanation: Chlorophyll is the green pigment found in plants' chloroplasts that captures sunlight, which is then used to convert carbon dioxide and water into glucose during photosynthesis. This process also releases oxygen as a byproduct.

24. B) Mitochondria. Explanation: The mitochondria are often called the "powerhouses" of the cell because they generate ATP (energy) through cellular respiration, which is essential for the cell's energy needs in both plants and animals.

25. B) Photosynthesis. Explanation: Photosynthesis is the process by which plants convert sunlight, carbon dioxide, and water into glucose (a form of sugar) and oxygen. This glucose serves as food for the plant, providing energy for growth and development.

Physical Science Practice Set 2

This set includes questions related to Electronics, Circuits, and Mechanical Comprehension. Select the correct answer from the choices given. For practice only. This practice set does not reflect the actual number of questions in the test.

1. A circuit breaker with a rating higher than recommended for a circuit

A) ensures longer device lifespan
B) provides no protection against overcurrent
C) prevents any circuit malfunction
D) is ideal for all loads

2. In a residential circuit, the hot wire is usually

A) green
B) black
C) blue
D) whitish

3. To measure electrical current in a circuit, you would use a(n)

A) ohmmeter
B) ammeter
C) voltmeter
D) wattmeter

4. A capacitor whose capacitance can be adjusted is called a

A) fixed capacitor
B) variable capacitor
C) variable resistor
D) potentiometer

5. A step-up transformer is used to

A) increase voltage
B) decrease resistance
C) reduce voltage
D) increase capacitance

6. Which of the following circuit components is least affected by a direct current?

A) 

B) 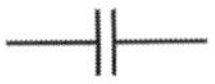

C) 

D) 

7. An inductor stores

A) electric charge
B) magnetic energy
C) heat
D) light

8. Which of the following describes a conductor?

A) Prevents the flow of current
B) Amplifies voltage
C) Allows easy flow of electrical current
D) Stores magnetic energy

9. Electrons have a ____ charge, and neutrons are ____ charged.

A) negative, neutral
B) positive, negative
C) neutral, positive
D) positive, neutral

10. What type of circuit has multiple paths for current to flow?

A) open circuit
B) series circuit
C) parallel circuit
D) short circuit

11. In the power formula, $P = V \times I$, the V represents

A) current
B) resistance
C) voltage
D) power

12. The N-type material in a diode is associated with

A) holes
B) free electrons
C) positive charge carriers
D) current amplification

13. In the given circuit, a 24V battery is connected across two parallel elements: a resistor of 10 ohms and a light bulb of 10 ohms. What is the total current supplied by the battery?

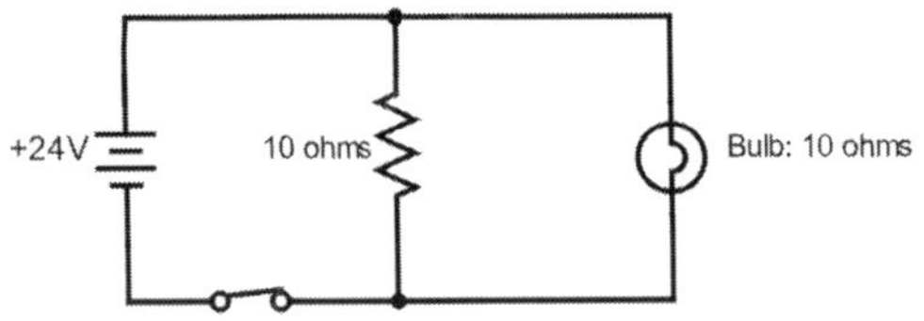

A) 1.2 A
B) 2.4 A
C) 3.6 A
D) 4.8 A

14. The addition of impurities to germanium to improve conductivity is called

A) annealing
B) doping
C) diffusion
D) oxidation

15. What type of circuit does the following show?

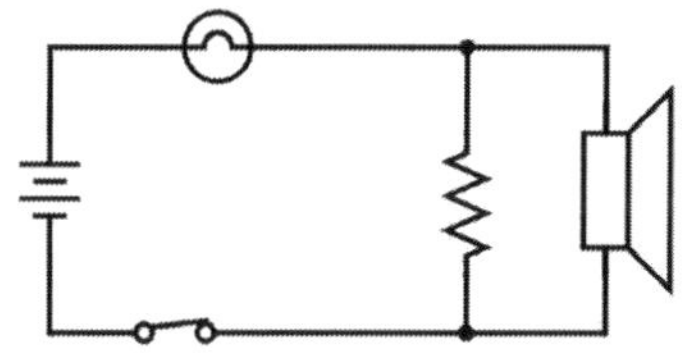

A) Series Circuit
B) Parallel Circuit
C) Series-Parallel Circuit
D) Short Circuit

16. Gear A, B, C, D, E drives Gear B, C, D, E, F respectively, as shown. If Gear A rotates clockwise, what is the rotational direction of Gear E?

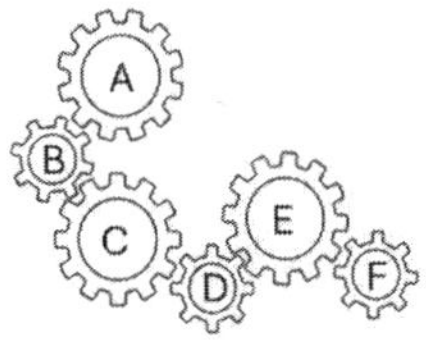

A) Clockwise
B) Counterclockwise
C) Both directions
D) Does not rotate

17. To calculate the efficiency of a machine, you should:

A) Multiply the input force by the output force.
B) Divide the output work by the input work.
C) Multiply the output work by the input work.
D) Subtract the input force from the output force.

18. In the pulley system shown, how much effort is needed to lift a 100-pound load?

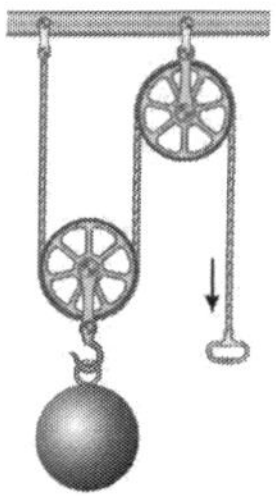

A) 50 pounds
B) 100 pounds
C) 200 pounds
D) 25 pounds

19. If Gear 1 has 50 teeth and Gear 2 has 20 teeth, how many revolutions does Gear 2 make for every 10 revolutions Gear 1 makes?

A) 5
B) 12
C) 25
D) 30

20. Two cars with different weights collide head-on while moving at the same speed. What most likely will happen to the heavier car upon impact?

(A) It will be thrown backward.
(B) It will stay in place.
(C) It will keep moving forward.
(D) It will flip over the smaller car.

21. A person is pulling a cart weighing 400 pounds up a ramp that is 20 feet long. If 100 pounds of force is used, how tall is the ramp?

(A) 4 feet
(B) 5 feet
(C) 6 feet
(D) 10 feet

22. The force required to sustain pushing a box across a rough surface is the same as

(A) the weight of the box times the static friction coefficient
(B) the force required to overcome friction while the box is being pushed
(C) the potential energy of the box
(D) the mass of the box divided by gravity

23. How much force is required to lift a 60-pound load using a lever system where the effort arm is 3 times longer than the resistance arm?

(A) 20 lbs
(B) 30 lbs
(C) 60 lbs
(D) 180 lbs

24. A car stuck in the sand requires 1,500 N of force to be pushed 6 m. How much work is done to push the car?

(A) 250 J
(B) 9,000 J
(C) 1,500 J
(D) 6,000 J

25. A machine lifts parts with a weight of 30 N from the ground to a height of 4 meters. If it takes 3 seconds to lift 10 parts, how much power is used?

(A) 120 watts
(B) 400 watts
(C) 40 watts
(D) 160 watts

26. A block and tackle pulley system has 4 segments supporting the load. Ignoring friction, what is the mechanical advantage of the system?

(A) 4
(B) 3
(C) 2
(D) 1

27. A hydraulic lift has a smaller piston area of 10 square inches and a larger piston area of 50 square inches. If you apply 100 pounds of force to the smaller piston, how much force is exerted on the larger piston?

(A) 500 pounds
(B) 1,000 pounds
(C) 2,000 pounds
(D) 10 pounds

28. If a copper rod, a plastic rod, and a wooden rod are left in the sun, which will heat up the fastest?

A) Copper
B) Plastic
C) Wood
D) All will heat up at the same rate

29. In a playground seesaw, one child weighing 40 pounds sits 4 feet on the left side of the fulcrum, and another child weighing 80 pounds sits 2 feet on the left side of the fulcrum. How far must an 80-pound child sit on the right-hand side of the fulcrum to balance the seesaw?

(A) 3 feet
(B) 4 feet
(C) 6 feet
(D) 2 feet

30. According to Ohm's law, which formula correctly defines the relationship between current, voltage, and resistance?

A) Resistance = Voltage / Current
B) Current = Resistance × Voltage
C) Power = Current + Resistance
D) Voltage = Current - Resistance

Answers and Explanations

1. B) provides no protection against overcurrent. Explanation: Using a higher-rated circuit breaker means it may not trip when necessary, failing to protect against overcurrent. A), C), and D) are incorrect because they misstate the effect of a higher-rated breaker.

2. B) black. Explanation: The hot wire is typically black in residential wiring. A) is incorrect; green represents ground. C) is incorrect; blue is used for specific phases. D) is incorrect; whitish represents neutral.

3. B) ammeter. Explanation: An ammeter measures current flowing through a circuit. A) measures resistance. C) measures voltage. D) measures power.

4. B) variable capacitor. Explanation: A variable capacitor allows the capacitance to be adjusted. A) is incorrect as a fixed capacitor has a set value. C) and D) are unrelated to capacitors.

5. A) increase voltage. Explanation: A step-up transformer is used to increase voltage from the primary winding to the secondary winding. B), C), and D) are incorrect, as they do not describe the function of a step-up transformer.

6. A) . Explanation:

A) Resistor: A resistor opposes the flow of current and will behave the same regardless of whether the current is AC (alternating current) or DC (direct current) or the direction of the DC. It simply converts electrical energy into heat, and there is no significant change in its operation with DC, making it the least affected.

B) Capacitor: A capacitor stores energy in an electric field. With DC, it will charge up to the applied voltage and then block further current flow once fully charged. Its behavior is significantly different with DC compared to AC, where it continuously charges and discharges.

C) Inductor: An inductor stores energy in a magnetic field and resists changes in current. With DC, once the initial change occurs, the inductor will allow the steady flow of current, but its behavior depends on current change, making it more affected by DC than a resistor.

D) Diode: A diode allows current to flow in only one direction. It significantly changes its behavior with DC depending on the direction of the current, either allowing it to pass or blocking it entirely.

Thus, the resistor is the component least affected by direct current, making option A the correct answer.

7. B) magnetic energy. Explanation: An inductor stores magnetic energy when current flows through it. A) is incorrect, as electric charge is stored by capacitors. C) and D) are unrelated to an inductor's function.

8. C) Allows easy flow of electrical current. Explanation: A conductor is a material that allows current to flow easily. A) and B) are incorrect; they describe an insulator and an amplifier, respectively. D) refers to an inductor.

9. A) negative, neutral. Explanation: Electrons are negatively charged, and neutrons are neutral (having no charge). B) is incorrect; electrons are not positively charged. C) and D) are incorrect as they mix up the charges.

10. C) parallel circuit. Explanation: A parallel circuit has multiple paths for current to flow. A) does not allow current to flow. B) has only one path. D) indicates a faulty condition where the current bypasses the load.

11. C) voltage. Explanation: In the formula $P = V \times I$, V represents voltage.

12. B) free electrons. Explanation: N-type material has free electrons that contribute to conductivity. A) describes P-type material. C) is incorrect because N-type carries negative charges. D) is not directly related to diode conductivity.

13. D) 4.8 A. Explanation:

1. The two elements (resistor and bulb) are in parallel, each with 10 ohms of resistance.

2. The equivalent resistance of the two parallel resistances R_{eq} can be found using the formula for parallel resistors: $\frac{1}{R_{eq}} = \frac{1}{R_1} + \frac{1}{R_2} \Rightarrow \frac{1}{R_{eq}} = \frac{1}{10} + \frac{1}{10} = \frac{2}{10} = \frac{1}{5}$. Hence, $R_{eq} = 5\,\Omega$.

3. Using Ohm's Law to find the total current supplied by the battery: $I = \frac{V}{R_{eq}} = \frac{24\,V}{5\,\Omega} = 4.8\,A$

14. B) doping. Explanation: Doping is the process of adding impurities to a semiconductor to enhance conductivity. A) and C) are unrelated to the introduction of impurities. D) is a process for creating an insulating layer, not for improving conductivity.

15. C) Series-Parallel Circuit. Explanation: The given diagram shows a combination of both series and parallel elements: The battery, switch, and lamp are in a series connection. The resistor and speaker are connected in parallel with each other. This combination of series and parallel components within the same circuit indicates that it is a series-parallel circuit.

16. A) Clockwise. Explanation: In systems of gears, when one rotates, it causes the gear it drives to rotate in the opposite direction. If Gear A turns clockwise, Gear B will turn counterclockwise, and Gear C will again rotate opposite to Gear B, which means Gear C rotates clockwise. Hence, Gear A, C, E all rotate clockwise, while Gear B, D, F rotates counterclockwise.

17. B) Divide the output work by the input work. Explanation: Efficiency is determined by dividing the output work by the input work and is often expressed as a percentage.

18. A) 50 pounds. Explanation: This is a moveable pulley system with two support segments. The mechanical advantage is equal to the number of support segments, which is 2. Therefore, the effort needed to lift the load is the load divided by the mechanical advantage: Effort $= \frac{\text{Load}}{\text{Mechanical Advantage}} = \frac{100\text{ pounds}}{2} = 50$ pounds.

19. C) 25. Explanation: The number of revolutions Gear 2 makes is inversely proportional to the number of teeth. Gear 1 has 50 teeth, and Gear 2 has 20 teeth. For every revolution of Gear 1, Gear 2 makes $\frac{5}{2}$ revolutions. Therefore, for 10 revolutions of Gear 1, Gear 2 will make: $10 \times \frac{5}{2} = 25$ revolutions.

20. (C) It will keep moving forward. Explanation: when two cars collide, the one with more mass/weight will continue moving forward because it has greater momentum. The smaller car, having less mass/weight, will likely be pushed backward. The laws of motion, particularly the conservation of momentum, dictate that the car with more mass/weight retains more of its forward motion.

21. (B) 5 feet. Explanation: The mechanical advantage of the inclined plane is determined by the length of the ramp divided by its height. With 100 pounds of force used to move a 400-pound object, the mechanical advantage is 4. Therefore, the height of the ramp is 20/4=5 feet.

22. (B) the force required to overcome friction while the box is being pushed. Explanation: When pushing an object across a rough surface, the force that needs to be applied is primarily to overcome the friction between the object and the surface. Friction depends on the surface roughness and the normal force, which is typically the weight of the object. The correct force is the one that counteracts the friction, allowing the object to move. The other choices are not correct because the force needed to push the box does not directly depend on the object's potential energy or mass in isolation. The weight of the object and the static coefficient of friction together define the force needed to make the object move initially; it is usually bigger than the force is needed to actually move the object across a surface once the movement is under way.

23. (A) 20 lbs. Explanation: The mechanical advantage (MA) of a lever is calculated as the ratio of the length of the effort arm to the length of the resistance arm. In this case, the effort arm is 3 times longer than the resistance arm, so: MA = 3. The mechanical advantage tells us how much the force is reduced. Therefore, the effort required to lift the 60-pound load is:

$$\text{Effort} = \frac{\text{Load}}{\text{MA}} = \frac{60\text{ lbs}}{3} = 20\text{ lbs}$$

This means only 20 lbs of force is needed to lift the 60-lb load, thanks to the lever's mechanical advantage.

24. (B) 9,000 J. Explanation: Work is the product of force and distance, calculated using the formula: $W = F \times d$

In this case, the force F is 1,500 N and the distance d is 6 m. Hence: $W = 1{,}500\ N \times 6\ m = 9{,}000\ J$

25. (A) 120 watts. Explanation: Power is calculated using the formula: $P = \frac{W}{t}$, where W is the work done and t is the time taken. Work $W = F \times d = 30\text{ N} \times 4\text{ m} = 120\text{ J}$. Since the machine lifts 10 parts, the total work is $120\text{ J} \times 10 = 1{,}200\text{ J}$. The time is 3 seconds, so the power is:

$$P = \frac{1{,}200\text{ J}}{3\text{ s}} = 400\text{ W}$$

26. (A) 4. Explanation: The mechanical advantage in a block and tackle pulley system is equal to the number of rope segments supporting the load. Here, 4 segments give a mechanical advantage of 4.

27. (B) 500 pounds. Explanation: The force on the larger piston is proportional to the area difference between the two pistons. Using Pascal's Law, the formula to calculate the force exerted on the larger piston is: $F_2 = F_1 \times \frac{A_2}{A_1}$

Where: F_1 is the force applied to the smaller piston, A_1 is the area of the smaller piston, A_2 is the area of the larger piston, F_2 is the force exerted on the larger piston.

Given: $F_1 = 100$ pounds, $A_1 = 10$ square inches, $A_2 = 50$ square inches

Substitute the values into the formula: $F_2 = 100 \times \frac{50}{10} = 100 \times 5 = 500$ pounds. So, the force exerted on the larger piston is 500 pounds.

28. A) Copper. Explanation: Copper is a metal and an excellent conductor of heat, meaning it absorbs and transfers heat faster than other materials.

B (Plastic) and C (Wood) are poor heat conductors compared to metals, so they will heat up more slowly.

D is incorrect because the materials have different thermal properties and will not heat up at the same rate.

29. Correct (B) 4 feet. Explanation: To balance the seesaw, the moments (weight × distance) on both sides must be equal. The side with the 40 and 80-pound children has a total moment of $(40 \times 4) + (80 \times 2) = 320\ foot - pounds$. To balance this with a 80-pound child, we set $80 \times \text{distance} = 320$, which gives a distance of 4 feet.

30. A) Resistance = Voltage / Current. Explanation: Ohm's Law defines the relationship between voltage (V), current (I), and resistance (R) as $V = I \times R$. Rearranging this formula gives us $R = V / I$. This makes Option A correct.

Chapter 11. Self-Description Inventory

To complete the Self-Description Inventory subtest, you will have 45 minutes to answer 240 questions.

This section is not scored as part of the AFOQT composite scores, but it is an integral part of the test. It serves as a personality gauge to help assess your characteristics and preferences. For each question, you will read a statement and select which option best fits how the statement describes you based on the following scale:

A. Strongly Disagree

B. Moderately Disagree

C. Neither Agree nor Disagree

D. Moderately Agree

E. Strongly Agree

Here are a few examples of the types of statements you can expect to see in the Self-Description Inventory questions:

1. I enjoy working with others.
2. I am assertive.
3. I believe something can always be improved.
4. I often drive over the speed limit.
5. I love being in nature.
6. I get annoyed when someone is late.
7. I frequently procrastinate.
8. I manage stress well.
9. I am sometimes too reserved.
10. I am comfortable with public speaking.

Your answers should simply reflect the way you perceive yourself. You have about 10 seconds per question, which is just enough time to read it, ponder for a few seconds, and then move on. You should not agonize over any one question, nor should you try to give the answers you think the Air Force wants to hear. Just answer honestly, and do not waste time as you go through the questions.

CHAPTER 12. SITUATIONAL JUDGMENT

Situational judgment is a way to measure your baseline ability to effectively lead by asking you to determine what you consider the MOST effective and LEAST effective means of resolving an issue or problem. These situations are intended to resemble what you might encounter in real life as an officer. You will be presented with a short paragraph, typically about interpersonal conflict, and then given five possible ways to address that conflict or issue.

A plausible example would be the following statement:

"One of your Airmen is not performing to standards but is the favorite person of the group, and everyone likes him/her. You have been pressured by your superiors to move that Airman elsewhere, but that would have a detrimental effect on morale for the other airmen remaining in your unit." The possible answers would look something like the following choices:

A. Just comply with your superiors; they know what is right, and there is no reason to risk your reputation.

B. Ignore your superiors; they are too far removed and do not understand the dynamics.

C. Counsel Airman on how to increase performance and give 30 days to improve.

D. Do the Airman's work for him/her to cover up the low standards.

E. Take a "wait and see" approach; maybe the Airman will improve, and you can avoid the confrontation.

As you can see, there is no objectively "right" or "wrong" answer here. However, your choices will reflect your decision-making style and alignment with **Air Force leadership standards**. While it is true that the issues presented are nuanced and subjective, certain patterns in responses may better align with the leadership qualities valued by the Air Force.

This subtest primarily evaluates your leadership style, not specific abilities. Effective leaders are often assertive in decision-making, able to make timely decisions while maintaining fairness and a balanced perspective. Key qualities measured include assertiveness, ethical decision-making, and adaptability. These traits are critical for success as an officer and are subtly revealed in how you approach complex situations.

The best tactic is to avoid viewing this section as a "test." There are no definitively right or wrong choices. For example, you won't encounter extreme scenarios like "You see an airman steal weapons ordinance to sell to the enemy." Instead, the issues presented are more nuanced, resembling the real-life dilemmas faced by leaders. It's important to demonstrate balance: assertiveness in making decisions when necessary, patience to assess the situation, and a focus on the mission above all else.

CHAPTER 13: PRACTICE TESTS

TEST 1. INSTRUMENT COMPREHENSION PRACTICE TEST

For each question, select the aircraft diagram that matches the corresponding altimeter heading and compass direction. For practice only. This practice set does not reflect the actual number of questions in the test.

1.

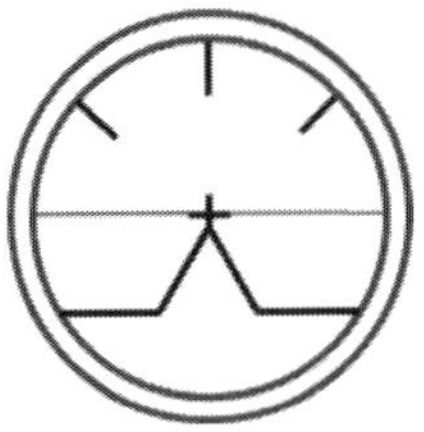

Artificial Horizon

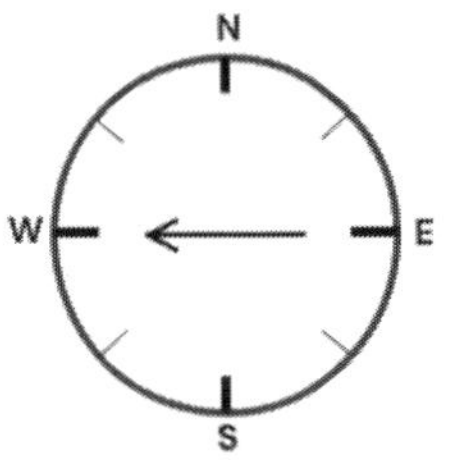

Compass

A B C D

2.

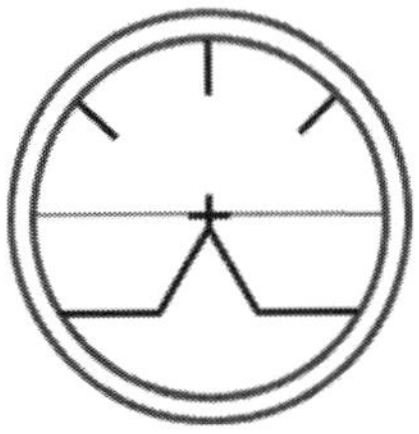

Artificial Horizon

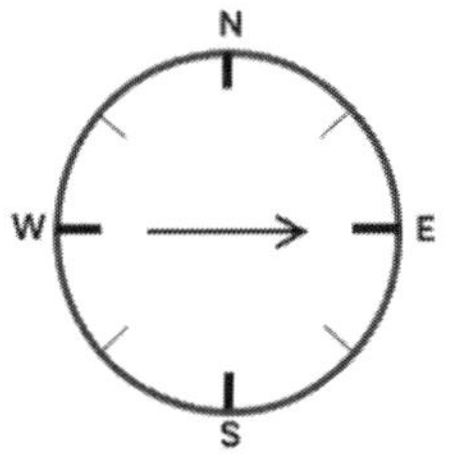

Compass

A B C D

3.

Artificial Horizon

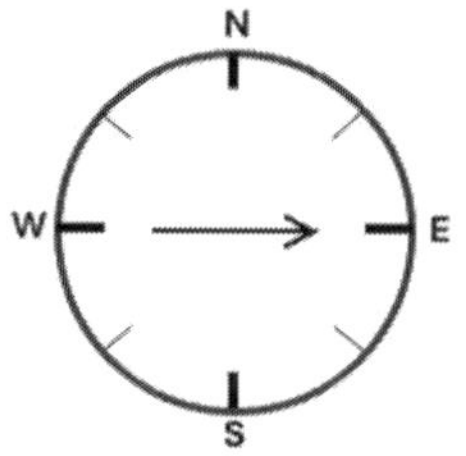

Compass

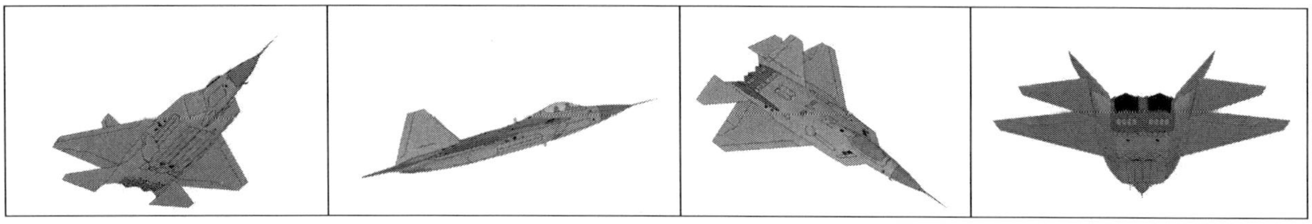

A B C D

4.

Artificial Horizon

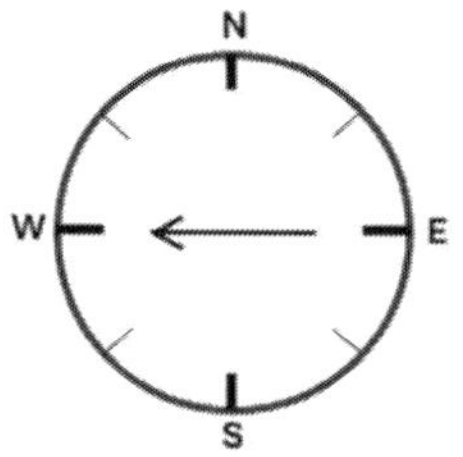

Compass

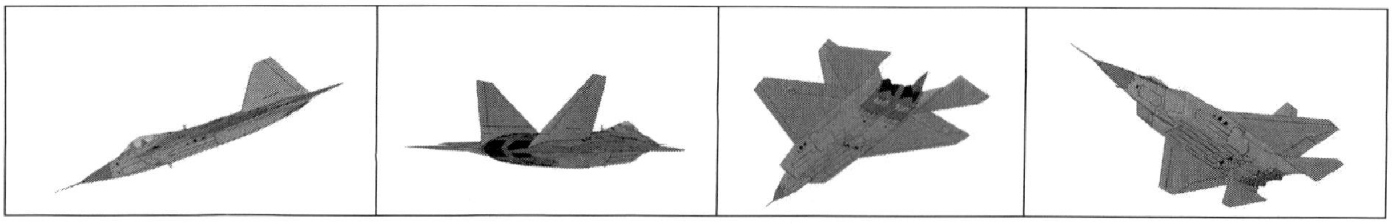

A B C D

5.

Artificial Horizon

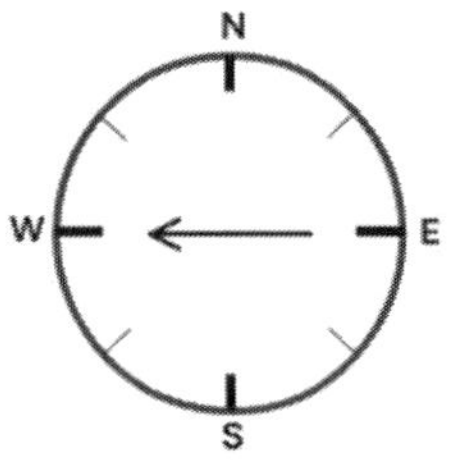

Compass

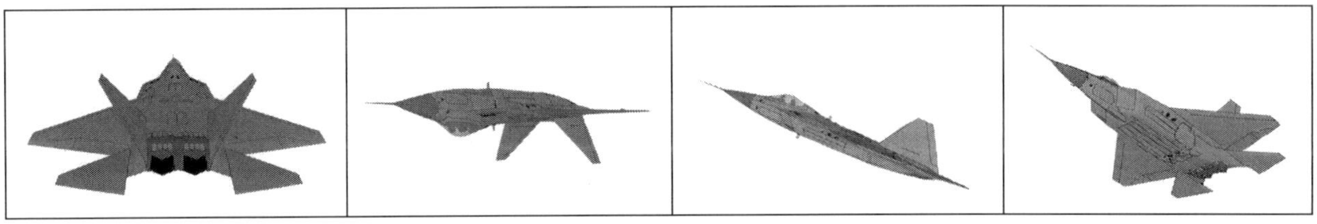

A B C D

6.

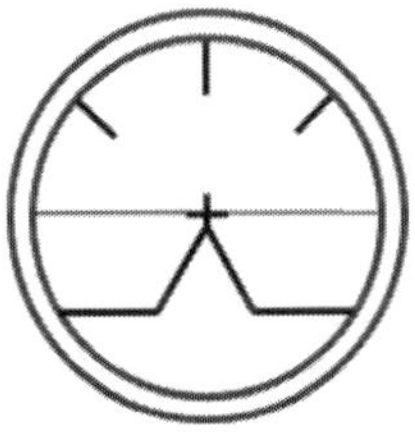

Artificial Horizon

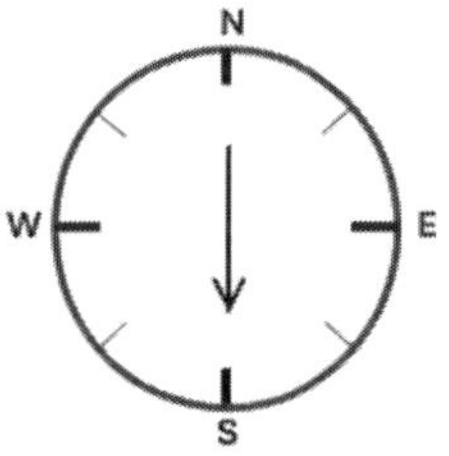

Compass

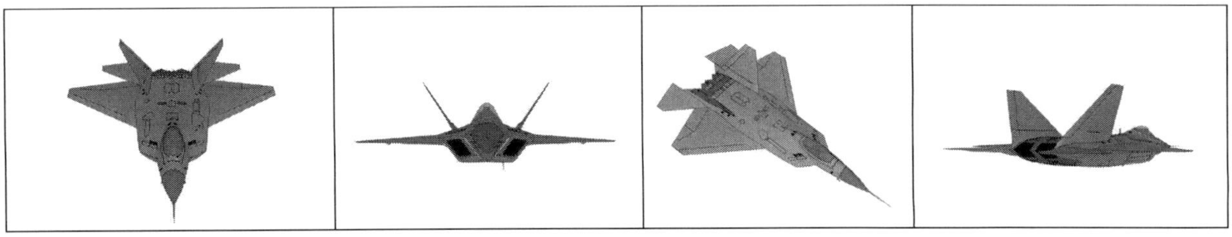

A B C D

7.

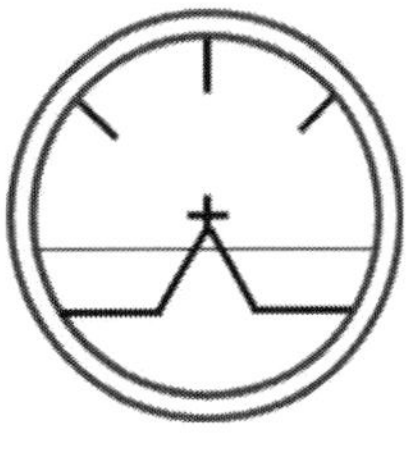

Artificial Horizon

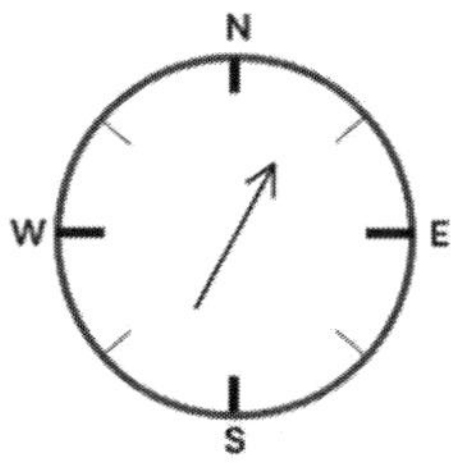

Compass

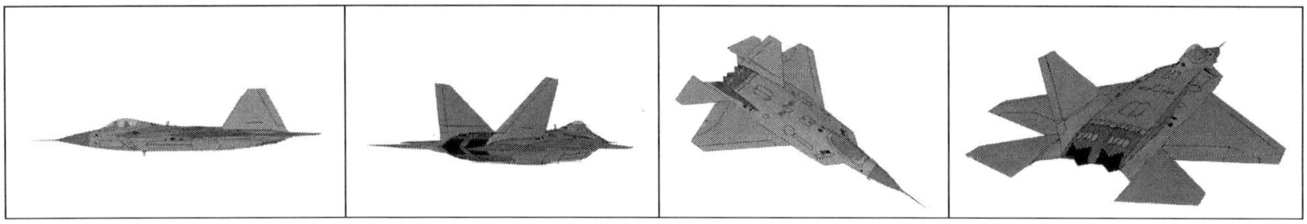

A B C D

8.

Artificial Horizon

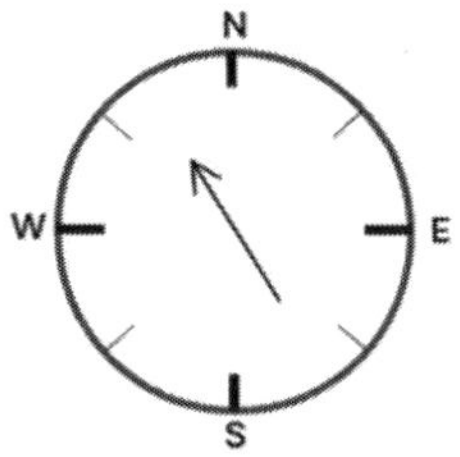

Compass

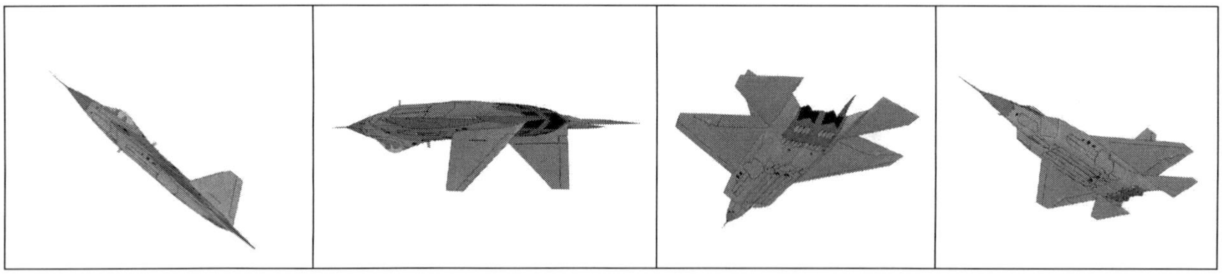

A B C D

9.

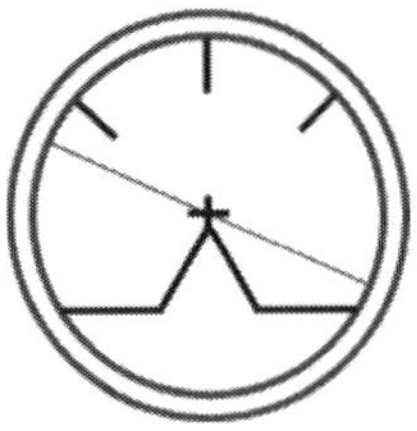

Artificial Horizon

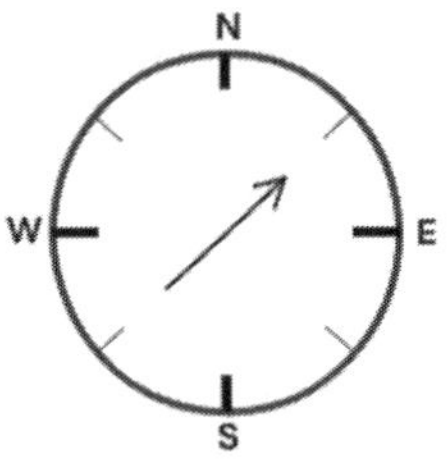

Compass

A B C D

10.

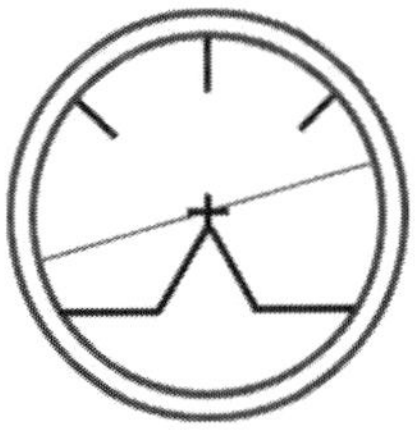

Artificial Horizon

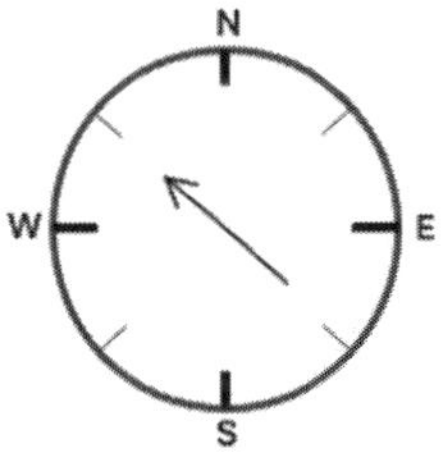

Compass

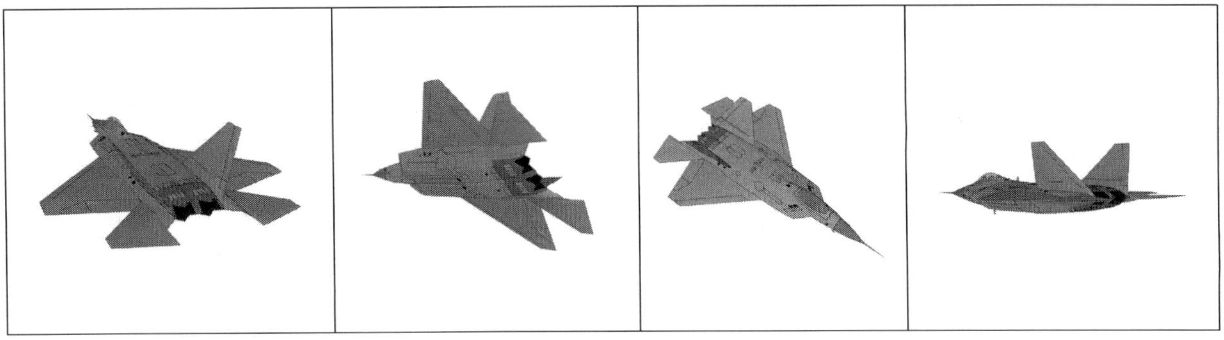

A B C D

11.

Artificial Horizon

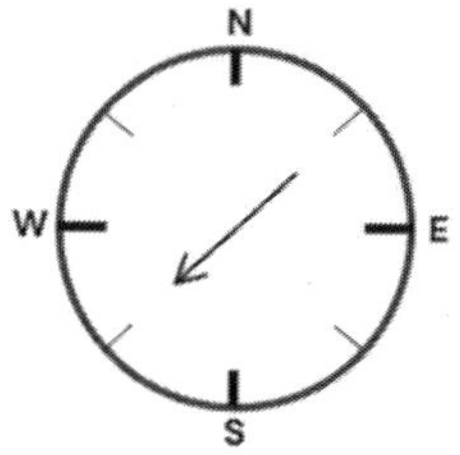

Compass

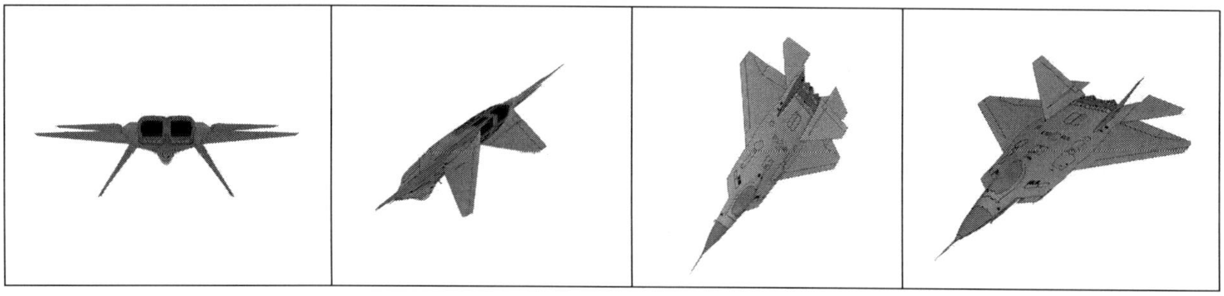

A B C D

12.

Artificial Horizon

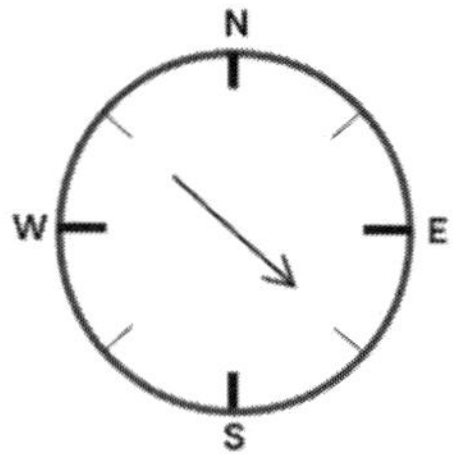

Compass

A B C D

13.

Artificial Horizon

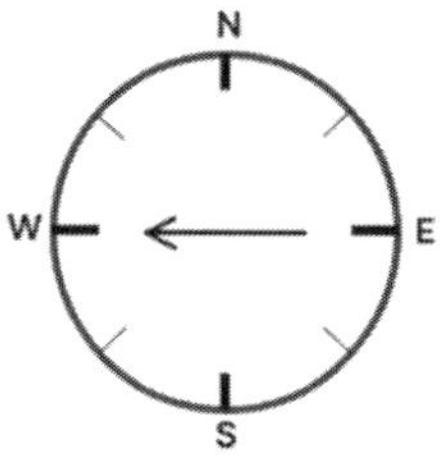

Compass

A B C D

14.

Artificial Horizon

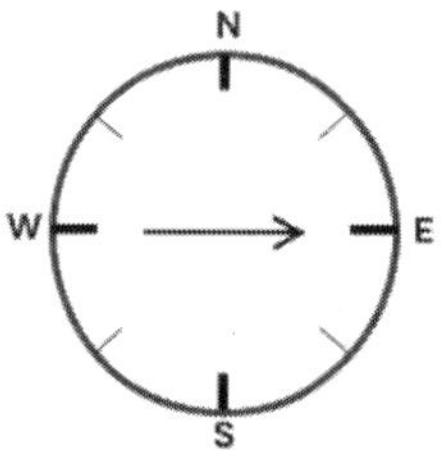

Compass

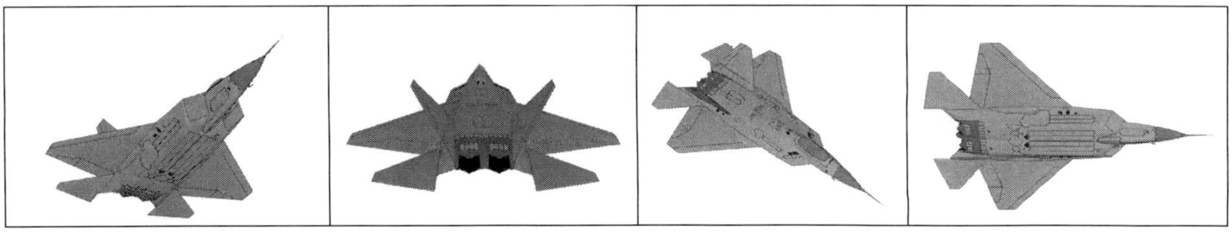

A B C D

15.

Artificial Horizon

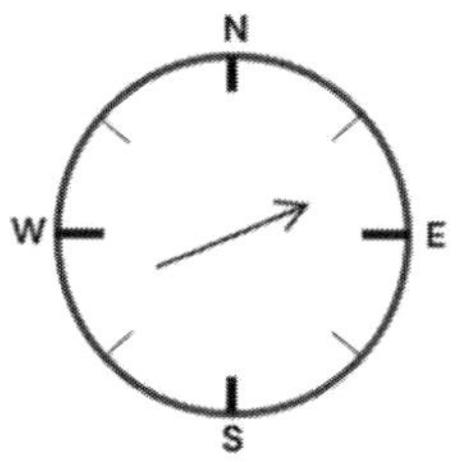

Compass

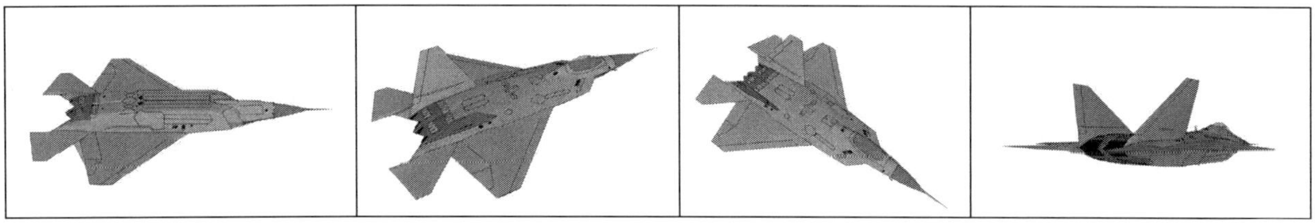

A B C D

16.

Artificial Horizon

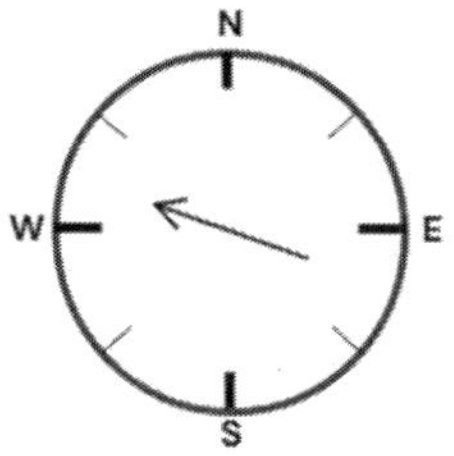

Compass

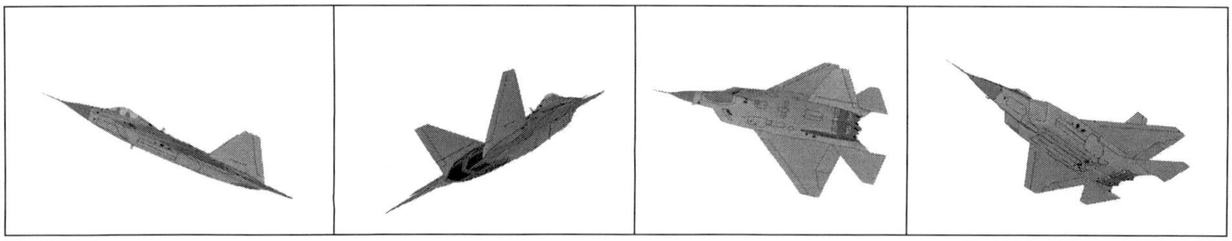

A B C D

17.

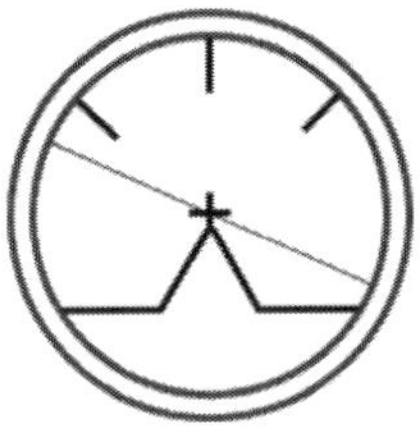

Artificial Horizon

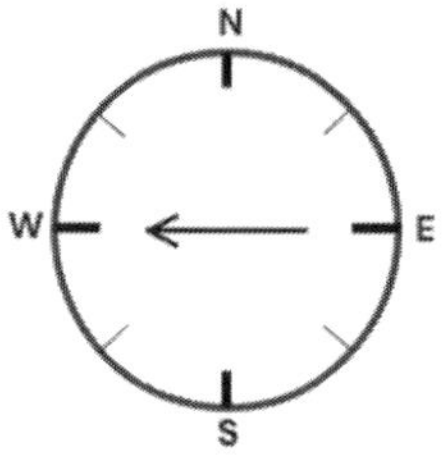

Compass

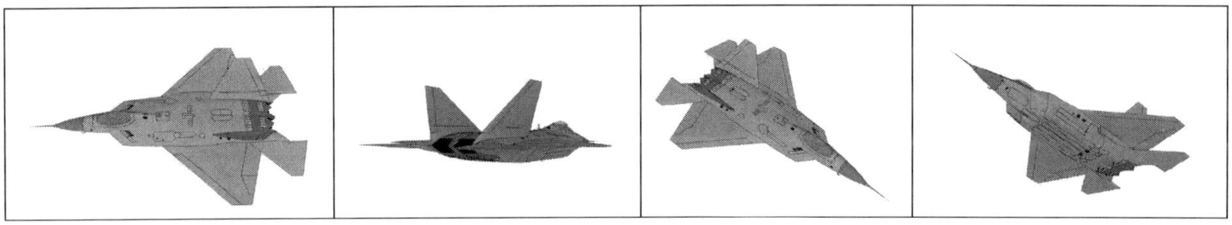

A B C D

18.

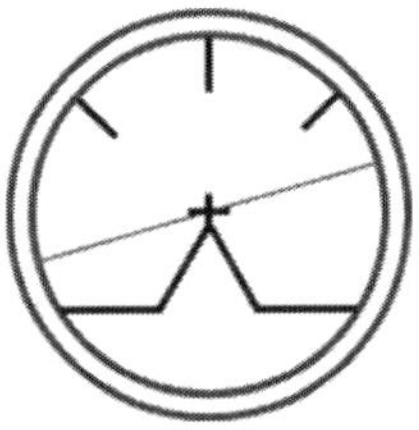

Artificial Horizon

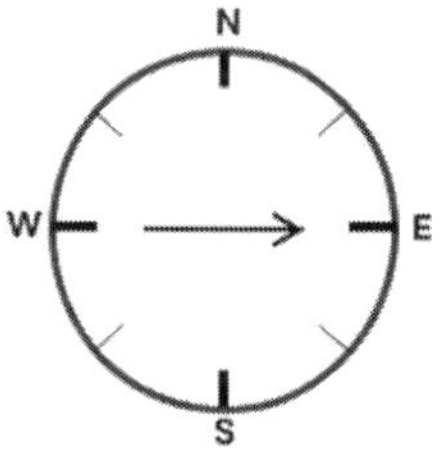

Compass

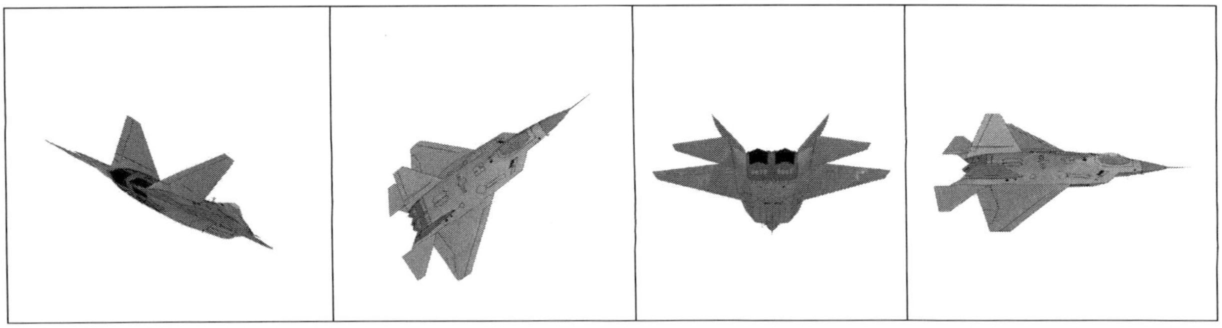

A B C D

19.

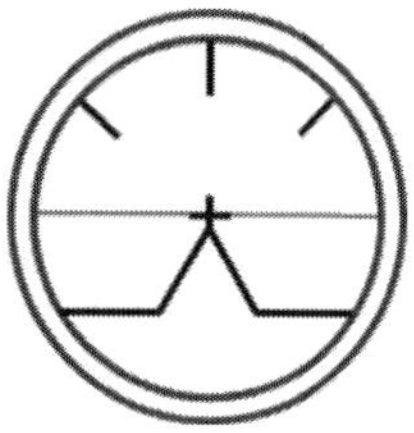

Artificial Horizon

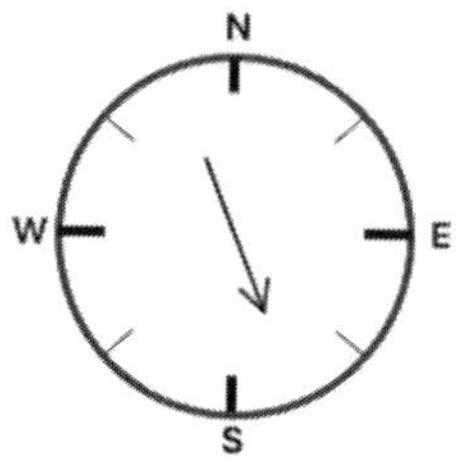

Compass

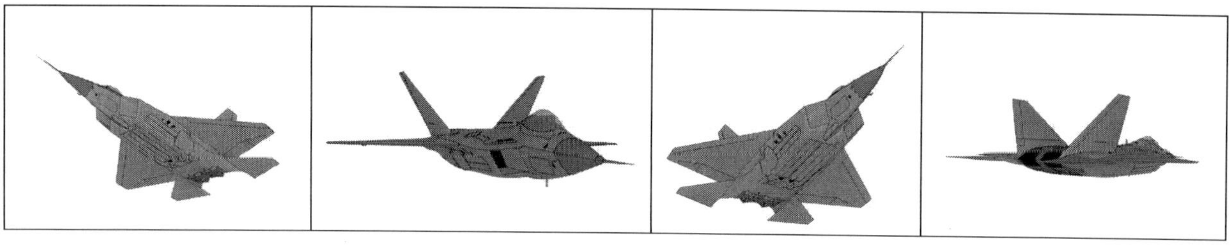

A B C D

20.

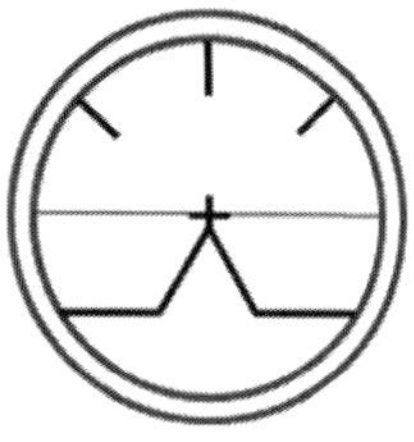

Artificial Horizon

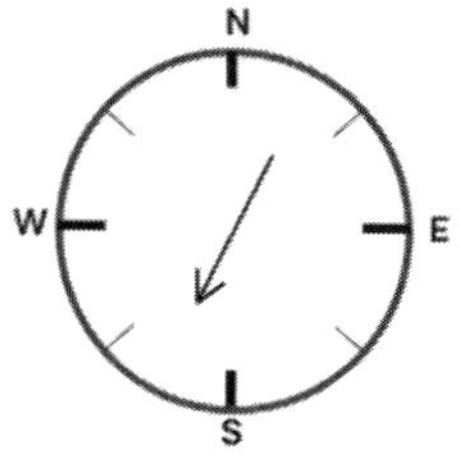

Compass

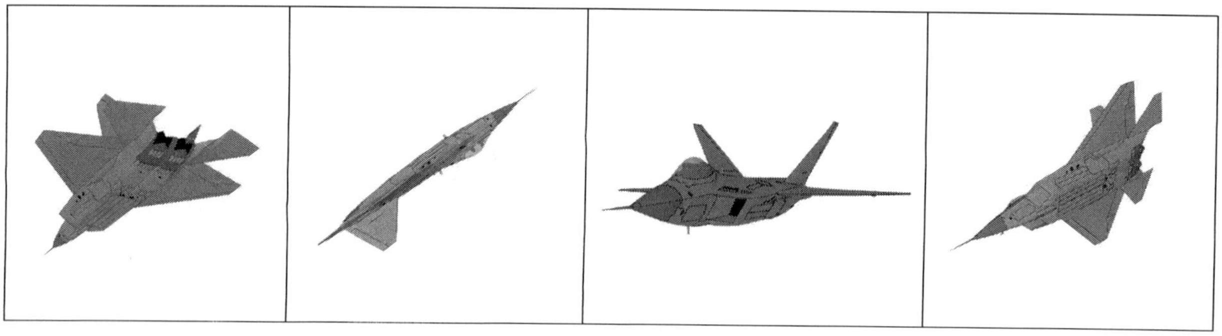

A B C D

21.

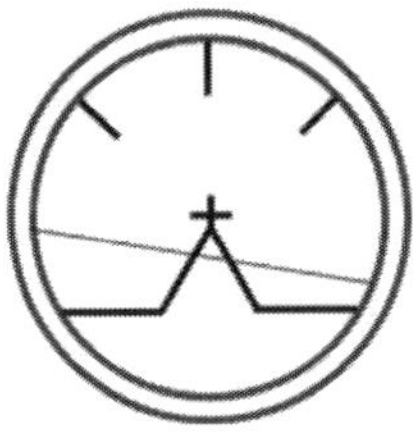

Artificial Horizon

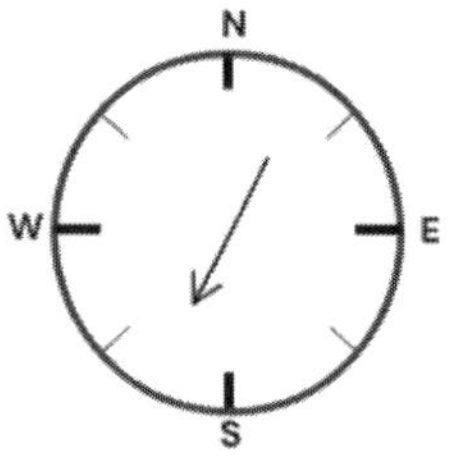

Compass

A B C D

22.

Artificial Horizon

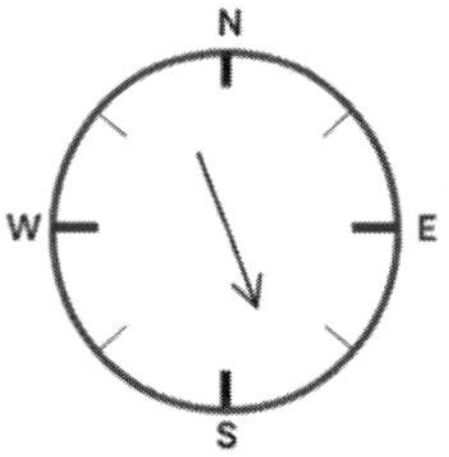

Compass

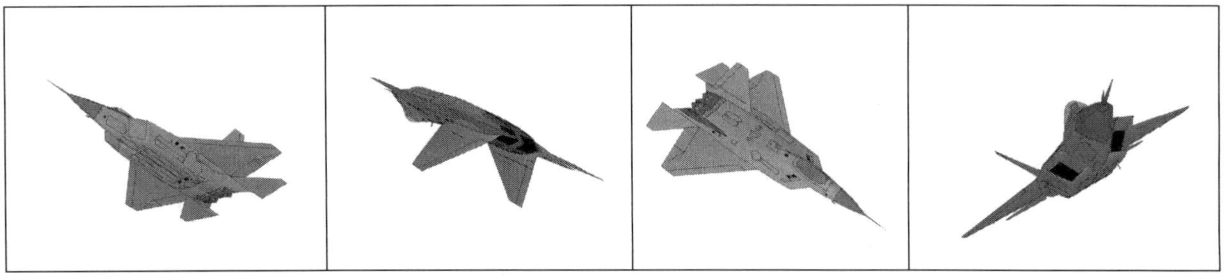

A B C D

23.

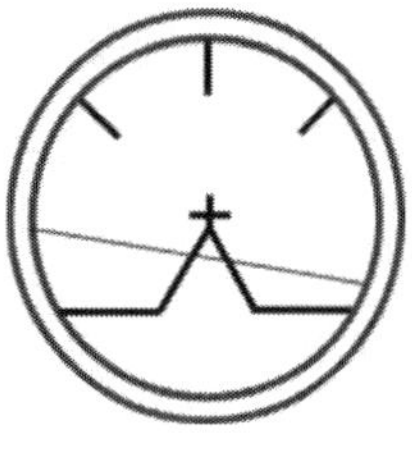
Artificial Horizon

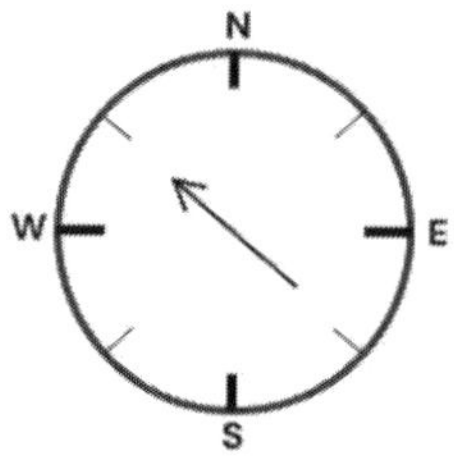

Compass

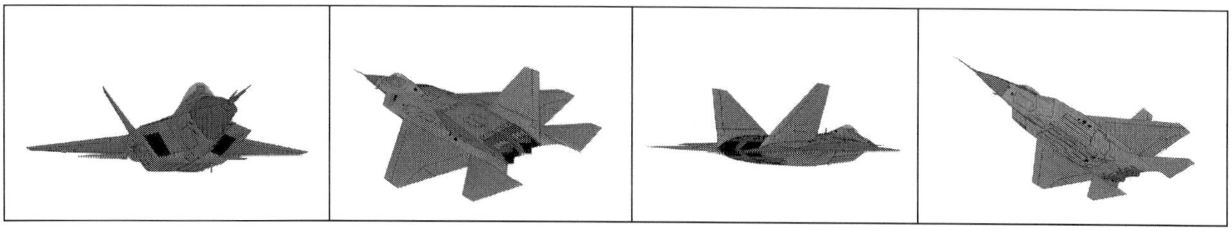

A B C D

24.

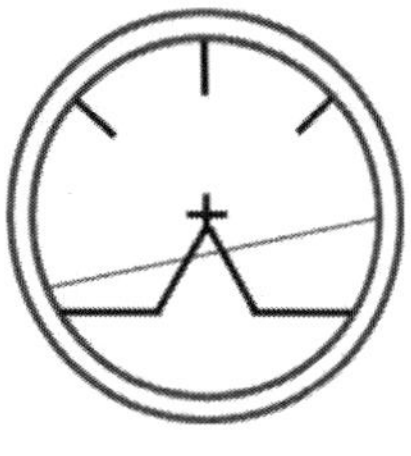
Artificial Horizon

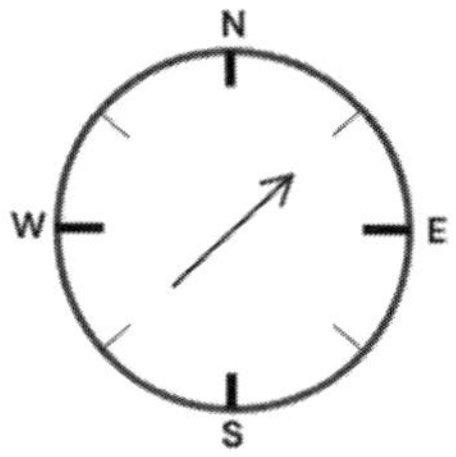

Compass

A B C D

25.

Artificial Horizon

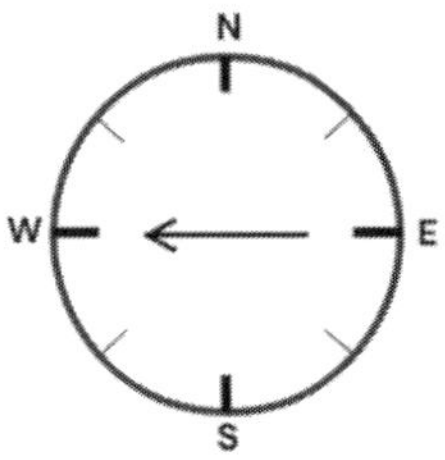

Compass

A

B

C

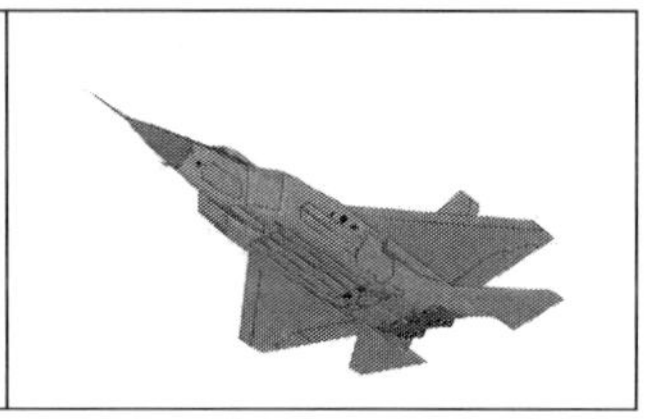

D

TEST 2. VERBAL ANALOGIES PRACTICE TEST

For each question, select the answer choice that most closely completes the analogy. To give you plenty of opportunities to practice, this set has more questions than in the actual test.

1. DOLL is to TOY as BASEBALL CAP is to ________:
   A. Baseball
   B. Hat
   C. Head
   D. Sports

2. LEMON is to FRUIT as ________:
   A. Peel is to potato
   B. Potato is to peel
   C. Carrot is to vegetable
   D. Apple is to vegetable

3. PLANE is to PILOT as CAR is to ________:
   A. Chauffeur
   B. Cashier
   C. Telemarketer
   D. Doctor

4. SHOES are to FEET as ________:
   A. Feet are to legs
   B. Glasses are to eyes
   C. Feet are to dancing
   D. Glasses are to seeing

5. NOTE is to MUSIC as ________:
   A. Dancing is to music
   B. Flute is to music
   C. English is to language
   D. Letter is to writing

6. TWO is to FOUR as EIGHT is to ________:
   A. Sixteen
   B. Twelve
   C. Seven
   D. Twenty

7. CAT is to KITTEN as ________:
   A. Bee is to larva
   B. Caterpillar is to butterfly
   C. Dog is to wolf
   D. Larva is to bee

8. CUBA is to HAVANA as RUSSIA is to ________:
   A. Japan
   B. Dublin
   C. Moscow
   D. Seattle

9. MOUSE is to CAT as ________:
   A. Cat is to bear
   B. Fly is to spider
   C. Goat is to sheep
   D. Dog is to wolf

10. ONION is to TEARS as ________:
   A. Pepper is to sneeze
   B. Salt is to taste
   C. Salt is to pepper
   D. Pepper is to spice

11. GLOVES are to COLD as SUNGLASSES are to ________:
   A. Rain
   B. Summer
   C. Spring
   D. Sun

12. OBJECT is to SEEING as ________:
   A. Music is to dancing
   B. Sound is to song
   C. Noise is to hearing
   D. Hearing is to noise

13. VACUUM is to DYSON as COMPUTER is to ________:
   A. Mouse
   B. Phone
   C. Dell
   D. Laptop

14. TRIED is to TRY as ________:
   A. Climbed is to climb
   B. Write is to wrote
   C. Fish is to fishing
   D. Thought is to thinking

15. LAMP is to LIGHT as AIR CONDITIONER is to ________:
   A. Electricity
   B. Cool
   C. Heat
   D. Gas

16. ADDITION is to MATH as ________:
   A. Subtraction is to division
   B. Language is to English
   C. Spanish is to English
   D. Painting is to Art

17. SALMON is to FISH as PARAKEET is to ________:
   A. Rodent
   B. Bird
   C. Tree
   D. Flower

18. CLOTH is to DRESS as ________:
   A. Second is to minute
   B. Needle is to thread
   C. Hour is to minute
   D. Thread is to needle

19. ONE HUNDRED is to TEN as FORTY-NINE is to ________:
   A. Twelve
   B. Seven
   C. Six
   D. Two hundred

20. COW is to CALF as ________:
   A. Kitten is to cat
   B. Goose is to gander
   C. Chicken is chick
   D. Fish is to lake

21. THUNDER is to LIGHTNING as ________:
   A. Rain is to snow
   B. Spark is to flame
   C. Bud is to leaf
   D. Heat is to fire

22. WISCONSIN is to MINESOTA as TEXAS is to ________:
   A. Maine
   B. California
   C. Oklahoma
   D. Florida

23. CHEF is to KNIFE as ________:
   A. Carpenter is to hammer
   B. Plumber is to hatchet
   C. Screwdriver is to carpenter
   D. Car is to mechanic

24. MOTORCYCLE is to VEHICLE as NECKLACE is to ________:
   A. Wear
   B. Neck
   C. Locket
   D. Jewelry

25. TULIP is to BULB as ________:
   A. Leaf is to branch
   B. Oak is to acorn
   C. Blossom is to flower
   D. Pinecone is to pine tree

26. PIG is to BACON as is to COW is to ________:
   A. Mutton
   B. Eggs
   C. Ham
   D. Steak

27. SPICE is to PEPPER as ________:
    A. Organ is to heart
    B. Grapes are to wine
    C. Water is to life
    D. Soda is to drink

28. SWIM is to SWAM as CATCH is to ________:
    A. Catching
    B. Caught
    C. Has caught
    D. Did catch

29. SPAGHETTI is to PASTA as ________:
    A. Bird is to Robin
    B. Batman is to Robin
    C. Bird is to Batman
    D. Robin is to bird

30. SCREEN is to TELEVISION as HINGE is to ________:
    A. Chair
    B. Table
    C. Door
    D. Window

31. FIVE is to TWENTY-FIVE as ________:
    A. Seven is to forty-nine
    B. Four is to forty
    C. Six is to twelve
    D. Ten is to sixty

32. CAIRO is to EGYPT as LONDON is to ________:
    A. France
    B. England
    C. Ireland
    D. Spain

33. KITCHEN is to COOK as ________:
    A. Pool is to swim
    B. Hot dog is to grill
    C. Slide is to park
    D. Hair is to cut

34. CAR is to ROAD as BOAT is to ________:
    A. River
    B. Pilot
    C. Dock
    D. Sky

35. FLU is to FEVER as ________:
    A. Pneumonia is to lungs
    B. Tumor is to cancer
    C. Brain is to stroke
    D. Cold is to congestion

36. BOUGHT is to TAUGHT as RUNNING is to ________:
   A. Will hold
   B. Playing
   C. Ran
   D. Jumps

37. DAISY is to FLOWER as ________:
   A. Maple is to tree
   B. Mind is to body
   C. Song is to music
   D. Dance is to song

38. COLD is to SHIVER as HEAT is to ________:
   A. Fire
   B. Spark
   C. Sweat
   D. Warm

39. BANK is to RIVER as ________:
   A. Ocean is to beach
   B. Beach is to coast
   C. Ocean is to waves
   D. Coast is to ocean

40. LEAF is to BUD as TREE is to ________:
   A. Seed
   B. Branch
   C. Oak
   D. Flower

41. PERSIAN is to CAT as ________:
   A. Dolphin is to whale
   B. Dalmatian is to dog
   C. Tree is to elm
   D. Flower is to seed

42. HURRICANE is to OCEAN as ________:
   A. Bird is to owl
   B. Tornado is to land
   C. Lightning is to thunderstorm
   D. Hail is to snow

43. RIDDLE is to SPHINX as SONG is to ________:
   A. Dance
   B. Sound
   C. Singer
   D. Music

44. HARP is to INSTRUMENT as ________:
   A. Smartphone is to phone
   B. Phone is to cell phone
   C. iPhone is to Android
   D. Android is to iPhone

45. HAMMER is to NAIL as SCREWDRIVER is to ________:
   A. Philips
   B. Screw
   C. Bolt
   D. Chisel

46. OVEN is to BAKER as ________:
   A. Salon is to stylist
   B. Plumber is to sink
   C. Jack is to mechanic
   D. Electrician is to wires

47. SIX is to EIGHTEEN as TWELVE is to ________:
   A. Seventeen
   B. Forty
   C. Five
   D. Thirty-six

48. MONTANA is to UNITED STATES as ________:
   A. Ontario is to Canada
   B. Canada is to Alberta
   C. Montreal is to Canada
   D. Montreal is to Quebec

49. NOODLE is to FOOD as PITBULL is to ________:
   A. Cat
   B. Poodle
   C. Fight
   D. Dog

50. APPLE is to CORE as ________:
   A. Petal is to stem
   B. Peach is to pit
   C. Peel is to banana
   D. Pork is to beans

TEST 3. TABLE READING PRACTICE TEST

For each question, select the correct coordinates from the table values.

X-Value

Y-Value	-15	-14	-13	-12	-11	-10	-9	-8	-7	-6	-5	-4	-3	-2	-1	0	1	2	3	4	5	6	7	8	9	10	11	12	13	14	15
15	64	2	8	74	64	46	42	74	74	95	60	97	68	25	10	43	84	44	87	89	9	52	71	45	1	78	13	69	25	13	62
14	55	13	14	42	43	44	82	15	46	93	53	78	47	56	77	83	40	89	5	4	71	15	80	86	51	48	2	95	93	51	32
13	92	72	18	4	51	90	22	14	79	27	53	53	3	86	78	94	59	76	30	39	97	94	3	64	8	20	86	30	37	89	60
12	80	25	12	67	14	78	27	16	19	47	99	61	50	61	75	78	3	62	72	59	37	85	83	85	13	45	31	50	75	19	50
11	62	44	78	54	54	95	76	5	66	83	52	81	93	83	24	89	53	32	29	37	12	4	85	16	39	32	38	19	90	79	66
10	41	90	35	57	35	97	22	24	58	16	88	17	4	29	51	18	37	5	45	42	28	45	2	25	37	43	97	28	20	99	54
9	41	49	26	43	50	56	60	21	72	74	95	7	50	52	45	1	37	23	68	98	25	86	3	37	67	80	53	15	35	92	15
8	53	50	6	44	7	75	16	78	23	49	33	47	7	55	60	50	61	82	72	6	74	61	11	75	76	75	9	16	91	81	56
7	14	48	60	22	6	33	46	79	48	72	29	89	76	56	14	81	35	7	53	55	49	60	56	92	45	10	62	52	64	71	33
6	90	94	52	80	34	26	96	1	22	45	29	96	50	70	71	8	37	10	33	47	13	26	29	48	57	58	74	34	10	77	92
5	3	60	89	47	90	85	49	56	1	44	10	93	31	19	69	52	34	20	91	65	44	93	40	37	59	99	92	86	10	85	23
4	37	50	90	83	64	18	69	28	40	53	24	77	16	10	19	54	70	47	72	5	70	10	77	14	93	63	42	61	76	49	29
3	43	77	16	73	18	30	51	70	2	82	87	39	47	72	90	38	21	61	95	34	57	60	72	99	76	17	64	67	4	98	36
2	6	71	36	70	14	88	8	18	19	38	56	24	13	52	70	79	26	76	16	91	71	8	9	44	95	76	43	94	9	86	87
1	69	4	27	63	68	3	55	3	33	16	6	16	4	52	26	82	71	6	11	18	79	30	55	35	37	3	47	15	68	4	18
0	47	83	5	48	91	37	71	83	82	93	68	42	74	6	44	26	64	76	16	74	22	6	84	70	34	42	11	29	38	64	34
-1	99	46	76	95	43	6	50	70	87	72	89	60	23	28	41	88	23	17	43	4	48	73	59	53	89	81	77	84	86	17	44
-2	55	50	16	85	61	41	10	79	36	46	90	13	21	60	98	72	20	55	66	37	22	41	12	13	1	26	18	44	78	74	78
-3	81	85	56	82	8	56	87	84	23	47	49	59	90	53	99	65	15	83	78	32	77	97	40	18	90	65	18	5	25	62	9
-4	28	89	83	14	17	75	56	51	14	77	41	57	10	82	9	94	21	89	35	70	48	18	17	13	64	63	31	74	12	98	43
-5	62	92	9	4	49	17	86	90	75	8	48	75	86	45	54	60	86	6	47	62	34	36	98	38	69	47	66	93	61	14	60
-6	14	14	26	91	50	23	66	23	29	39	17	38	72	44	8	95	13	36	90	45	44	18	36	88	3	11	76	61	74	58	86
-7	12	93	94	72	38	95	12	51	56	88	9	71	76	99	10	23	78	2	76	85	97	19	69	19	21	30	72	66	66	16	70
-8	71	97	45	37	62	4	93	81	47	43	40	83	90	92	59	84	96	84	51	61	47	67	65	6	72	71	37	75	67	80	78
-9	24	89	81	46	85	18	52	25	10	97	26	20	49	70	46	11	94	70	86	96	97	3	27	63	31	9	53	72	60	53	25
-10	41	97	4	15	33	79	37	74	38	93	56	8	86	3	55	93	28	41	96	67	7	16	65	89	82	5	36	18	76	2	94
-11	27	95	94	82	20	30	31	81	59	94	95	99	45	8	7	94	11	59	23	62	3	36	62	90	79	71	13	4	64	83	7
-12	19	5	84	33	76	11	17	5	11	35	85	3	49	23	1	35	71	42	42	55	85	63	47	98	75	82	98	71	55	55	5
-13	11	62	19	31	92	61	12	66	12	72	99	75	87	13	10	39	37	3	52	49	41	8	95	92	75	45	30	86	91	96	28
-14	20	33	24	76	89	51	80	14	29	55	7	13	59	66	72	30	52	26	59	59	94	44	91	1	24	89	14	60	51	5	62
-15	72	25	55	75	16	98	73	85	77	21	39	84	85	78	59	24	47	32	65	74	13	58	55	90	35	9	43	5	9	28	63

	X	Y	A	B	C	D	E
1	1	-9	94	3	19	1	98
2	0	-6	8	37	77	95	91
3	13	3	67	98	4	14	45
4	-8	-7	14	51	56	78	22
5	1	-12	71	33	97	47	76
6	-13	10	12	31	55	18	35
7	10	-2	24	73	26	52	17
8	15	-7	14	41	75	90	70
9	-4	10	17	81	73	7	1
10	2	-3	14	4	99	83	15
11	-3	6	51	50	8	14	77
12	-14	13	19	15	72	11	10
13	10	-4	77	75	81	22	34
14	8	4	21	52	97	35	14
15	7	-7	66	1	69	34	25
16	0	-9	4	65	33	11	41
17	5	-8	47	99	28	79	17
18	10	11	11	32	33	14	15
19	15	8	55	42	56	13	37
20	9	-3	14	17	15	87	90
21	-7	7	48	88	74	84	2
22	-11	-10	45	86	43	33	41
23	-4	-11	53	23	73	84	99
24	-13	12	3	12	53	22	64
25	-9	-6	66	4	9	19	67
26	-4	9	65	7	53	74	27
27	4	5	15	77	65	31	12
28	8	-1	9	12	33	77	53
29	6	-10	69	29	48	54	16
30	0	0	26	11	2	42	79
31	-7	-2	18	99	36	3	61
32	-15	2	15	6	21	29	45
33	15	-4	27	55	43	83	87
34	9	-14	13	15	89	92	24
35	-7	-11	5	59	27	39	43
36	9	-2	97	39	64	18	1
37	2	1	6	29	4	13	81
38	10	-7	59	15	33	30	65
39	12	14	13	95	47	62	59
40	1	5	63	14	34	78	86

TEST 4. AVIATION INFORMATION PRACTICE TEST

For each question, select the best answer.

1. The aircraft barometric altimeter determines altitude by measuring which of the following?
   A. Radar signal
   B. Density
   C. Calibrated
   D. Above Ground Level
   E. Air pressure

2. What occurs because of a forward center of gravity?
   A. Difficulty applying downward pitch
   B. Decreased stability
   C. Longer range
   D. Difficulty applying upward pitch
   E. Decreased roll responsiveness

3. What type of environmental condition(s) yields the greatest aircraft performance?
   A. Sea level and cold
   B. High altitude and cold
   C. High altitude and hot
   D. Sea level and hot
   E. Humid

4. Describe how to recover from a stall...
   A. Decrease power and increase pitch
   B. Increase power and increase pitch
   C. Decrease pitch and increase power
   D. Decrease pitch and decrease power
   E. Increase power and add left or right rudder

5. Dead Reckoning navigation involves...
   A. Using landmarks to pinpoint position
   B. Using VORs and TACANs to pinpoint position
   C. Utilizing a previously known position and estimating present position based on course, time, and distance calculations
   D. GPS to estimate position
   E. Determining position by only using a compass

6. In flight, the rudder controls...
   A. Pitch
   B. Yaw
   C. Roll
   D. Power
   E. Speed

7. What effect do flaps have on landing distance?
   A. Decrease
   B. None
   C. Increase
   D. Decrease on wet runways only
   E. Increase on wet runways only

8. When the aircraft yoke or control wheel is moved to the right, what happens to the right aileron?
   A. Remains neutral
   B. Moves upward
   C. Moves downward
   D. Initially moves upwards then downwards
   E. Initially moves downwards then upwards

9. How does ice accumulated on the aircraft's wings affect the stall speed?
   A. No effect
   B. Decrease
   C. Decrease at high altitude
   D. Decrease at low altitude
   E. Increase

10. What conditions are conducive to fog formation?
    A. High temperature-dew point spread and high winds
    B. Dry desert areas and high winds
    C. Thunderstorms nearby and calm winds
    D. Low temperature-dew point spread and calm winds, especially in low lying valleys and coastal areas
    E. Hurricanes and high winds

11. In straight and level flight with the autopilot disengaged, after adding power or thrust the aircraft will...
    A. Descend
    B. Remain level
    C. Climb
    D. Roll to the right
    E. Roll to the left

12. What type of aircraft has the most severe wake turbulence?
    A. Heavy, slow, gear up, flaps up
    B. Light, slow, gear up, flaps up
    C. Heavy, fast, gear down, flaps deployed
    D. Light, fast, gear down, flaps deployed
    E. Heavy, fast, gear down, flaps up

13. What must be done to maintain altitude in a turn?
    A. Nothing
    B. Decrease pitch
    C. Decrease power
    D. Add rudder
    E. Increase pitch

14. What flight control is used to mitigate adverse yaw in a turn?
    A. Aileron
    B. Spoilers
    C. Flaps
    D. Rudder
    E. Aileron trim

15. Air Traffic Control orders a pilot-discretion descent from 12000 feet to meet a 3,000-feet crossing restriction. Using a 3-degree descent profile, how far away from the 3,000-feet constraint must the descent be initiated?

A. 5 miles
B. 10 miles
C. 20 miles
D. 30 miles
E. 38 miles

16. In calm winds, what happens to fuel consumption as the aircraft climbs?

A. No change
B. Decreases
C. Increases
D. Increases up to 10,000 feet then decreases
E. Decreases up to 5,000 feet then increases

17. What effect does a tailwind have on takeoff distance?

A. None
B. Decreases
C. Increases
D. Decreases on contaminated runways only
E. Increases on contaminated runways only

18. What is the best airspeed for maximum fuel endurance?

A. Maximum lift and lowest drag known as L/D Max
B. Minimum lift and highest drag known as L/D Max
C. Maximum lift and highest drag known as V_{ne}
D. Just above aircraft stall speed

19. When is the magnetic compass most accurate?

A. Climb
B. Descent
C. Straight and level flight
D. Right turn
E. Left turn

20. An aircraft has a ground speed of 120 knots. How long will it take to travel 20 nautical miles?

A. 1 minute
B. 2 minutes
C. 5 minutes
D. 10 minutes
E. 30 minutes

TEST 5. BLOCK COUNTING PRACTICE TEST

For each group, select the best answer for each of the 5 related questions.

GROUP 1

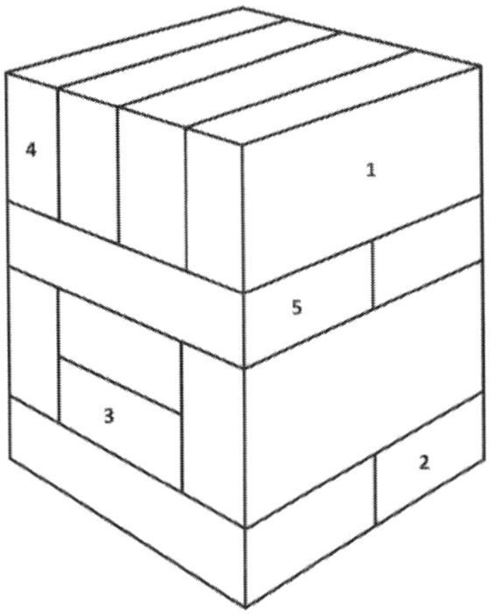

1. Block 1 is touched by______blocks.
   A. 1
   B. 2
   C. 3
   D. 4
   E. 5

2. Block 2 is touched by______blocks.
   A. 4
   B. 2
   C. 6
   D. 3
   E. 1

3. Block 3 is touched by______blocks.
   A. 4
   B. 5
   C. 1
   D. 3
   E. 2

4. Block 4 is touched by______blocks.
   A. 2
   B. 4
   C. 6
   D. 2
   E. 3

5. Block 5 is touched by______blocks.
   A. 7
   B. 8
   C. 3
   D. 4
   E. 5

GROUP 2

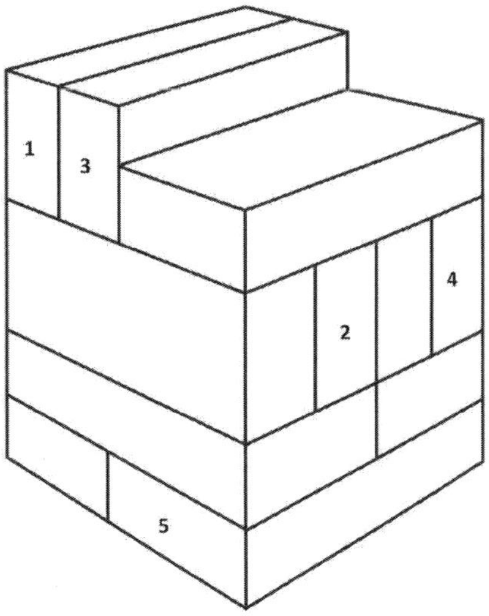

1. Block 1 is touched by_____blocks.
   A. 2
   B. 3
   C. 4
   D. 5
   E. 6

2. Block 2 is touched by_____blocks.
   A. 5
   B. 8
   C. 2
   D. 1
   E. 6

3. Block 3 is touched by_____blocks.
   A. 5
   B. 6
   C. 3
   D. 2
   E. 7

4. Block 4 is touched by_____blocks.
   A. 5
   B. 4
   C. 1
   D. 8
   E. 6

5. Block 5 is touched by_____blocks.
   A. 1
   B. 2
   C. 3
   D. 5
   E. 6

GROUP 3

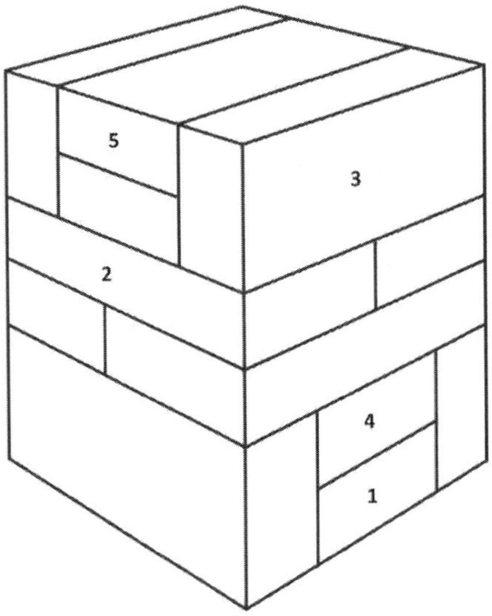

1. Block 1 is touched by______blocks.
   A. 4
   B. 3
   C. 1
   D. 2
   E. 5

2. Block 2 is touched by______blocks.
   A. 5
   B. 7
   C. 6
   D. 2
   E. 3

3. Block 3 is touched by______blocks.
   A. 4
   B. 3
   C. 2
   D. 1
   E. 5

4. Block 4 is touched by______blocks.
   A. 4
   B. 3
   C. 5
   D. 1
   E. 7

5. Block 5 is touched by______blocks.
   A. 1
   B. 2
   C. 3
   D. 4
   E. 5

GROUP 4

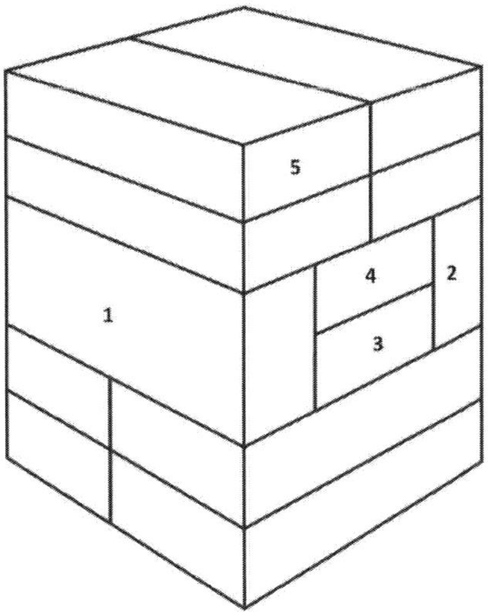

1. Block 1 is touched by______blocks.
   A. 5
   B. 3
   C. 4
   D. 7
   E. 2

2. Block 2 is touched by______blocks.
   A. 3
   B. 4
   C. 5
   D. 6
   E. 2

3. Block 3 is touched by______blocks.
   A. 4
   B. 6
   C. 3
   D. 5
   E. 2

4. Block 4 is touched by______blocks.
   A. 1
   B. 2
   C. 4
   D. 6
   E. 5

5. Block 5 is touched by______blocks.
   A. 4
   B. 6
   C. 8
   D. 1
   E. 2

GROUP 5

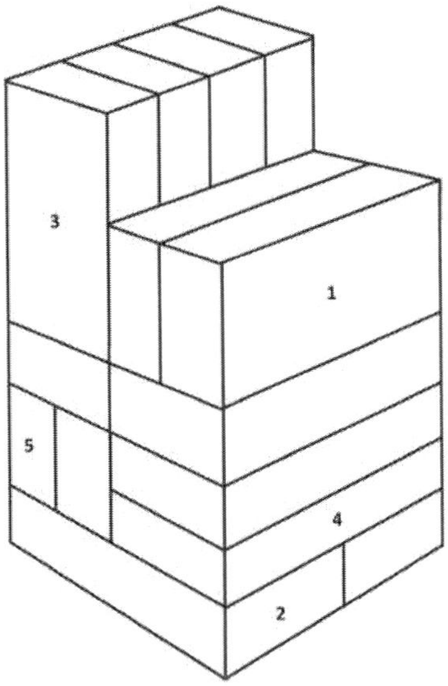

1. Block 1 is touched by______blocks.
   - A. 1
   - B. 3
   - C. 2
   - D. 4
   - E. 5

2. Block 2 is touched by______blocks.
   - A. 4
   - B. 3
   - C. 5
   - D. 1
   - E. 2

3. Block 3 is touched by______blocks.
   - A. 2
   - B. 6
   - C. 5
   - D. 3
   - E. 1

4. Block 4 is touched by______blocks.
   - A. 3
   - B. 5
   - C. 2
   - D. 6
   - E. 4

5. Block 5 is touched by______blocks.
   - A. 4
   - B. 3
   - C. 5
   - D. 6
   - E. 2

GROUP 6

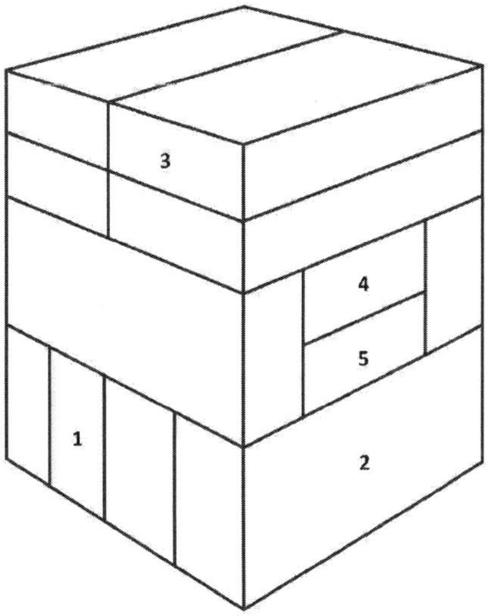

1. Block 1 is touched by______blocks.
   - A. 4
   - B. 3
   - C. 2
   - D. 6
   - E. 5

2. Block 2 is touched by______blocks.
   - A. 3
   - B. 4
   - C. 5
   - D. 2
   - E. 6

3. Block 3 is touched by______blocks.
   - A. 4
   - B. 6
   - C. 5
   - D. 2
   - E. 1

4. Block 4 is touched by______blocks.
   - A. 5
   - B. 4
   - C. 3
   - D. 1
   - E. 6

5. Block 5 is touched by______blocks.
   - A. 6
   - B. 4
   - C. 3
   - D. 2
   - E. 7

TEST 6. ARITHMETIC REASONING PRACTICE TEST

For each question, select the best answer. To give you plenty of opportunities to practice, this set has more questions than in the actual test.

1. The number of students enrolled at Two Rivers Community College increased from 3,450 in 2010 to 3,864 in 2015. What was the percent increase?
   A. 9%
   B. 17%
   C. 12%
   D. 6%

2. Amy drives her car until the gas gauge is down to $^1/_8$ full. Then she fills the tank to capacity by adding 14 gallons. What is the total capacity of the gas tank?
   A. 16 gallons
   B. 18 gallons
   C. 20 gallons
   D. 22 gallons

3. Jean buys a textbook, a flash drive, a printer cartridge, and a ream of paper. The flash drive costs three times as much as the ream of paper. The textbook costs three times as much as the flash drive. The printer cartridge costs twice as much as the textbook. The ream of paper costs $10. How much does Jean spend in total for all supplies?
   A. $250
   B. $480
   C. $310
   D. $180

4. In the graduating class at Emerson High School, 52% of the students are girls and 48% are boys. There are 350 students in the class. Among the girls, 98 plan to attend college. How many girls do not plan to attend college?
   A. 84 girls
   B. 48 girls
   C. 66 girls
   D. 72 girls

5. A cell phone on sale for 30% off costs $210 after the discount. What was the original price of the phone?
   A. $240
   B. $273
   C. $300
   D. $320

6. Seven added to four-fifths of a number equals 15. What is the number?
   A. 10
   B. 15
   C. 20
   D. 25

7. The sum of two numbers is 360, and their ratio is 7:3. What is the smaller number of the two?
   A. 72
   B. 105
   C. 98
   D. 108

8. Alicia must earn a score of 75% to pass an 80-question test. How many total questions can she miss and still pass the test?

   A. 20 questions
   B. 25 questions
   C. 60 questions
   D. 15 questions

9. Which number below is the largest?

   A. $^5/_8$
   B. $^3/_5$
   C. $^2/_3$
   D. 0.72

10. It took Charles four days to write a history paper. He wrote 5 pages on the first day, 4 pages on the second day, and 8 pages on the third day. If Charles ended up writing an average of 7 pages per day, how many pages did he write on the fourth day?

    A. 11 pages
    B. 8 pages
    C. 12 pages
    D. 9 pages

11. Danvers is 8 miles due south of Carson and 6 miles due west of Baines. If a driver could drive in a straight line from Carson to Baines, how many miles would the trip be?

    A. 8 miles
    B. 10 miles
    C. 12 miles
    D. 14 miles

12. Four friends plan to share equally the cost of a retirement gift. If one person drops out of the arrangement, the cost per person for the remaining three would increase by $12. How much does the gift cost?

    A. $144
    B. $136
    C. $180
    D. $152

13. Amanda makes $14 an hour as a bank teller, and Oscar makes $24 an hour as an auto mechanic. Both work 8 hours a day, 5 days a week. Which equation below can be used to calculate how much money Amanda and Oscar make collectively in a 5-day week?

    A. $(14 + 24) \times 8 \times 5$
    B. $\frac{14+24}{(8)(5)}$
    C. $(14 + 24)(8 + 5)$
    D. $14 + 24 * 8 * 5$

14. The population of Mariposa County in 2015 was 90% of the 2010 population. The population in 2010 was 145,000 people. What was the population in 2015?

    A. 160,000 people
    B. 142,000 people
    C. 120,500 people
    D. 130,500 people

15. What is the sum of $^1/_3$ and $^3/_8$?

A. $^3/_{24}$

B. $^4/_{11}$

C. $^{17}/_{24}$

D. $^{15}/_{16}$

16. In four years, Tom will be twice as old as Serena was 3 years ago. Tom is 3 years younger than Serena. How old are Tom and Serena now?

A. Serena is 28, Tom is 25

B. Serena is 7, Tom is 4

C. Serena is 18, Tom is 15

D. Serena is 21, Tom is 18

17. Marisol's score on a standardized test was ranked in the 78th percentile. If 660 students took the test, approximately how many students scored lower than Marisol?

A. 582 students

B. 515 students

C. 612 students

D. 486 students

18. Sam worked 40 hours at *d* dollars per hour and received a bonus of $50. His total earnings were $530. What was his hourly wage?

A. $18

B. $16

C. $14

D. $12

19. When you add two numbers, the sum is 480. If the ratio of the two numbers is 5:1, what is the smaller number?

A. 60

B. 70

C. 72

D. 80

20. In a high school French class, 45% of the students are sophomores, and 9 sophomores are in the class. How many total students are in the class?

A. 16 students

B. 18 students

C. 20 students

D. 22 students

21. At a lunch cart, there are 2 orders of diet soda for every 5 orders of regular soda. If the owner of the lunch cart sells 112 sodas a day, how many are diet and how many are regular?

A. 28 diet, 84 regular

B. 32 diet, 80 regular

C. 34 diet, 82 regular

D. 36 diet, 84 regular

22. Two rectangles are proportional. In other words, the ratio of length to width is the same for both rectangles. The smaller rectangle has a length of 8 inches and a width of 3 inches. The larger rectangle has a length of 12 inches. What is the width of the larger rectangle?

A. 4 inches
B. 4.5 inches
C. 6 inches
D. 8.5 inches

23. The average weight of five friends (Al, Bob, Carl, Dave, and Ed) is 180 pounds. Al weighs 202 pounds, Bob weighs 166 pounds, Carl weighs 190 pounds, and Dave weighs 192 pounds. How much does Ed weigh?

A. 180 pounds
B. 172 pounds
C. 186 pounds
D. 150 pounds

24. Alan commutes 18 miles to work. Bob's commute is 4 miles shorter. Ted's commute is 6 miles shorter than Bob's. Rebecca's commute is shorter than Alan's but longer than Bob's. Which option below could be the length of Rebecca's commute?

A. 12 miles
B. 14 miles
C. 15 miles
D. 18 miles

25. Of the patients admitted to an ER over a one-week period, 14 had heart attacks, 15 had workplace injuries, 24 were injured in auto accidents, 12 had respiratory problems, 21 were injured in their homes, and 34 had other medical problems. What percent of these patients had respiratory problems?

A. 10%
B. 12%
C. 15%
D. 18%

26. The ratio of female to male nurses in a hospital is 9:1. If 144 nurses are female, how many nurses are male?

A. 12 nurses are male
B. 14 nurses are male
C. 16 nurses are male
D. 18 nurses are male

27. A committee studying an economic issue includes 3 state legislators, 6 state employees, and several members of the public. If one person is selected at random from the committee, the probability that the person will be a state legislator is $^1/_5$. How many of the members of the committee are members of the public?

A. 5 members
B. 6 members
C. 8 members
D. 9 members

28. At Pleasantville College, the ratio of female to male students is exactly 5 to 4. Which option below could be the number of students at the college?

A. 8,200 students
B. 2,955 students
C. 3,500 students
D. 3,105 students

29. The perimeter of a rectangle is 24 inches, and the ratio of the length to the width is 2:1. What is the area of the rectangle?

A. 60 square inches
B. 18 square inches
C. 32 square inches
D. 48 square inches

30. The three teams with the best records in the division are the Bulldogs, the Rangers, and the Statesmen. The Bulldogs have won 9 games and lost 3. The Rangers have won 10 games and lost 2. The Statesmen have also won 10 games and lost 2. Each team has 1 game left before the playoffs. The Bulldogs will be playing the Black Sox, and the Rangers will be playing the Statesmen. The team with the best record will win a spot in the playoffs. Which of the following statements is true?

A. The Statesmen will be in the playoffs.
B. The Bulldogs will not be in the playoffs.
C. The Rangers will not be in the playoffs.
D. The Statesmen will not be in the playoffs.

31. Brian pays 15% of his gross salary in taxes. If he pays $7,800 in taxes, what is his gross salary?

A. $52,000
B. $48,000
C. $49,000
D. $56,000

32. In her retirement accounts, Janet has invested $40,000 in stocks and $65,000 in bonds. If she wants to rebalance her accounts so that 70% of her investments are in stocks, how much money will she have to move from bonds?

A. $33,500
B. $35,000
C. $37,500
D. $40,000

33. Eight identical machines can produce 96 parts per minute. How many parts could 12 of these machines produce in 3 minutes?

A. 144 parts
B. 288 parts
C. 256 parts
D. 432 parts

34. There are 3 more men than women on the board of directors of the Big Box Retail Company. The board has 13 members. How many are women?

A. 3 are women
B. 4 are women
C. 5 are women
D. 6 are women

35. Emma borrowed a total of $1,200 with simple interest. She took the loan for as many years as the rate of interest. If she paid $432 in interest at the end of the loan period, what was the rate of simple interest on the loan?

A. 5%
B. 15%
C. 9%
D. 6%

36. In mathematics class, you have taken five tests and your average test grade is 91%. On the next test, your grade is 78%. What is your new test average?

A. 84.5

B. 90.5

C. 87.5

D. 88.8

37. Jorge and his younger sister Alicia have ages that combine to a total of 42. If their ages are separated by eight years, how old is Alicia?

A. 25 years old

B. 32 years old

C. 17 years old

D. 11 years old

38. You are making a budget for your money very carefully. Buying a smoothie each day costs $3.59 during the week and $3.99 on weekends. How much does your weekly budget allow, if you have a smoothie each workday and one day on the weekend?

A. $22.74

B. $23.54

C. $23.94

D. $21.94

39. Four out of twenty-eight students in your class must go to summer school. What is the ratio of the classmates who do not go to summer school in lowest terms?

A. 6/7

B. 1/7

C. 4/7

D. 3/7

40. Gourmet cookies are regularly priced at 89 cents each. Approximately how much is each cookie if one and a half dozen cookies sell for $12.89?

A. 65 cents

B. 82 cents

C. 72 cents

D. 80 cents

41. Which of the following is not an integer?

A. 0

B. 1

C. -45

D. All the answer choices are integers

42. Subtracting a negative number is the same as adding a ____________ number.

A. Positive

B. Negative

C. Zero

D. Irregular

43. What are the factors of 128?

A. 2

B. 2, 64

C. 2, 4, 8, 16

D. 1, 2, 4, 8, 16, 32, 64, 128

44. What are the two even prime numbers?

A. 0, 2

B. -2, 2

C. Cannot answer with the information given

D. There is only one even prime number

45. What are the prime factors of 128?

A. 2

B. 2, 3

C. 2, 4, 8, 16, 32, 64

D. Cannot answer with the information given

46. What is the proper "name" for the following: $[(52 + 25) + 3] / 58x$

A. An equation

B. An expression

C. A polynomial

D. An exponent

47. What is the greatest common factor (GCF) of 16 and 38?

A. 2

B. 16

C. 19

D. Cannot determine with the information given

48. What is the least common multiple (LCM) of 5 and 8?

A. 13

B. 40

C. 80

D. Cannot determine with the information given

49. What is the value of 7! ?

A. 127

B. 3,490

C. 5,040

D. 12,340

50. Which of the following is an irrational number?

A. $\sqrt{4}$

B. $\sqrt{9}$

C. $\sqrt{17}$

D. All the above

TEST 7. MATHEMATICS KNOWLEDGE PRACTICE TEST

For each question, select the best answer. To give you plenty of opportunities to practice, this set has more questions than in the actual test.

1. Find the median in the following series of numbers: 80, 78, 73, 69, 100
   A. 69
   B. 73
   C. 78
   D. 80

2. Solve the following equation: $x = \sqrt{11 \times 44}$
   A. $x = 36$
   B. $x = 24$
   C. $x = 18$
   D. $x = 22$

3. Which number below is the largest?
   A. -345
   B. 42
   C. -17
   D. 3^4

4. Which number below is a prime number?
   A. 81
   B. 49
   C. 59
   D. 77

5. The area of a triangle equals one-half the base times the height. Which equation below shows the correct way to calculate the area of a triangle that has a base of 6 and a height of 9?
   A. $(6+9)/2$
   B. $\frac{1}{2}(6+9)$
   C. $2(6 \times 9)$
   D. $\frac{(6 * 9)}{2}$

6. Evaluate (i.e., calculate the value of) the following expression: $2 + 6 \times 3 \times (3 \times 4)^2 + 1$
   A. 2,595
   B. 5,185
   C. 3,456
   D. 6,464

7. If $x \geq 9$, what is a possible value of x?
   A. 2^3
   B. 9
   C. -34
   D. 8.5

8. Solve the following equation: $x = (-9) \times (-9)$
   A. $x = 18$
   B. $x = 0$
   C. $x = 81$
   D. $x = -81$

9. Solve the following equation: $x = a^2 \times a^3$

   A. $x = a$

   B. $x = a^5$

   C. $x = 1$

   D. $x = 0$

10. What is the smallest possible integer value of x in the following equation: $x > 3^2 - 4$

   A. 3

   B. 5

   C. 6

   D. 7

11. Carmen has a box that is 18 inches long, 12 inches wide, and 14 inches high. What is the volume of the box?

   A. 44 cubic inches

   B. 3,024 cubic inches

   C. 216 cubic inches

   D. 168 cubic inches

12. The lines in the diagram below are ________.

   A. Parallel

   B. Perpendicular

   C. Acute

   D. Obtuse

13. Find the area of the shape below.

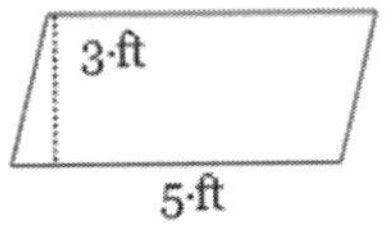

   A. 16 square feet

   B. 7.5 square feet

   C. 15 square feet

   D. 30 square feet

14. Which option below is equivalent to the following equation: $\frac{5}{mp} \div \frac{p}{4}$

   A. $\frac{5p}{4mp}$

   B. $\frac{20}{mp^2}$

   C. $\frac{20mp}{4mp}$

   D. $\frac{4m}{5p}$

15. What is the approximate area of the portion of the square shown that is not covered by the circle?

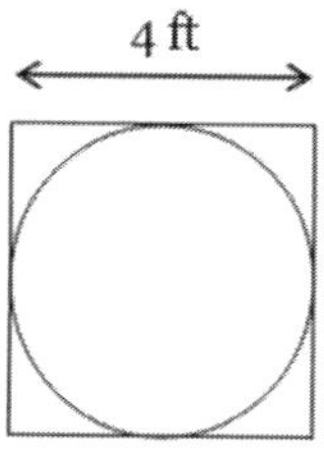

A. 4.14 square feet
B. 3.44 square feet
C. 6.25 square feet
D. 5.12 square feet

16. Lourdes rolls a pair of 6-sided dice. What is the probability that the result (the sum of the numbers on the top of the dice) will equal 10?

A. $^1/_{36}$
B. $^2/_{36}$
C. $^3/_{36}$
D. $^4/_{36}$

17. What option below shows the factorial of 5?

A. 25
B. 5 and 1
C. 120
D. 125

18. Which digit is in the thousandths place in the following number: 1,234.567

A. 1
B. 2
C. 6
D. 7

19. What is the mode in the following set of numbers: 4, 5, 4, 8, 10, 4, 6, 7

A. 6
B. 4
C. 8
D. 7

20. Which number below is a perfect square?

A. 5
B. 15
C. 49
D. 50

21. A rectangle's length is three times its width. The area of the rectangle is 48 square feet. How long are the sides?

A. Length = 12, width = 4
B. Length = 15, width = 5
C. Length = 18, width = 6
D. Length = 24, width = 8

22. Which option below shows the prime factorization of 24?

A. $24 = 8 \times 3$
B. $24 = 2 \times 2 \times 2 \times 3$
C. $24 = 6 \times 4$
D. $24 = 12 \times 2$

23. x is a positive integer. Dividing x by a positive number less than 1 will yield ________.

A. A number greater than x

B. A number less than x

C. A negative number

D. An irrational number

24. Solve the following equation: $x = 8 - (-3)$

A. $x = 5$

B. $x = -5$

C. $x = 11$

D. $x = -11$

25. Which statement below is true?

A. The square of a number is always less than the number.

B. The square of a number may be either positive or negative.

C. The square of a number is always a positive number.

D. The square of a number is always greater than the number.

26. Solve for r in the following equation: $p = 2r + 3$

A. $r = 2p - 3$

B. $r = p + 6$

C. $r = \frac{p+3}{2}$

D. $r = \frac{p-3}{2}$

27. What is the least common multiple of 8 and 10?

A. 80

B. 40

C. 18

D. 72

28. Solve the following equation: $x = -12 \div -3$

A. $x = -4$

B. $x = -15$

C. $x = 9$

D. $x = 4$

29. Convert the improper fraction $\frac{17}{6}$ to a mixed number.

A. $2\frac{5}{6}$

B. $3\frac{1}{6}$

C. $\frac{6}{17}$

D. $3\frac{5}{6}$

30. What value of q is a solution to the following equation: $130 = q(-13)$

A. 10

B. −10

C. 1

D. 10^2

31. Find the value of $a^2 + 6b$ when $a = 3$ and $b = 0.5$

A. 12

B. 6

C. 9

D. 15

32. Which of the following is equal to half a billion?

A. 50,000,000
B. 500,000,000
C. 500,000
D. 50,000,000,000

33. If the radius of the circle in the diagram below is 4 inches, what is the perimeter of the square?

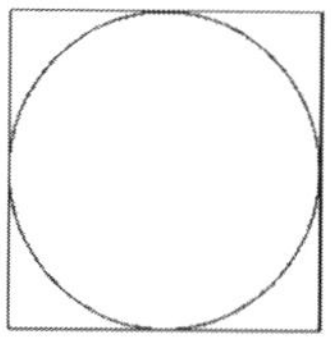

A. 24 inches
B. 32 inches
C. 64 inches
D. 96 inches

34. Solve for x in the following equation: $x = \frac{3}{4} \times \frac{7}{8}$

A. $x = \frac{7}{8}$
B. $x = \frac{9}{8}$
C. $x = \frac{10}{12}$
D. $x = \frac{21}{32}$

35. Which answer represents the relationship between x and y in the below table?

x	y
0	7
3	13
5	17
7	21
8	23

A. $y = x + 7$
B. $y = 4x + 1$
C. $y = 2x + 10$
D. $y = 2x + 7$

36. What is the perimeter of the figure shown?

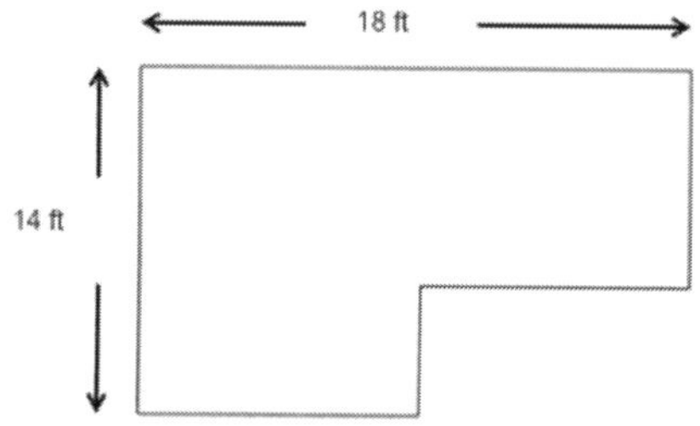

A. 64 feet
B. 72 feet
C. 84 feet
D. 96 feet

37. What exponent should replace the question mark in the following equation: $15{,}200 = 1.52 * 10^{?}$

A. 2
B. 3
C. 4
D. 5

38. Which option below is equal to 0.0065?

A. 6.5×10^{-2}
B. 6.5×10^{-3}
C. 6.5×10^{-4}
D. 6.5×10^{-5}

39. What is the greatest common factor of 48 and 64?

A. 4
B. 8
C. 16
D. 32

40. Find the value of the expression $x^2 + y^3$ when $x = -3$ and $y = -5$.

A. −116
B. 134
C. −134
D. 116

41. A square with 8-inch sides is cut into smaller squares with 1-inch sides. How many of the smaller squares can be made?

A. 8 squares
B. 16 squares
C. 24 squares
D. 64 squares

42. Which statement below is true?

A. 2 is the only even prime number.
B. The largest prime number less than 100 is 89.
C. The greatest common factor of 24 and 42 is 4.
D. The least common multiple of 8 and 6 is 48.

43. Which option below falls between $^2/_3$ and $^3/_4$?

A. $^3/_5$
B. $^4/_5$
C. $^7/_{10}$
D. $^5/_8$

44. What is the value of the following expression if $a = 10$ and $b = -4$: $\sqrt{2a + b^2}$?

A. 6
B. 7
C. 8
D. 9

45. Kevin has his glucose levels checked monthly. The results are below:

January	February	March	April	May	June	July
98	102	88	86	110	92	90

In which month was his glucose level equal to the median level over the 7-month period?

A. January
B. March
C. April
D. June

46. Mark will randomly choose 2 different letters from the word MINUTE. If the first letter he chooses is an M or an N, what is the probability that the next letter he chooses will be an M or an N?

A. $^1/_4$

B. $^1/_5$

C. $^1/_6$

D. $^1/_3$

47. Which option below is equivalent to 60% of 90?

A. 0.6×90

B. $90 \div 0.69$

C. $^3/_5$

D. $^2/_3$

48. Which option below is equivalent to 1.34×10^5?

A. 134,000

B. 13,400

C. 1,340,000

D. 13,400,000

49. If $a = -4$, what is the value of $a^3 - a - 2$?

A. 66

B. −62

C. 68

D. −66

50. Which expression below is not equivalent to the others?

A. $4^2 \times 5 \times 3$

B. $2^3 \times 15 \times 2$

C. $3^3 \times 5 \times 2$

D. $2^3 \times 10 \times 3$

TEST 8. WORD KNOWLEDGE PRACTICE TEST

For each question, select the answer that best fills in the blank. To give you plenty of opportunities to practice, this set has more questions than in the actual test.

1. The word <u>expansive</u> most closely means ________:
   A. costly
   B. vast
   C. sensible
   D. competitive

2. The word <u>credible</u> most closely means ________:
   A. enthusiastic
   B. dishonest
   C. reliable
   D. professional

3. The word <u>devastation</u> most closely means ________:
   A. Continuity
   B. Restoration
   C. Clarity
   D. Destruction

4. The word <u>vague</u> most closely means ________:
   A. Unclear
   B. Specific
   C. Pessimistic
   D. Gloomy

5. The word <u>irreverent</u> most closely means ________:
   A. Religious
   B. Disrespectful
   C. Humorous
   D. Boring

6. The word <u>aversion</u> most closely means ________:
   A. Attraction
   B. Inclination
   C. Optimism
   D. Distaste

7. The word <u>laborious</u> most closely means ________:
   A. Difficult
   B. Laid-back
   C. Noisy
   D. Lonely

8. The word <u>interminable</u> most closely means ________:
   A. Dull
   B. Valuable
   C. Endless
   D. Rushed

9. The word achromatic most closely means ________:
    A. Time-related
    B. Non-romantic
    C. Without color
    D. Before history

10. The word cursory most closely means ________:
    A. Ungodly
    B. Rapid
    C. Not smooth
    D. In a circular motion

11. The word hearsay most closely means ________:
    A. Bovine
    B. Unverified secondhand information
    C. Taking place in arid regions
    D. Communicative

12. The word magnanimous most closely means ________:
    A. Latin
    B. Large quantities of liquid
    C. Forgiving
    D. Antipathy

13. The word terrestrial most closely means ________:
    A. Of the earth
    B. Ordinary
    C. Existing in a "miniature" environment
    D. Foreign

14. The word nonchalant most closely means ________:
    A. Magnanimous
    B. Hurried
    C. Indifferent
    D. Positive

15. The word palpable most closely means ________:
    A. Tangible
    B. Unconcerned
    C. Capable of being manipulated
    D. Easygoing

16. The word daub most closely means ________:
    A. Suave
    B. Plaster
    C. Heinous
    D. Muddy

17. The word distend most closely means ________:
    A. Enforce
    B. Soften
    C. Swell
    D. Indemnify

18. The word gaffe most closely means ________:
    A. Taciturn
    B. Mistake
    C. Fishlike
    D. Cane

19. The word papal most closely means ________:
    A. Lightweight
    B. Regal
    C. Downtrodden
    D. Of or relating to leader of Catholicism

20. The word Pernicious most closely means ________:
    A. Harmful
    B. Fable
    C. Mossy
    D. Cold

21. The word hoopla most closely means ________:
    A. Cavernous
    B. Heinous
    C. Commotion
    D. Sweet

22. The word doctrinaire most closely means ________:
    A. Negative
    B. Dogmatic
    C. Insipid
    D. Diffident

23. The word plumose most closely means ________:
    A. Shy
    B. Sly
    C. Furry
    D. Feathery

24. The word supplant most closely means ________:
    A. Amplify
    B. Clarify
    C. Uproot
    D. Stabilize

25. The word vagary most closely means ________:
    A. Limitless
    B. Wispy
    C. Capable
    D. Caprice

26. The word calamari most closely means ________:
    A. Shamu
    B. Digit
    C. Squid
    D. Matrimony

27. The word lethargic most closely means ________:
    A. Apathetic
    B. Cozy
    C. Bouncy
    D. Deadly

28. The word teensy most closely means ________:
    A. Adolescent
    B. Immature
    C. Tiny
    D. Feminine

29. The word bland most closely means ________:
    A. Common
    B. Solid
    C. Rakish
    D. Dull

30. The word fulsome most closely means ________:
    A. Cruel
    B. Copious
    C. Graceful
    D. Handy

31. The word nomenclature most closely means ________:
    A. Jealousy
    B. Name
    C. Steadiness
    D. Aggression

32. The word reluctant most closely means ________:
    A. Casual
    B. Timid
    C. Intense
    D. Unwilling

33. The word shabby most closely means ________:
    A. Feline
    B. Sophisticated
    C. Dilapidated
    D. Miserly

34. The word Ubiquitous most closely means ________:
    A. Aspiring
    B. Burning
    C. Omnipresent
    D. Contemptuous

35. The word defer most closely means ________:
    A. Remove
    B. Cancel
    C. Quiet
    D. Delay

36. The soldier was undaunted by the challenge ahead.
    A. Fearful
    B. Discouraged
    C. Brave
    D. Confused

37. The judge's decision was immutable, and there was no chance of an appeal.
    A. Flexible
    B. Final
    C. Changeable
    D. Undecided

38. The witness gave a lucid description of the events.
    A. Vague
    B. Confused
    C. Clear
    D. Incomprehensible

39. The word enervate most nearly means:
    A. Energize
    B. Weaken
    C. Strengthen
    D. Support

40. The word cognizant most nearly means:
    A. Unaware
    B. Aware
    C. Distracted
    D. Forgetful

41. The word sagacious most nearly means:
    A. Foolish
    B. Wise
    C. Reckless
    D. Uninformed

42. The word benign most nearly means:
    A. Harmful
    B. Gentle
    C. Aggressive
    D. Severe

43. His tenacity in pursuing his goals impressed everyone.
    A. Determination
    B. Laziness
    C. Hesitation
    D. Confusion

44. The manager had to mitigate the negative effects of the company's policy change.
    A. Worsen
    B. Reduce
    C. Emphasize
    D. Hide

45. The manager was known for his meticulous attention to detail.
    A. Careless
    B. Lazy
    C. Precise
    D. Forgetful

46. The word assiduous most nearly means:
    A. Lazy
    B. Diligent
    C. Indifferent
    D. Inconsistent

47. The word most opposite in meaning to sagacious is:
    A. Foolish
    B. Wise
    C. Prudent
    D. Thoughtful

48. The word ominous most nearly means:
    A. Threatening
    B. Optimistic
    C. Cheerful
    D. Encouraging

49. The word capacious most nearly means:
    A. Small
    B. Spacious
    C. Crowded
    D. Cramped

50. The word precarious most nearly means:
    A. Secure
    B. Dangerous
    C. Stable
    D. Reliable

Test 9. Reading Comprehension Practice Test

For each passage, read the passage, and then select the best answer for each of the 5 related questions.

Questions 1-5 are based on the following passage:

His obsessions and wealth led him to own the most magnificent collection of specimens ever compiled by a single man. He had committed to and invested in the widespread slaughter of some of the world's most exotic bird species. His greatest passion had destroyed nature's innocence in a maelstrom of cruel arrogance. He had robbed nature of its riches, while filling his coffers with its spoils. The ravenous hysteria shown for the rarest of feathers to adorn the hats of the fashionable rich showed no signs of abating. Parisian women were festooned in the feathers, sometimes the whole skins, of snowy egrets, birds of paradise, ospreys, and toucans. Concerning nature, humankind is rarely content to witness its spectacle without desiring to possess it.

1. What is the main topic of this paragraph?

   A) Fashion
   B) Nature
   C) Death
   D) Wealth

2. What is the writer's main point about this topic?

   A) It is more important than fashion
   B) Humans can destroy it
   C) It can cause man to be greedy
   D) All the above

3. The final sentence is ________:

   A) The main topic of the paragraph
   B) A sentence that contrasts the main topic to another topic
   C) The beginning of a new topic unrelated to the paragraph
   D) None of the above

4. In the passage, the writer refers to the destruction of "nature's innocence". What literary device is the writer using here?

   A) Personification
   B) Alliteration
   C) Metaphor
   D) Simile

5. The writer views the destruction of the natural world:

   A) As inevitable
   B) Sarcastically
   C) Cheerfully
   D) None of the above

Questions 6-10 are based on the following passage:

The working-class gambler is the lowest and vilest player ever to walk among God's blessed Earth. Frequenting the "hells" of London by night, the nefarious gamester sustains on a meagre diet of bread and quarter gin measures doled out by the degenerate owners of gambling dens to encourage play. The more desperate of these characters bully the weaker ones to play and prey upon them like hungry parasites when their luck has run its course. The impoverished gamblers—profligates and blackguards at heart—blight our great city with the moral disease of the destitute class. In contrast, the aristocratic gambler begins play at the respectable hour of three and partakes in only the most superior forms of gambling. With morality on his side, he chooses to squander inherited wealth with the graceful recklessness expected of his class. The pleasant, well-lit comfort of his surroundings is reflected in his affable, gentlemanly countenance. But the working-class gambler and the aristocratic gambler share a common destiny: being left in poverty and despair.

6. What is the main topic of the paragraph?

A) Social class
B) Gambling
C) Morality
D) None of the above

7. What is the main point of the paragraph?

A) Aristocratic gamblers are immoral
B) There are more working-class gamblers than aristocratic gamblers
C) Gambling is beneficial for society
D) All gamblers will become poor in the end if they continue to gamble

8. What is the purpose of comparing the "working-class" to the "aristocratic gambler"?

A) To emphasize the differences between the working class and the aristocracy
B) To show the immorality of working-class gamblers
C) To highlight how, regardless of class (i.e., rich or poor), all gamblers end up destitute
D) None of the above

9. What word best describes the writer's tone in this paragraph?

A) Ironic
B) Uncertain
C) Resentful
D) None of the above

10. The writer refers to some gamblers as being "like hungry parasites." What literary device is the writer using?

A) Onomatopoeia
B) Foreshadowing
C) Metaphor
D) Simile

Questions 11-15 are based on the following passage:

What had I been waiting for—a sign from above, a notice from the bank, a list of instructions telling me step-by-step how to live life? I had thought for all those years that something or someone would tell me that I had been living. But in that moment, while staring fear directly in the face, I realized that something else had gone. I'd gone. Me. I'd simply disappeared. How had I spent forty-five years on this planet and never even realized? I couldn't have been living because I was already too busy dying. I understood this reality now with perfect clarity. My death was not an event to be feared; it had already begun. I was already dead.

11. What is the main topic of the paragraph?

A) Death
B) Fear
C) Love
D) Growing up

12. What is the writer's main point about this topic?

A) It is something to be resented
B) It is something to be feared
C) It is something to be laughed at
D) None of the above

13. Which literary device does the writer use to present his/her position?

A) Rhetorical questions
B) Metaphors
C) Pathetic fallacy
D) Similes

14. The writer views death as ________:

A) Inevitable
B) Transient
C) Joyful
D) None of the above

15. What word best describes the writer's tone in this paragraph?

A) Didactic
B) Mocking
C) Light-hearted
D) Poignant

Questions 16-20 are based on the following passage:

Technology has fundamentally transformed the landscape of modern warfare, making conflicts faster, more precise, and far-reaching. From unmanned aerial vehicles (UAVs) to advanced cyber warfare tools, innovations in technology have shifted the focus from brute force to intelligence and precision. For instance, drones have become a staple in military operations, offering the ability to conduct surveillance and targeted strikes without endangering human operators. Additionally, artificial intelligence (AI) is increasingly used to analyze vast amounts of data, predict enemy movements, and even assist in autonomous decision-making during missions.

Another critical area of technological advancement is cyber warfare. Modern militaries now engage in battles not only on land, sea, and air but also in the digital realm. Cyberattacks can disrupt communication systems, disable critical infrastructure, and steal classified information. The introduction of space-based technologies, like satellite-guided missiles and GPS, further demonstrates how technology has expanded the battlefield into new domains.

However, while technology offers unparalleled advantages, it also introduces new vulnerabilities. Reliance on interconnected systems makes militaries susceptible to cyberattacks and technical malfunctions. As a result, defense strategies increasingly emphasize securing digital networks alongside traditional military tactics.

In conclusion, technology's role in modern warfare is both a game-changer and a challenge. While it allows militaries to fight smarter and with greater precision, it also necessitates new strategies to mitigate emerging risks in an increasingly interconnected world.

16. What is the main idea of the passage?

A) Technology has introduced vulnerabilities to modern militaries.
B) Technology has made modern warfare faster and more precise but also more vulnerable.
C) Modern warfare has moved away from traditional strategies entirely.
D) Cyber warfare is now the most critical aspect of modern conflict.

17. Which of the following is an example of a technological advancement mentioned in the passage?

A) Submarine warfare
B) Satellite-guided missiles
C) Naval blockades
D) Radio communication

18. What can be inferred about the future of warfare based on the passage?

A) Traditional military strategies will no longer be relevant.
B) Militaries will rely entirely on artificial intelligence for decision-making.
C) Securing digital networks will become increasingly critical.
D) Cyber warfare will replace all other forms of conflict.

19. Based on the passage, what does the term "autonomous decision-making" most likely mean?

A) Decisions made by commanders in real-time.
B) Decisions made by AI systems without human input.
C) Decisions based on past military campaigns.
D) Decisions influenced by political leaders.

20. What can be inferred about the challenges of cyber warfare?

A) Cyberattacks are only effective against small militaries.
B) Cyber warfare is less important than traditional battles.
C) Cyber warfare can cause significant disruptions without physical conflict.
D) Cyber warfare requires the use of satellite technology.

Questions 21-25 are based on the following passage:

The history of military aviation dates back to the early 20th century, when nations first recognized the strategic potential of flight in warfare. The Wright brothers' first powered flight in 1903 paved the way for the development of aircraft as military tools. By World War I, aviation played a critical role in reconnaissance, allowing armies to gather intelligence about enemy positions. Fighter planes were later introduced, leading to dogfights that became an iconic feature of the war.

Between the World Wars, advancements in technology revolutionized military aviation. Aircraft became faster, more durable, and capable of carrying heavier payloads, such as bombs. These innovations were prominently showcased during World War II, where airpower was a decisive factor. The introduction of long-range bombers, such as the B-29 Superfortress, and advanced fighters like the P-51 Mustang underscored the growing importance of aerial dominance.

In the post-war period, the Cold War spurred rapid advancements in jet propulsion, stealth technology, and missile systems. Military aviation evolved to include unmanned systems such as drones, which have become essential tools in modern warfare. Today, airpower remains a cornerstone of military strategy, enabling precision strikes, rapid troop deployments, and global surveillance.

From the fragile biplanes of World War I to modern stealth fighters, the history of military aviation reflects humanity's relentless drive to master the skies. Each technological leap has redefined the nature of warfare, making military aviation an enduring symbol of innovation and strategic power.

21. What is the main idea of the passage?

A) Military aviation originated during World War II.
B) Airpower has consistently redefined the nature of warfare.
C) Drones are the pinnacle of military aviation innovation.
D) The history of military aviation began with the Wright brothers' first flight.

22. Which of the following is an example of an aircraft introduced during World War II?

A) Wright Flyer
B) P-51 Mustang
C) F-22 Raptor
D) MQ-9 Reaper

23. What can be inferred about the impact of drones on modern military aviation?

A) They have replaced traditional aircraft entirely.
B) They have made airpower more efficient and less risky for pilots.
C) They were primarily developed during the Cold War.
D) They are less significant than manned fighter jets.

24. What does the phrase "relentless drive to master the skies" most likely mean?

A) Humanity's obsession with controlling the weather.
B) The continuous effort to improve airpower and aviation technology.
C) The need to build aircraft for civilian use.
D) The focus on defeating enemies in aerial combat.

25. What can be inferred about military aviation during the Cold War?

A) Aircraft technology stagnated due to a lack of conflict.
B) Stealth technology and jet propulsion became priorities.
C) Unmanned systems were abandoned in favor of manned aircraft.
D) Bombers were the only aircraft developed during this period.

TEST 10. PHYSICAL SCIENCE PRACTICE TEST

For each question, select the best answer. To give you plenty of opportunities to practice, this set has more questions than in the actual test.

1. What is NOT one of the ways that scientists differentiate between the concentric spherical layers of the earth?
   A. The material's state of matter
   B. The type of rocks
   C. Fossil records
   D. Chemical formation

2. What is the layer of the earth that extends approximately 40-60 miles down from the surface?
   A. Mantle/asthenosphere
   B. Outer core
   C. Onion
   D. Lithosphere

3. What causes the phenomena known as the Aurora Borealis?
   A. Weather patterns
   B. Charged particles from the sun interacting with the earth's magnetic field
   C. High levels of iron and silicon in the "axis" of the earth
   D. The igneous rock throughout earth's crust

4. Which of the following is NOT a type of sedimentary rock?
   A. Basalt
   B. Shale
   C. Coal
   D. Sandstone

5. How much more energetically powerful is a seismic event measuring 7 on the Richter scale compared to an event measuring 5 on the Richter scale?
   A. 4 times more energetically powerful
   B. 8 times more energetically powerful
   C. 1,000 times more energetically powerful
   D. 1,000,000 times more energetically powerful

6. In what atmospheric layer will you be traveling for the majority of the flight distance on a commercial airline flight from London, England to Frankfurt, Germany?
   A. Troposphere
   B. Stratosphere
   C. Mesosphere
   D. Exosphere

7. If you are driving through a thick layer of fog, at or near sea level, you are actually passing through what type of cloud?
   A. Cirrus
   B. Stratus
   C. Cumulonimbus
   D. Cumulus

8. Which of the following statements is true if an atom has an atomic number of 17?
   A. Its valence quantity is above the critical level
   B. It likely has an atomic mass that is less than 17
   C. It has 17 protons
   D. The sum of its protons and neutrons is at least 17

9. Which pair of physical properties is NOT matched with its appropriate SI symbol?
   A. Luminous intensity: Candela
   B. Electric current: Watt
   C. Pressure: Pascal
   D. Temperature: Kelvin

10. Using the common conversion factor between meters and feet, approximately how many miles would an 8K race be?
   A. 4 miles
   B. 4 1/2 miles
   C. 5 miles
   D. 6 1/2 miles

11. What is the composition of Neptune?
   A. Metamorphic rock with a thin atmosphere of methane
   B. A relatively tiny rock and mineral core with an enormous thick outer layer of gas
   C. An iron core covered with a thin layer of ice
   D. A solid, frozen methane and frozen water surface, a liquid water middle layer and a rock and mineral core.

12. The Asteroid Belt is located between which two planets?
   A. Venus and Earth
   B. Earth and Mars
   C. Mars and Jupiter
   D. Jupiter and Saturn

13. What is the correct order of the taxonomy that biologists have established to describe Kingdom Animalia (i.e., animals)?
   A. Phylum, order, class, family, genus, species
   B. Order, phylum, class, genus, family, species
   C. Phylum, class, order, genus, family, species
   D. Phylum, class, order, family, genus, species

14. Which of the following characteristic(s) do all organisms in kingdom Animalia share? (I) Cell walls, (II) ability to actively move on their own from place to place at least at some point in their life cycle, (III) a nervous system, (IV) a vertebrae.
   A. I and III
   B. II and III
   C. III only
   D. II only

15. What organism or category of organism is NOT included in Kingdom Fungi?
   A. Yeast
   B. Molds
   C. Sponges
   D. Mushrooms

16. What does the vertical axis of a taxonomy tree represent?
   A. Phyla
   B. Degree of specialization
   C. Degree of intelligence
   D. Time

17. What is the main goal of biological evolution?
   - A. To change an organism so that it is more likely to produce viable offspring
   - B. To create a "better" organism
   - C. To make the organism's "society" more complex
   - D. To produce new species of organisms

18. Nature "designs" organisms to improve through the process of natural ____________?
   - A. Selection
   - B. Programming
   - C. Randomness
   - D. Intention

19. What type of heavenly body is most likely to have a highly elliptical orbit that "slingshots" itself as it passes around the sun?
   - A. Quasars
   - B. Comets
   - C. Black holes
   - D. Asteroids

20. What are the phases of water?
   - A. Evaporation, condensation and sublimation
   - B. Solid, liquid and gas/vapor
   - C. A solute, a solvent and a solution
   - D. None of the above

21. In which era did "dinosaurs roam the earth"?
   - A. Paleozoic
   - B. Mesozoic
   - C. Cenozoic
   - D. Mesoproterozoic

22. When does geographic speciation occur?
   - A. When a population is separated by geographic change
   - B. When two species meet in a common place and interbreed
   - C. When one population drives another in the same geographic locale to extinction
   - D. When biomes lose diversity to human activity

23. What is the classification of rocks formed by the hardening of molten magma from deep within the earth?
   - A. Sedimentary
   - B. Metamorphic
   - C. Igneous
   - D. None of the above

24. How do fungi reproduce?
   - A. Sexually
   - B. Asexually
   - C. Sometimes A, sometimes B, sometimes both A and B
   - D. None of the above

25. Which brain lobe receives sensory information from the body?
   - A. Parietal
   - B. Temporal
   - C. Occipital
   - D. Frontal

26. What is the formula for determining the weight of an object (with g = gravitational acceleration, v = velocity, m = mass, and a = acceleration)?

A. $w = m \times v$

B. $w = m^2 \times a$

C. $w = m \times g$

D. $w = a \times g^2$

27. What is the acceleration of a train given the following facts: At time = 0, the train is moving at 43 miles/hour; at time = 3 hours 30 minutes, the train is moving at 64 miles/h.

A. 3 miles/h^2

B. 6 miles/h^2

C. 21 miles/h

D. 21 miles/h^2

28. Three different objects (choices A, B, and C) are dropped from three different heights. Which object has the most potential energy? Assume that the gravitational acceleration is a constant: 32.174 ft/s^2.

A. A 10-pound object dropped from 70 ft

B. A 58-pound object dropped from 4 yards

C. A 0.01-ton object dropped from 300 inches

D. All answer choices are equal

29. What is the unit for work, according to the International System of Units, the internationally recognized standard metric system?

A. Newton

B. Joule

C. Watt

D. Pascal

30. What is the "tradeoff" when using an inclined plane?

A. Less "output" force

B. Less distance the object must travel

C. Greater distance the object must travel

D. Both A and C

31. If you want to lift a very heavy load using a simple lever (in the form of a "teeter-totter") so that the lever arm becomes horizontal, what would be an effective way to accomplish your goal?

A. Increase the applied force

B. Move the fulcrum such that the resistance arm and the effort arm are balanced

C. Increase the height of the fulcrum

D. Move the fulcrum in the direction of the load to be lifted

32. What is the name of a simple machine that is designed to amplify force by increasing and/or decreasing the amount of torque applied at one of the interacting parts in its "system" of parts?

A. Gear

B. Pump

C. Vacuum

D. None of the above

33. What is the definition of mechanical advantage?

A. Force: Direction

B. Output Force: Input Force

C. Direction: Magnitude

D. All the above

34. If a pulley is an example of a simple machine, which of the following is always true?
    A. A pulley changes the magnitude of the applied force
    B. A pulley changes the direction of the applied force
    C. Both A and B
    D. Neither A nor B

35. How does a wedge create mechanical advantage?
    A. Amplifies applied force by transmitting it from a larger surface to a smaller surface
    B. Increasing the pressure along the length or the wedge
    C. Converting output force to applied force
    D. None of the above

36. What is the benefit of having well-greased automobile bearings?
    A. Increase mechanical advantage
    B. Reduces the coefficient of friction
    C. Increases the acceleration of the automobile
    D. None of the above

37. Which object (O) has the least momentum?
    A. moves at 3 ft/h and weighs 1 ton
    B. moves at 5,000 ft/h and weighs 2 pounds
    C. moves at 1,000 ft/min and weighs 25 pounds
    D. moves at 300,000 ft/s and weighs 0.001 pounds

38. What "things" can change either the magnitude or the direction of a force?
    A. Inclined plane
    B. Lever
    C. Wedge
    D. All the above

39. Why do bicyclists use their low gear when going uphill?
    A. It increases the torque produced
    B. It decreases the number of revolutions of the pedals
    C. It reduces the need to buy lot of bicycle accessories
    D. It both increases the torque and allows you to move your legs slower

40. Which of the following "simple machines" is NOT an example of a wedge?
    A. Knife
    B. Axe
    C. Door stop
    D. All the examples listed ARE wedges

41. A 5 kg mass traveling at 15 m/s needs to be stopped in 3 seconds. How much force is required?
    A. 15 newtons
    B. 75 newtons
    C. 25 newtons
    D. 9 newtons

42. Accelerating a 3 kg mass to 12 m/s in 4.5 seconds requires how much force?
    A. 15 newtons
    B. 0.9 newtons
    C. 18 newtons
    D. 8 newtons

43. Find your potential energy if you are on a roller coaster at the top of a 25-meter drop and if your mass is 95 kg.
    A. 23,275 joules
    B. 37.25 joules
    C. 242 joules
    D. 931 joules

44. How much kinetic energy is required to launch a 1.3 kg toy rocket to a height of 1500 meters?
    A. 11,307 joules
    B. 14,700 joules
    C. 24,305 joules
    D. 19,110 joules

45. Find the kinetic energy for your car traveling at 80 km/h. Assume your car is 1400 kg in mass.
    A. 1,097,600 joules
    B. 172,840 joules
    C. 345,679 joules
    D. 7,856 joules

46. What do you call an atom that has an electron stripped off or added?
    A. Ionized
    B. Charged
    C. One fundamental unit—1 amu
    D. Neutral

47. What happens when electricity passes through an inductor in a circuit board?
    A. A temperature drop or rise
    B. Rectification
    C. A magnetic field is generated
    D. Two parallel plates with a dielectric (non-conducting) material are "activated"

48. What is the main difference between alternating current (AC) and direct current (DC)?
    A. The presence or absence of a magnetic field
    B. Ionization
    C. The direction electrons move
    D. All the above

49. Why is copper the metal most commonly used to make wires that carry electricity?
    A. Low resistance, flexible, common
    B. High resistance, rigid, common
    C. Low resistance, flexible, rare
    D. None of the above

50. How does a resistor change the current and voltage in an electrical circuit?
    A. Accumulating charge
    B. Operating as a switch for the passage of electrons
    C. Generating a magnetic field
    D. Reducing the rate that electrons can flow

51. In a circuit, the greater the resistance, the lower the ______________.
    A. Voltage
    B. Diodes
    C. Current
    D. Ohms

52. What is the "driving force" behind the flow of current?
   A. Voltage
   B. Amps
   C. Resistance
   D. None of the above

53. If you have a 12-volt battery that is powering a circuit with a specified internal resistance of 120 ohms, how many amps is the circuit drawing?
   A. 0.064 amps
   B. 0.10 amps
   C. 0.12 amps
   D. 10 amps

54. What is the measurement of how much electrical current is moving through a circuit?
   A. Volt (voltage)
   B. Amp (ampere)
   C. Bandwidth
   D. Alternating current (AC)

55. What is the measurement of the EMF exerted on electrical charge in a circuit?
   A. Volt (voltage)
   B. Amp (ampere)
   C. Current
   D. Direct current (DC)

56. If Ohm's law says that $V = I \times R$, what current results when a 6-volt battery is connected to a circuit with a resistance of 150 ohms?
   A. 900 amps
   B. 25 amps
   C. 40 milliamperes
   D. 156 amps

57. If Ohm's law says that $V = I \times R$, what voltage must be applied to create a 10.5 ampere current in a circuit with a resistance of 50 ohms?
   A. 4.8 volts
   B. 525 volts
   C. 0.21 volts
   D. 156 volts

58. If $V = I \times R$, what is the resistance of a circuit with a 7.5 ampere current and a 6-volt power source?
   A. 0.80 ohms
   B. 45 ohms
   C. 800 ohms
   D. ohms

59. Current in a wire is a result of ________ mobility.
   A. Nuclear
   B. Neutron
   C. Proton
   D. Electron

60. Electric charge motion is a result of ________.
   A. Resistance
   B. Voltage
   C. Heat
   D. Electron

Chapter 14: Practice Test Answers

Test 1. Instrument Comprehension Practice Test Answers

1. A	11. D	21. A
2. D	12. C	22. D
3. B	13. A	23. B
4. A	14. D	24. C
5. C	15. B	25. C
6. B	16. C	
7. D	17. A	
8. C	18. D	
9. A	19. B	
10. B	20. C	

Test 2. Verbal Analogies Practice Test Answers

1. B	11. D	21. D	31. A	41. B
2. C	12. C	22. C	32. B	42. B
3. A	13. C	23. A	33. A	43. C
4. B	14. A	24. D	34. C	44. A
5. D	15. B	25. B	35. D	45. B
6. A	16. D	26. D	36. B	46. C
7. A	17. B	27. A	37. A	47. D
8. C	18. A	28. B	38. C	48. A
9. B	19. B	29. D	39. D	49. D
10. A	20. C	30. C	40. A	50. B

Test 3. Table Reading Practice Test Answers

1. A	11. B	21. A	31. C
2. D	12. C	22. D	32. B
3. C	13. B	23. E	33. C
4. B	14. E	24. B	34. E
5. A	15. C	25. A	35. B
6. E	16. D	26. B	36. E
7. C	17. A	27. C	37. A
8. E	18. B	28. E	38. D
9. A	19. C	29. E	39. B
10. D	20. E	30. A	40. C

Test 4. Aviation Information Practice Test Answers

1. E	11. C
2. D	12. A
3. A	13. E
4. C	14. D
5. C	15. D
6. B	16. B
7. A	17. C
8. B	18. A
9. E	19. C
10. D	20. D

Test 5. Block Counting Practice Test Answers

Group 1

1. C
2. A
3. B
4. E
5. B

Group 2

1. D
2. E
3. B
4. A
5. C

Group 3

1. B
2. C
3. A
4. C
5. C

Group 4

1. A
2. C
3. D
4. E
5. E

Group 5

1. C
2. A
3. D
4. E
5. A

Group 6

1. E
2. B
3. D
4. A
5. E

Test 6. Arithmetic Reasoning Practice Test Answers

1. **Answer:** C. 12%

   Rationale: To find the percent increase, you first need to know the amount of the increase. Enrollment went from 3,450 in 2010 to 3,864 in 2015. This change shows an increase of 414 students. Now, to find the percent of the increase, divide the amount of the increase by the original amount:

   $$414 \div 3{,}450 = 0.12$$

   To convert a decimal to a percent, move the decimal point two places to the right:

   $$0.12 = 12\%$$

   When a question instead asks for the percent of a decrease, you also divide the amount of the decrease by the original amount.

2. **Answer:** A. 16 gallons

   Rationale: Amy drives her car until the gas tank is $^1/_8$ full. This means that it is $^7/_8$ empty. She fills it by adding 14 gallons. In other words, 14 gallons is $^7/_8$ of the tank's capacity. Draw a simple diagram to represent the gas tank.

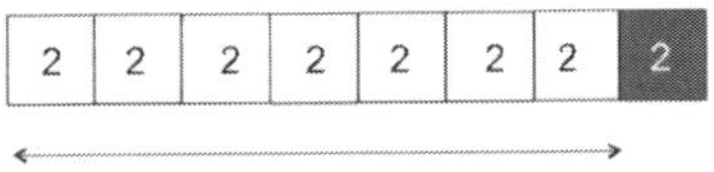

   $$^7/_8 = 14$$

   Each eighth of the tank is 2 gallons. Thus, the capacity of the tank is 2×8, or 16 gallons.

3. **Answer:** C. $310

   Rationale: The individual costs of all these items can be expressed in terms of the cost of the ream of paper. Use x to represent the cost of a ream of paper. The flash drive costs three times as much as the ream of paper, so it costs $3x$. The textbook costs three times as much as the flash drive, so it costs $9x$. The printer cartridge costs twice as much as the textbook, so it costs $18x$. Those values can then be plugged into the follow equation:

   $$x + 3x + 9x + 18x = 31x$$

   The ream of paper costs $10, so $31x$ (the total cost) is $310.

4. **Answer:** A. 84 girls

   Rationale: First, find the number of girls in the class. First, convert 52% to a decimal by moving the decimal point two places to the left

   $$52\% = .52$$

   Now, multiply .52 times the number of students in the class:

   $$.52 \times 350 = 182$$

   Of the 182 girls, 98 plan to go to college, so 84 do not plan to go to college.

   $$182 - 98 = 84$$

5. **Answer:** C. $300

   Rationale: If the phone was on sale at 30% off, the sale price was 70% of the original price. Therefore, the answer can be obtained by solving the following equation for x:

   $$\$210 = 70\% \text{ of } x$$

   where x is the original price of the phone. When you convert 70% to a decimal, you get the following equation:

$$\$210 = .70 \times x$$

To isolate x on one side of the equation, divide both sides of the equation by .70 to find the following:

$$x = \$300$$

6. **Answer:** A. 10
Rationale: The facts provided can be used to write the following equation:

$$7 + 45n = 15$$

First, subtract 7 from both sides of the equation to get the following equation:

$$45n = 8$$

Then, divide both sides of the equation by $\frac{4}{5}$. To divide by a fraction, invert the fraction so that $\frac{4}{5}$ becomes $\frac{5}{4}$, then multiply:

$$n = \frac{8}{1} \times \frac{5}{4}$$
$$n = \frac{40}{4}$$
$$n = 10$$

7. **Answer:** D. 108

Rationale: The ratio of the two numbers is 7:3. Therefore, the larger number is $\frac{7}{10}$ of 360 and the smaller number is $\frac{3}{10}$ of 360. To find the smaller number, multiply the smaller fraction by the total of the two numbers, as shown in the equation below:

$$\frac{3}{10} \times \frac{360}{1} = \frac{1{,}080}{10} = 108$$

8. **Answer:** A. 20 questions

Rationale: Alicia must have a score of 75% on a test with 80 questions. To find how many questions she must answer correctly, first convert 75% to a decimal by moving the decimal point two places to the left: 75% = .75. Now multiply 75 times 80:

$$.75 \times 80 = 60$$

Alicia must answer 60 questions correctly to pass the test, but the question asks how many questions she can miss. If she must answer 60 correctly, then she can miss 20 questions and still pass the test.

9. **Answer:** D. 0.72

Rationale: The simplest way to answer this question is to convert the fractions to decimals. To convert a fraction to a decimal, divide the numerator (the top number) by the denominator (the bottom number). The questions below show the conversions for answers A, B, and C:

$$^5/_8 = 0.625$$
$$^3/_5 = 0.6$$
$$^2/_3 = 0.67$$

The largest number is therefore 0.72, answer D.

10. **Answer:** A. 11 pages

Rationale: If Charles wrote an average of 7 pages per day for four days, he wrote a total of 28 pages. He wrote a total of 17 pages on the first three days, so he must have written 11 pages on the fourth day.

11. **Answer:** B. 10 miles
Rationale: If you made a simple map with the three cities, it would look like the following diagram:

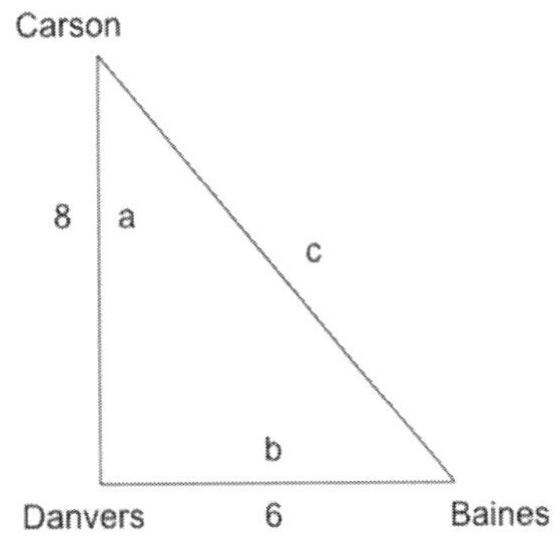

The preceding figure is a right triangle. The longest side of a right triangle is called the hypotenuse. The two legs of the triangle are labeled a and b. The hypotenuse is labeled c. You can find the length of the hypotenuse (the distance between Caron and Baines) with the following equation:

$$a^2 + b^2 = c^2$$

In this case, the equation would be calculated as follows:

$$\begin{aligned} 8^2 + 6^2 &= c^2 \\ 64 + 36 &= c^2 \\ 100 &= c^2 \end{aligned}$$

To find c, ask the following question: What number times itself equals 100? The answer is 10.

12. **Answer:** A. $144

Rationale: When one person dropped out of the arrangement, the cost for the remaining three went up by $12 per person for a total of $36. This means that each person's share was originally $36. There were four people in the original arrangement, so the cost of the gift was 4 × $36, or $144.

Or alternatively, let $4x$ equal the original cost of the gift. If the number of shares decreases to 3, then the total cost is $3(x + 12)$. Those expressions must be equal, so:

$$\begin{aligned} 4 &= 3(x + 12) \\ 4x &= 3x + 36 \end{aligned}$$

Subtracting $3x$ from both sides, obtains the following:

$$x = 36$$

Then the original price of the gift was 4 times 36, or $144.

13. **Answer:** A. $(14 + 24) \times 8 \times 5$

Rationale: In the correct answer, $(14 + 24) \times 8 \times 5$, the hourly wages of Amanda and Oscar are first combined, and then this amount is multiplied by 8 hours in a day and 5 days in a week. One of the other choices, $14 + 24 \times 8 \times 5$, looks like the correct choice, but it is incorrect because the hourly wages must be combined before they can be multiplied by 8 and 5.

14. **Answer:** D. 130,500 people

Rationale: The population of Mariposa County in 2015 was 90% of its population in 2010. Convert 90% to a decimal by moving the decimal point two places to the left: 90% = .90. Now multiply .90 times 145,000 (the population in 2010), as shown in the following equation:

$$.90 * 145{,}000 = 130{,}500$$

15. **Answer:** C. $^{17}/_{24}$

Rationale: To add $^{1}/_{3}$ and $^{3}/_{8}$, you must find a common denominator. The simplest way to do this is to multiply the denominators: $3 \times 8 = 24$. Therefore, 24 is a common denominator. (This method will not always give you the <u>lowest</u> common denominator, but does in this case.)

Once you have found a common denominator, convert both fractions in the problem to equivalent fractions that have that same denominator. To do this conversion, multiply each fraction by an equivalent of 1, as shown in the following:

$$\frac{1}{3} \times \frac{8}{8} = \frac{8}{24}$$

$$\frac{3}{8} \times \frac{3}{8} = \frac{9}{24}$$

Now you can add $^{8}/_{24}$ and $^{9}/_{24}$ to solve the problem.

16. **Answer:** B. Serena is 7, Tom is 4

 Rationale: Use S to represent Serena's age. Tom is 3 years younger than Serena, so his age is S – 3. In 4 years, Tom will be twice as old as Serena was 3 years ago. Thus, you can write the equation:

$$Tom + 4 = 2(Serena - 3)$$

 Now substitute S for Serena and S – 3 for Tom:

$$(S - 3) + 4 = 2(S - 3)$$

 Simplify the equation, as shown in the following equation:

$$S + 1 = 2S - 6$$

 Subtract S from both sides of the equation:

$$1 = S - 6$$

 Add 6 to both sides of the equation. You get the following:

$$7 = S, \text{Serena's age}$$

$$4 = S - 3, \text{Tom's age}$$

17. **Answer:** B. 515 students

 Rationale: Marisol scored higher than 78% of the students who took the test. First, convert 78% to a decimal by moving the decimal point two places to the left: 78% = .78. Now multiply .78 times the number of students who took the test:

$$.78 \times 660 = 514.8 \text{ or } 515 \text{ students (must be whole numbers)}$$

18. **Answer:** D. $12
 Rationale: Use the information given to write an equation, as shown in the following equation:

$$530 = 40d + 50$$

 When you subtract 50 from both sides of the equation, you get the following:

$$480 = 40d$$

 Divide both sides of the equation by 40 to complete the equation:

$$12 = d, \text{Sam's hourly wage}$$

19. **Answer:** D. 80

 Rationale: If we call the smaller number x, then the larger number is $5x$. The sum of the two numbers is 480, so the following equation can be used to solve for x:

$$\begin{aligned} x + 5x &= 480 \\ 6x &= 480 \\ x &= 80 \end{aligned}$$

20. **Answer:** C. 20 students

Rationale: To solve this problem, first convert 45% to a decimal by moving the decimal point two places to the left: $45\% = .45$. Use x to represent the total number of students in the class and then set up the following equation.

$$.45x = 9$$

Solve for x by dividing both sides of the equation by .45. Since 9 divided by .45 is 20, the correct answer is C.

21. **Answer:** B. 32 diet, 80 regular

    Rationale: If the owner sells 2 diet sodas for every 5 regular sodas, then $^2/_7$ of the sodas sold are diet and $^5/_7$ are regular. Multiply these fractions by the total number of sodas sold, as follows:

    $$\frac{2}{7} \times \frac{112}{1} = \frac{224}{7} = 32\ diet\ sodas$$

    $$\frac{5}{7} \times \frac{112}{1} = \frac{560}{7} = 80\ regular\ sodas$$

    Remember: When multiplying a fraction by a whole number, it is usually simpler to dividing by the denominator first is usually simpler. Then multiply by the numerator.

22. **Answer:** B. 4.5 inches

    Rationale: The length of the larger rectangle is 12 inches, and the length of the smaller rectangle is 8 inches. Therefore, the length of the larger rectangle is 1.5 times the length of the smaller rectangle. Since the rectangles are proportional, the width of the larger rectangle must be 1.5 times the width of the smaller rectangle. The width can thus be calculated as follows:

    $$1.5 \times 3\ inches = 4.5\ inches$$

23. **Answer:** D. 150 pounds

    Rationale: The average weight of the five friends is 180 pounds, so the total weight of all five is 5×180, or 900 pounds. Add the weights of Al, Bob, Carl, and Dave. Together, they weigh 750 pounds. Subtract 750 from 900 to find Ed's weight of 150 pounds.

24. **Answer:** C. 15 miles

    Rationale: Rebecca's commute is shorter than Alan's but longer than Bob's. Alan's commute is 18 miles, and Bob's is 14 miles; thus, Rebecca's must be longer than 14 but shorter than 18 miles. Neither 14 nor 18 is correct since the distances cannot be equal to either of the examples. Therefore, 15 miles is the only correct answer.

25. **Answer:** A. 10%

    Rationale: First, find the total number of patients admitted to the ER by adding the number admitted for all the reasons given in the question. There were 120 patients admitted. To find what percent were admitted for respiratory problems, divide the number admitted for respiratory problems by the total number admitted, as shown in the equation below:

    $$12 \div 120 = .10$$

    Convert this decimal to a percent by moving the decimal point two places to the right:

    $$.10 = 10\%.$$

    Remember: When you are asked what percent of a total for a certain part, divide the part by the whole.

26. **Answer:** C. 16 are male nurses

    Rationale: There are 9 times as many female nurses as male nurses. To find the number of male nurses, divide the number of female nurses by 9, as shown in the following equation:

    $$144 \div 9 = 16\ male\ nurses$$

27. **Answer:** B. 6 members

Rationale: Three state legislators on the committee make up $^1/_5$ of the committee members. Therefore, 15 members total sit on the committee. Of these 15 members, 3 are state legislators and 6 are state employees; thus, 6 must be members of the public.

28. **Answer:** D. 3,105 students

    Rationale: The ratio of female to male students is exactly 5 to 4, so $^5/_9$ of the students are female and $^4/_9$ of the students are male. Therefore, the total number of students must be evenly divisible by 9, and 3,105 (answer D) is the only answer that fits this requirement.

29. **Answer:** C. 32 square inches

    Rationale: The perimeter of the rectangle is 24 inches. Therefore, the length plus the width must equal half of 24, or 12 inches. The ratio of length to width is 2:1; thus, the length is $^2/_3$ of 12, and the width is $^1/_3$ of 12. Based on those fractions, the length is 8 inches, and the width is 4 inches. The area $(length\ x\ width)$ is 32 square inches.

30. **Answer:** B. The Bulldogs will not be in the playoffs.

    Rationale: The Rangers are playing the Statesmen in the final game, so one of these teams will finish with a record of 11 wins and 2 losses. Even if the Bulldogs win their game, their final record will be 10 wins and 3 losses. Thus, the Bulldogs will not be in the playoffs.

31. **Answer:** A. $52,000

    Rationale: Convert 15% to a decimal by moving the decimal point two places to the left: $15\% = 0.15$. Using x to represent Brian's gross salary, the equation can be written as follows:

    $$0.15x = \$7{,}800$$

    To solve for x, divide both sides of the equation by 0.15. $7,800 divided by 0.15 is $52,000.

32. **Answer:** A. $33,500

    Rationale: Janet has a total of $105,000 in her accounts ($40,000 in stocks, $65,000 in bonds). 70% of that amount (her goal for her stock investments) is $73,500. To reach that goal, she would have to move $33,500 from bonds to stocks.

33. **Answer:** D. 432 parts

    Rationale: Each machine can produce 12 parts per minute $(96 \div 8)$. Multiply 12 by 12 (number of machines) by 3 (number of minutes) to get 432 parts every 3 minutes, as shown in the following equation:

    $$12 \times 12 \times 3 = 432\ parts\ every\ 3\ minutes$$

34. **Answer:** C. 5 are women

    Rationale: Use x to represent the number of women on the board. Then, the number of men is $x + 3$. The following equation is used to solve for x (i.e., number of women on the board):

    $$x + (x + 3) = 13$$
    $$2x + 3 = 13$$

    To isolate $2x$ on one side of the equation, subtract 3 from both sides to obtain the following:

    $$2x = 10$$

    $$x = 5\ women\ on\ the\ board$$

35. **Answer:** D. 6%

    Rationale: The loan was for $1,200, and the amount paid out was $432. You know that the number of years of the loan and the interest rate of the loan is the same number.

    There are the four possible scenarios in the multiple-choice answers:

    A. 5—$5\ years \times 0.05 = 0.25 \times 1{,}200 = \$300\ paid\ in\ interest$

    B. 15—$15\ years \times 0.15 = 2.25 \times 1{,}200 = \$2{,}700\ paid\ in\ interest$

C. $9 - 9\ years \times 0.09 = 0.81 \times 1{,}200 = \$972\ paid\ in\ interest$

D. $6 - 6\ years \times 0.06 = 0.36 \times 1{,}200 = \$432\ paid\ in\ interest$

This method of finding the correct answer is based on eliminating the incorrect ones as much as finding the correct one. Often the time spent trying to find an equation or formula is more than the time needed to just model the possible outcomes.

36. **Answer:** D. 88.8

Rationale: If you have taken 5 tests and your average grade was 91% (0.91), then you have earned a total of 455 points out of a possible 500 points thus far:

$$5 \times 0.91 = 4.55.$$

If you have earned a grade of 78% on the next test, you must have gotten 78 points out of a possible 100 points which can be added to the previous total points for your grade

Therefore, you have earned $455 + 78 = 533$ out of a possible 600 points. Since you have taken 6 tests, divide the number of points earned by the number of tests you have taken:

$$553 \div 600 = 0.8883 = 88.83\%.$$

37. **Answer:** C. 17 years old

Rationale: Rather than try figure out a proper formula, ask yourself, "What if Jorge and Alicia were the same age? If their combined ages are 42, then they would each be 21 years old. Since they are 8 years apart, Jorge has to "get older" by 4 years and Alicia has to "get younger" by 4 years. Therefore, add 4 years to Jorge to get 25 years of age, and subtract 4 years from Alicia to get 17 years of age. It might be good to check your results before basking in the glory of knowing how to do this problem! If Jorge + Alicia should equal 42, then $25 + 17 = 42$. Remember, the question is asking about Alicia's age, not Jorge's.

38. **Answer:** D. $21.94

Rationale: In this scenario, you will be buying a smoothie each day during the workweek—five times—and once on the weekend. The weekend smoothie will cost slightly more. Notice that you will NOT be buying a smoothie on one of the two weekend days. The weekly cost of your smoothie consumption (SC) can be determined in the following manner:

$$SC = (3.59 \times 5) + (3.99 \times 1)$$
$$SC = 17.95 + 3.99$$
$$SC = 21.94$$

39. **Answer:** A. 6/7

Rationale: If 4 out of 28 students ARE going to summer school, then 24 out of the 28 students ARE NOT going to summer school. Therefore, to find out the ratio, you divide 24 by 28, and then simplify your fraction by dividing both the numerator and the denominator by 4, which is the same thing as multiplying the fraction by 1 because 4/4 = 1:

$$\frac{24}{28} = \frac{\frac{24}{4}}{\frac{28}{4}} = \frac{6}{7}$$

40. **Answer:** C. 72 cents

Rationale: Reading the question carefully, you will note that the regular price for the cookies is irrelevant since the question is only about the price for each cookie if you buy one and a half dozen, i.e. 18, cookies. If 18 cookies cost $12.89, then 1 cookie costs $12.89/18 = 0.716$, or $0.716, and rounded up they each cost $0.72 or 72 cents.

41. **Answer:** D. All the answer choices are integers

Rationale: An integer is defined as a number that can be written without a fraction or decimal component. The set of integers includes zero (0), the natural numbers (1, 2, 3...), also called whole numbers or counting numbers. It also includes their additive inverses, the negative integers (-1, -2, -3 . . .).

42. **Answer:** A. Positive

Rationale: Within an equation, subtracting a negative number (-A), will give the same result as adding the corresponding positive number (A). Here's an example:

$$Z + Y = Z - (-Y)$$
$$43 + 6 = 43 - (-6)$$

43. **Answer:** D. 1, 2, 4, 8, 16, 32, 64, 128

Rationale: Factors are the set of numbers that can be multiplied to form a given number. 128 is created in the following ways:

$$128 = 2 \times 64$$
$$128 = 4 \times 32$$
$$128 = 8 \times 16$$
$$128 = 1 \times 128$$

Therefore, factors of 128 are simply the set of these factors: 1, 2, 4, 8, 16, 32, 64, 128.

44. **Answer:** D. There is only one even prime number

Rationale: By definition, there is only one even prime number; 2. Memorize it. Prime numbers are natural numbers greater than 1 that have no positive divisors other than 1 and itself. A composite number is a natural number greater than 1 that is not a prime number. For example, 7 is prime because no integer (natural number), other than 1 and itself, can be divided into it without remainder. The even number 10, for example, is a composite number because both 2 and 5 can be divided into it without remainder.

45. **Answer:** A. 2

Rationale: The prime factorization of 128 is found in the following way:

$$128 = 2 \times 64 = 2 \times 2 \times 32 = 2 \times 2 \times 2 \times 16 = 2 \times 2 \times 2 \times 2 \times 8 = 2 \times 2 \times 2 \times 2 \times 2 \times 4$$

The prime factorization of 128 is the last form in the list. The question asked what the prime factors of 128 are. In the final form two is the only number in the prime factorization of 128.

46. **Answer:** B. An expression

Rationale: By definition, an equation is a statement that two mathematical expressions are equal. Notice that an equation has, by definition, an equal sign. A polynomial is an expression of more than two algebraic terms, especially terms that contain different powers of the same variables. Notice that a polynomial is a specific type of expression. An exponent is a quantity representing the power to which a given number or expression is to be raised. The exponent is the superscript symbol beside the number or expression. There is no exponent in the given quantity. Since none of these mathematic terms can be used to describe the given information, it must be an expression. An expression is a collection of symbols that jointly express a quantity.

47. **Answer:** A. 2

Rationale: The greatest common factor (GCF) is found by identifying all the factors of the two or more numbers in your set, and then finding the largest number that they share.

Factors of 16: 1, 2, 4, 8, 16

Factors of 38: 1, 2, 19, 38

The greatest (largest) common factor these two numbers share is 2.

48. **Answer:** B. 40

Rationale: The least common multiple (LCM) is found by listing numbers that are integer multiples of the original number. Multiplying a given number by all the integers (1, 2, 3, 4, 5, 6 etc.):

For the number 5 : 5, 10, 15, 20, 25, 30, 35, 40, 45, 50, etc.

For the number 8: 8, 16, 24, 32, 40, 48, 56, 64, etc.

The LCM is the smallest number that appears in both sets of multiples; in this case the number 40.

49. **Answer:** C. 5,040

Rationale: The use of the exclamation sign with a number simply means that the number is multiplied by all the integers smaller than that number. In this example:

$$7! = 1 \times 2 \times 3 \times 4 \times 5 \times 6 \times 7 = 5{,}040$$

50. **Answer:** C. $\sqrt{17}$

Rationale: If a square root is not expressed as an integer, it is an irrational number. Since $\sqrt{4} = 2$ and $\sqrt{9} = 3$, they are not irrational numbers. However, $\sqrt{17}$ cannot be expressed as an integer because it is irrational. The most notable irrational number is probably "pi," which is very useful in geometry; it is equal to approximately 3.14159. It is important to note that this is an approximate value.

TEST 7. MATHEMATICS KNOWLEDGE PRACTICE TEST ANSWERS

1. **Answer:** C. 78

   Rationale: To find the median in a series of numbers, arrange the numbers in order from smallest to largest:

   69, 73, 78, 80, 100

   The number in the center is the median. Thus, the answer is C.

   If there are an even number of numbers in the series—for example, 34, 46, 52, 54, 67, 81—then the median will be the average of the two numbers in the center. In this example, the median will be 53 (the average of 52 and 54). Remember: The median is different from the average.

2. **Answer:** D. 22

   Rationale: This question asks you to find the square root of 11 times 44. First, do the multiplication:

   $$11 \times 44 = 484$$

   Now multiply each of the possible answers by itself to see which one is the square root of 484. Since $22 * 22 = 484$, the answer is D.

3. **Answer:** D. 3^4

   Rationale: Positive numbers are larger than negative numbers, so this fact limits the possible answers to 42 and 3^4, the latter of which equals $81(3 \times 3 \times 3 \times 3)$. Thus, the answer is D.

4. **Answer:** C. 59

   Rationale: A prime number is a whole number greater than 1 that can be divided evenly only by itself and 1. In this question, only 59 fits that definition. Thus, the answer is C.

   Conversely, 81 is not prime because it can be divided evenly by 9; 49 is not prime because it can be divided evenly by 7; and 77 is not prime because it can be divided evenly by 7 and 11.

5. **Answer:** D. $\frac{(6 \times 9)}{2}$

   Rationale: The area of a triangle is one-half the product of the base and the height. Choices A and B are incorrect because they add the base and the height instead of multiplying them. Choice C is incorrect because it multiplies the product of the base and the height by 2 instead of dividing it by 2. Thus, the answer is D.

6. **Answer:** A. 2,595

   Rationale: The steps in evaluating a mathematical expression must be carried out in a certain order, which is called "the order of operations." The following list shows the steps in order:

   - **Step 1:** Parentheses: First do any operations in parentheses.
   - **Step 2:** Exponents: Then do any steps that involve exponents.
   - **Step 3:** Multiply and Divide: Multiply and divide from left to right.
   - **Step 4:** Add and Subtract: Add and subtract from left to right.

   One way to remember this order is to use the following mnemonic sentence:

   Please **E**xcuse **M**y **D**ear **A**unt **S**ally.

   To evaluate the expression in this question, follow the below steps:

   - **Step 1:** Multiply the numbers in Parentheses: $3 \times 4 = 12$
   - **Step 2:** Apply the Exponent 2 to the number in parentheses: $12^2 = 144$
   - **Step 3:** Multiply: $6 \times 3 \times 144 = 2{,}592$
   - **Step 4:** Add: $2 + 2{,}592 + 1 = 2{,}595$

7. **Answer:** B. 9

Rationale: The symbol ≥ means "greater than or equal to." The only answer that is greater than or equal to 9 is answer B.

8. **Answer:** C. $x = 81$

Rationale: When two numbers with the same sign (both positive or both negative) are multiplied, the answer is a positive number. Thus, the answer is C.

Conversely, when two numbers with different signs (one positive and the other negative) are multiplied, the answer is negative.

9. **Answer:** B. $x = a^5$

Rationale: When multiplying numbers that have the same base and different exponents, keep the base the same and add the exponents. In this case, $a^2 \times a^3$ becomes $a^{(2+3)}$ or a^5.

10. **Answer:** C. 6

Rationale: Solve the following equation:

$$x > 3^2 - 4$$

$$x > 9 - 4$$

$$x > 5$$

We know that x is greater than 5, so the answer could be either 6 or 7. However, the question asks for the smallest possible value of x, so the correct answer is 6.

11. **Answer:** B. 3,024 cubic inches

Rationale: The formula for the volume of a rectangular solid is $length \times width \times height$. Therefore, the volume of this box is calculated using the following formula:

$$18 \times 12 \times 14 = 3{,}024 \ cubic\ inches$$

12. **Answer:** B. Perpendicular

Rationale: Two lines that form a right angle are called perpendicular.

13. **Answer:** C. 15 square feet

Rationale: This shape is called a parallelogram. The area of a parallelogram equals the base times the height, as calculated in the following equation:

$$5\ feet \times 3\ feet = 15\ square\ feet$$

14. **Answer:** B. $\frac{20}{mp^2}$

Rationale: To divide by a fraction, invert the second fraction and multiply:

$$\text{so } \frac{5}{mp} \div \frac{p}{4} \text{ becomes } \frac{5}{mp} \times \frac{4}{p}$$

$$\text{and } \frac{5}{mp} \times \frac{4}{p} = \frac{20}{mp^2}$$

15. **Answer:** B. 3.44 square feet

Rationale: First, find the area of the square by multiplying the length of a side by itself. The area of the square is 16 square feet. Then, find the area of the circle by using the following formula:

$$A = \pi r^2$$

The symbol π equals approximately 3.14. The letter r is the radius of the circle. In this case, r is half the width of the square (or 2), so $r^2 = 4$. The area of the circle is calculated as follows:

$$A = 3.14 \times 4$$
$$A = 12.56$$

To find the area of the square not covered by the circle, subtract 12.56 square feet from 16 square feet, which equals 3.44 square feet.

16. **Answer:** C. $^3/_{36}$

Rationale: When you roll a pair of 6-sided dice, there are 36 possible combinations of numbers (or outcomes). There are six numbers on each of the dice, so there are 6×6 possible combinations. Only three of those combinations will yield a total of 10: $4 + 6$, $5 + 5$, and $6 + 4$. Much of the time, probability answers will be given in the form of simplified fractions. In this case, the correct answer could also have been $^1/_{12}$.

17. **Answer:** C. 120

Rationale: The factorial of a number is the product of all the integers less than or equal to the number. The factorial of 5 is $5 \times 4 \times 3 \times 2 \times 1 = 120$. The factorial is written this way: $5!$

18. **Answer:** D. 7

Rationale: In this number, the places are as follows:

- 1 is in the thousands place.
- 2 is in the hundreds place.
- 3 is in the tens place.
- 4 is in the ones place
- 5 is in the tenths place.
- 6 is in the hundredths place.
- 7 is in the thousandths place.

19. **Answer:** B. 4

Rationale: The mode is the number that appears most often in a set of numbers. Since 4 appears three times, it is the "mode."

20. **Answer:** C. 49

Rationale: A perfect square is the product of an integer times itself. In this question, 49 is a perfect square because it is the product of 7×7.

21. **Answer:** A. Length = 12, width = 4

Rationale: Use w to represent the width of the rectangle. The length is three times the width, so the length is $3w$. The area of the rectangle is the length times the width, so the area is $w \times 3w$, or $3w^2$:

$$3w^2 = 48$$

Divide both sides of the equation by 3 to get the following:

$$w^2 = 16$$

Therefore, $w = 4$, the width of the rectangle, and $3w = 12$, the length of the rectangle

22. **Answer:** B. $24 = 2 \times 2 \times 2 \times 3$

Rationale: The prime factors of a number are the prime numbers that divide that number evenly. The prime factorization of a number is a list of the prime factors that must be multiplied to yield that number.

The simplest method of finding the prime factorization is to start with a simple multiplication fact for that number. In this case, the following is a good choice:

$$24 = 6 \times 4$$

The prime factorization of 24 includes the prime factorization of both 6 and 4. Therefore, since $6 = 2 \times 3$ and $4 = 2 \times 2$, the prime factorization of 24 must be the following:

$$24 = 2 \times 2 \times 2 \times 3$$

23. **Answer:** A. A number greater than x

Rationale: When a positive number is divided by a positive number less than 1, the quotient will always be larger than the number being divided. For example, $5 \div 0.5 = 10$. If we solve this as a fraction, $5 \div (1/2)$ is the same as $5 \times (2/1)$, which equals 10, since dividing by a fraction is the same as multiplying by the reciprocal.

24. **Answer:** C. $c = 11$

Rationale: Subtracting a negative number is the same as adding a positive number. Therefore, $8 - (-3)$ is the same as $8 + 3$, both of which equal 11.

25. **Answer:** C. The square of a number is always a positive number.

Rationale: When two numbers with the same sign (positive or negative) are multiplied, the product is always positive. When a number is squared, it is multiplied by itself, so the numbers being multiplied have the same sign. Therefore, the product is always positive.

The square of a number greater than 1 is always greater than the original number. But the square of a positive number less than one (e.g., 0.5) is always less than the original number.

26. **Answer:** D. $r = \frac{p-3}{2}$

Rationale: Begin by subtracting 3 from both sides of the equation to get the following:

$$p - 3 = 2r$$

Now, to isolate r on one side of the equation, divide both sides of the equation by 2 to get the following:

$$r = \frac{p - 3}{2}$$

27. **Answer:** B. 40

Rationale: The least common multiple is used when finding the lowest common denominator. The least common multiple is the lowest number that can be divided evenly by both numbers.

Here is a simple method to find the least common multiple of 8 and 10: (1) write 8 on the left side of your paper, (2) then add 8 and write the result, (3) then add another 8 to that number, and (4) write the result. Keep going until you have a list that looks something like the following list:

8 16 24 32 40...

This partial list shows some multiples of 8. (If you remember your multiplication tables, these numbers are the column or row that go with 8.) Now do the same thing with 10.

10 20 30 40...

This partial list shows multiples of 10. Eventually, similar numbers will appear in both rows. The smallest of these numbers is the least common multiple. There will always be more multiples that are found in both rows (if you were to continue listing multiples), but the smallest number is the least common multiple.

28. **Answer:** D. $x = 4$

Rationale: When you multiply or divide numbers that have the same sign (both positive or both negative), the answer will be positive. When you multiply or divide numbers that have different signs (one positive and the other negative), the answer will be negative. In this case, both numbers have the same sign, so the answer is a positive number, 4 (answer D).

29. **Answer:** A. $2\frac{5}{6}$

Rationale: To convert an improper fraction to a mixed number, divide the numerator (in this case, 17) by the denominator (in this case 6). In this case, you get 2 with a remainder of 5. 2 becomes the whole number portion of the mixed number, and 5 becomes the numerator of the new fraction.

30. **Answer:** B. –10

Rationale: To find the value of q, divide both sides of the equation by –13. When a positive number is divided by a negative number, the answer is negative.

31. **Answer:** A. 12

Rationale: Replace the letters with the numbers they represent, and then perform the necessary operations, as shown in the following solution:

$$3^2 + 6(0.5)$$

$$9 + 3 = 12$$

32. **Answer:** B. 500,000,000

Rationale: A billion is 1,000 million (1,000,000,000). Half a billion is 500 million (500,000,000).

33. **Answer:** B. 32 inches

Rationale: The radius of a circle is one half of the diameter, so the diameter of this circle is 8 inches. The diameter is a line that passes through the center of a circle and joins two points on its circumference. If you study this figure, you can see that the diameter of the circle is the same as the length of each side of the square. The diameter is 8 inches, so the perimeter of the square is 32 inches $(8 + 8 + 8 + 8)$.

34. **Answer:** D. $^{21}/_{32}$

Rationale: To multiply fractions, multiply the numerators and the denominators. In this case, multiply 3 by 7, and 4 by 8. The correct answer is therefore $^{21}/_{32}$.

35. **Answer:** D. $y = 2x + 7$

Rationale: The simplest way to answer this question is to determine which of the equations would work for all the values of x and y in the table. Choices A and D would work when $x = 0$ and $y = 7$, but A would not work for any other values of x and y. Choice B would work when $x = 3$ and $y = 13$, but it would not work for any other values of x and y. Choice C would not work for any of the values of x and y. Only choice D would be correct for each ordered pair in the table.

36. **Answer:** A. 64 feet

Rationale: The dimensions of the left side and the top side of this figure are provided, but how do you find the dimensions of the other sides? Look at the two horizontal lines on the bottom of the figure. Together, the lines are as wide as the top side of the figure, so they must be 18 ft when measured together. Now look at the two vertical lines on the right side of the figure. Together, the vertical lines are as tall as the left side of the figure, so the lines must be 14 ft when measured together. Thus, the perimeter of the figure is $14 + 18 + 14 + 18$, or $32 + 32 = 64\,ft$.

37. **Answer:** C. 4

Rationale: Make a list of the powers of 10.

- $10^2 = 100$
- $10^3 = 1{,}000$
- $10^4 = 10{,}000$
- $10^5 = 100{,}000$

When you divide 15,200 by 1.52, you get 10,000. Therefore, the correct exponent is 4. That exponent is also the number of places the decimal must be moved in the number 1.52 to make the number 15,200.

38. **Answer:** B. 6.5×10^{-3}

Rationale: Make a list of the negative powers of 10:

- $10^{-2} = 0.01$
- $10^{-3} = 0.001$

- $10^{-4} = 0.0001$
- $10^{-5} = 0.00001$

Now multiply each of these numbers by 6.5 to see which one gives you 0.0065. The decimal must be moved 3 places to the left in the number 6.5 to make the number 0.0065.

39. **Answer:** C. 16

    Rationale: The greatest common factor of two numbers is the largest number that can be divided evenly into both numbers. The simplest way to answer this question is to start with the largest answer (32) and determine whether it can be divided evenly into 48 and 64. It cannot.

    Now try the next largest answer (16). That number can be divided evenly into 48 and 64. Thus, 16 is the correct answer. The other answers are also factors, but the largest of them is 16.

40. **Answer:** A. −116

    Rationale: Calculate the value of x^2 and y^3, then add the results, as shown in the following solution:

$$x^2 = (-3)(-3) = 9$$
$$y^3 = (-5)(-5)(-5) = -125$$
$$9 + (-125) = -116$$

41. **Answer:** D. 64 squares

    Rationale: If you drew a diagram of the larger square, you would see that you can make 8 rows of 8 smaller squares, so you can make a total of 64 squares ($8 * 8$).

42. **Answer:** A. 2 is the only even prime number.

    Rationale: 2 is the only even prime number since all other even numbers are multiples of 2. The largest prime number less than 100 is 97, not 89. The greatest common factor of 24 and 42 is 6. The greatest common factor of two numbers is the largest number that can be divided evenly into both numbers. The least common multiple of 8 and 6 is 24. The least common multiple of two numbers is the smallest number that can be divided evenly by both numbers.

43. **Answer:** C. $^7/_{10}$

    Rationale: The simplest way to solve this problem is to convert the fractions to decimals, which is done by dividing the numerators by the denominators:

$$^2/_3 = 0.67 \text{ and } ^3/_4 = 0.75$$

Thus, the correct answer is a decimal that falls between 0.67 and 0.75.

$$^3/_5 = 0.6$$
$$^4/_5 = 0.8$$
$$^7/_{10} = 0.7$$
$$^5/_8 = 0.625$$

Thus, answer C is correct.

44. **Answer:** A. 6

    Rationale: If $a = 10$, then $2a = 20$. Now compute the value of b^2:

$$b^2 = (-b)(-b)$$
$$b^2 = (-4)(-4)$$
$$b^2 = 16$$

Now you have the following: $\sqrt{20 + 16}$ or $\sqrt{36}$

The square root of 36 is 6.

45. **Answer:** D. June

Rationale: To find the median in a series of numbers with an odd number, arrange the numbers in order from smallest to largest. The number in the center, 92 in this case, is the median. If there are an even number of numbers in the series—for example, 23, 45, 62, 64, 77, 82—then the median will be the average of the two numbers in the center. In this example, the median will be 63 (the average of 62 and 64). Remember: The median is different from the average.

46. **Answer:** B. $^1/_5$

Rationale: After Mark chooses the first letter, five letters will remain. Thus, the probability that the second letter he chooses will be an M or an N is $^1/_5$.

47. **Answer:** A. 0.6×90

Rationale: To find 60% of 90, first convert 60% to a decimal by moving the decimal point two places to the left, then multiply this decimal (0.6) by 90.

48. **Answer:** A. 134,000

Rationale: Make a list of the powers of 10.

- $10^2 = 100$
- $10^3 = 1{,}000$
- $10^4 = 10{,}000$
- $10^5 = 100{,}000$

If you multiply 1.34 by 100,000, the answer is 134,000.

49. **Answer:** B. −62

Rationale: If you substitute 4 for a, you get the following expression:

$$(-4) \times (-4) \times (-4) - (-4) - 2$$

The product $(-4) \times (-4) \times (-4) = -64$. Subtracting (−4) is the same as adding 4. So the expression is now as follows:

$$-64 + 4 - 2 = -62$$

50. **Answer:** C. $3^3 \times 5 \times 2$

Rationale: For each of the answers, first calculate the value of the numbers with exponents, and then multiply, as shown in the following calculations:

$$16 \times 5 \times 3 = 240$$
$$8 \times 15 \times 2 = 240$$
$$27 \times 5 \times 2 = 270$$
$$8 \times 10 \times 3 = 240$$

Test 8. Word Knowledge Practice Test Answers

1. **Answer:** B. Vast

   Rationale: Expansive means covering a wide area regarding space or scope; extensive; wide-ranging.

2. **Answer:** C. Reliable

   Rationale: Credible means able to be believed; convincing; plausible; tenable.

3. **Answer:** D. Destruction

   Rationale: Devastation means great destruction or damage; ruin, havoc, wreckage.

4. **Answer:** A. Unclear

   Rationale: Vague means of uncertain, indefinite, or unclear character or meaning; indistinct; ill-defined.

5. **Answer:** B. Disrespectful

   Rationale: Irreverent means showing a lack of respect for people or things that are generally taken seriously; disdainful; scornful; derisive; contemptuous.

6. **Answer:** D. Distaste

   Rationale: Aversion means a strong dislike or disinclination; abhorrence; antipathy.

7. **Answer:** A. Difficult

   Rationale: Laborious means a task, process, or journey requiring considerable effort or time; arduous; strenuous.

8. **Answer:** C. Endless

   Rationale: Interminable means unending; monotonously or annoyingly protracted or continued; unceasing; incessant.

9. **Answer:** C. Without color

   Rationale: Achromatic means free from color.

10. **Answer:** B. Rapid

    Rationale: Cursory means hasty and therefore not thorough or detailed; perfunctory; desultory; casual; superficial.

11. **Answer:** B. Unverified secondhand information

    Rationale: Hearsay means information received from other people that one cannot adequately substantiate; rumor; gossip.

12. **Answer:** C. Forgiving

    Rationale: Magnanimous means very generous or forgiving, especially toward a rival or someone less powerful than oneself; generous; charitable; benevolent.

13. **Answer:** A. Of the earth

    Rationale: Terrestrial means of, on, or relating to the earth.

14. **Answer:** C. Indifferent

    Rationale: Nonchalant means having an air of indifference or easy concern.

15. **Answer:** A. Tangible

    Rationale: Palpable means capable of being touched or felt; tangible.

16. **Answer:** B. Plaster

    Rationale: Daub means plaster; to cover or coat with soft adhesive matter; to apply crudely.

17. **Answer:** C. Swell

    Rationale: Distend means to enlarge from internal pressure; to swell; to become expanded.

18. **Answer:** B. Mistake

    Rationale: Gaffe means a social or diplomatic blunder; mistake; faux pas.

19. **Answer:** D. Of or relating to leader of Catholicism

    Rationale: Papal means of or relating to a pope or the Roman Catholic Church.

20. **Answer:** A. Harmful

    Rationale: Pernicious means having a harmful effect, especially in a gradual or subtle way. It describes something that is destructive, causing great damage over time.

21. **Answer:** C. Commotion

    Rationale: Hoopla means a noisy commotion; boisterous merrymaking.

22. **Answer:** B. Dogmatic

    Rationale: Doctrinaire means very strict in applying beliefs and principles; dogmatic; dictatorial.

23. **Answer:** D. Feathery

    Rationale: Plumose means having feathers or plumes; feathered.

24. **Answer:** C. Uproot

    Rationale: Supplant means to supersede another, especially by force or treachery; uproot; to eradicate and supply a substitute for; to take the place of and serve as a substitute, especially because of superior excellence or power; replace.

25. **Answer:** D. Caprice

    Rationale: Vagary means an erratic, unpredictable, or extravagant manifestation, action, or notice; caprice.

26. **Answer:** C. Squid

    Rationale: Calamari means squid used as food; the inky substance the squid secretes.

27. **Answer:** A. Apathetic

    Rationale: Lethargic means indifferent; apathetic; sluggish.

28. **Answer:** C. Tiny

    Rationale: Teensy means tiny.

29. **Answer:** D. Dull

    Rationale: Bland means smooth and soothing in manner or quality; exhibiting no personal concern or embarrassment; unperturbed; not irritating, stimulating, or invigorating; dull.

30. **Answer:** B. Copious

    Rationale: Fulsome means characterized by abundance; copious.

31. **Answer:** B. Name

    Rationale: Nomenclature means name, designation; a system of terms used in a particular science, discipline, or art.

32. **Answer:** D. Unwilling

    Rationale: Reluctant means holding back; averse; unwilling; disinclined.

33. **Answer:** C. Dilapidated

Rationale: Shabby means clothed with worn or seedy garments; ill-kept; dilapidated.

34. **Answer:** C. Omnipresent

Rationale: Ubiquitous means being present, everywhere at the same time. It indicates something that is widespread, prevalent.

35. **Answer:** D. Delay

Rationale: Defer means to put off; delay; postpone; suspend.

36. **Answer:** C. Brave

Rationale: Undaunted means not intimidated or discouraged by difficulty, making brave the best synonym.

37. **Answer:** B. Final

Rationale: Immutable means unchanging or permanent, so final is the closest synonym.

38. **Answer:** C. Clear

Rationale: Lucid means expressed clearly and easily understood, making clear the best choice.

39. **Answer:** B. Weaken

Rationale: Enervate means to sap or drain someone of energy, making weaken the correct choice.

40. **Answer:** B. Aware

Rationale: Cognizant refers to being aware or mindful of something.

41. **Answer:** B. Wise

Rationale: Sagacious describes someone who is wise, discerning, or knowledgeable.

42. **Answer:** B. Gentle

Rationale: Benign refers to something that is harmless and gentle, the opposite of harmful or aggressive.

43. **Answer:** A. Determination

Rationale: Tenacity means the quality of being determined or persistent.

44. **Answer:** B. Reduce

Rationale: Mitigate means to make something less severe or reduce its impact.

45. **Answer:** C. Precise

Rationale: Meticulous means very careful and precise, particularly about details.

46. **Answer:** B. Diligent

Rationale: Assiduous describes someone who is hardworking and persistent.

47. **Answer:** A. Foolish

Rationale: Sagacious means wise or showing good judgment, so the antonym is foolish.

48. **Answer:** A. Threatening

Rationale: Ominous suggests something bad or threatening is likely to happen.

49. **Answer:** B. Spacious

Rationale: Capacious describes something that has a lot of space or room.

50. **Answer:** B. Dangerous

Rationale: Precarious refers to something that is unstable or risky.

TEST 9. READING COMPREHENSION PRACTICE TEST ANSWERS

Questions 1-5

1. **Answer:** B) Nature

   Rationale: The subtopics of fashion, the destruction of nature, and the inherently greedy nature of humankind are all introduced through the lens of nature. Several text features elucidate the importance of the main topic with, for example, the personification of nature. The final sentence ties together the relevance of the preceding information to make a conclusive observation on the main topic of nature.

2. **Answer:** D) All the above

   Rationale: Fashion is presented as extravagant and frivolous in comparison to nature by using hyperbole; the "ravenous hysteria" of the fashionable women implies their inferiority regarding the contrasting theme of nature. The destructive nature of humans is directly expressed through the violent language used, through the "widespread slaughter" that occurs and the mere fact that a man's "cruel arrogance" has destroyed "nature's innocence." Humans are shown to be greedy about nature by the way the subject is presented as a thief, robbing "nature of its riches." Humans are guilty not only of pecuniary greed in relation to nature but also of the desirous need to "possess" and therefore own nature's beauty.

3. **Answer:** A) The main topic of the paragraph

   Rationale: The last sentence links the strands of the subtopics by making a judgment on the relationship of humankind to nature. Humans seek to destroy nature because it is inherent in human nature to own and "possess" it.

4. **Answer:** A) Personification

   Rationale: The writer personifies nature by capitalizing the word. This approach makes nature appear as one living entity, emphasized by nature being imbued with human qualities, such as "innocence."

5. **Answer:** A) As inevitable

   Rationale: The destruction of the natural world is painted as wholly inevitable by the final sentence, which states that humankind is "rarely" concerned with only viewing nature, but must "possess" it. To possess nature, humans must in some way interfere with the natural order of the world; nature can no longer conceivably remain in an untouched state.

Questions 6-10

6. **Answer:** B) Gambling

   Rationale: While social class and questions of morality do feature heavily throughout the extract, gambling is the overarching, ubiquitous theme throughout the passage. This theme is emphasized with the final sentence, which suggests that morality and class are irrelevant as the fate of a "gamester" will remain the one overriding factor.

7. **Answer:** D) All gamblers will become poor in the end if they continue to gamble

   Rationale: The revelatory nature of the final sentence elucidates the main point the writer is trying to make. The writer contrasts different types of gamblers to reveal that these differences are wholly irrelevant as all gamblers share the same "destiny," which is "poverty and despair."

8. **Answer:** C) To highlight how, regardless of class (i.e., rich or poor), all gamblers end up destitute

   Rationale: The comparison between the working-class gambler and the aristocratic gambler is purposely created to contrast their different modi operandi in terms of how gamblers gamble and live their lives. However, this contrast is established to prove, with the final sentence, that regardless of a gambler's social class, all gamblers will all end up destitute in the end as that is a gambler's "destiny."

9. **Answer:** D) None of the above

   Rationale: Initially, the writer's tone appears to be accusatory. The writer launches biting critiques of the moral degeneracy of the working-class gambler, who is, among other things, a "bully" of the "lowest and vilest" order; the writer also appears to take the side of the aristocratic gambler. However, the revelation in the final sentence that, regardless of wealth or social class, both gamblers share the same fate. This statement considerably changes the tone of the passage; what initially appears as a biting critique of the working class becomes a biting critique of gambling itself.

10. **Answer:** D) Simile

    Rationale: The writer compares the working-class people who frequent gambling dens to "hungry parasites." The simile works as hyperbole to convince the reader of the predatory nature of this subsection of society.

Questions 11-15

11. **Answer:** A) Death

    Rationale: The writer's realization revolves around the unfulfilled life he/she has lived. While still physically living, the writer comes to the realization that something had always been missing to the point where he/she was figuratively "already dead."

12. **Answer:** D) None of the above

    Rationale: The writer shows little feelings toward the subject of his/her death as the passage revolves around the epiphanic moment when the writer realizes that he/she is dead. Yet the mere fact that death "was not an event to be feared" implies that this realization had put the writer beyond feeling because he/she had already faced the worst fear imaginable.

13. **Answer:** A) Rhetorical questions

    Rationale: The writer uses several rhetorical questions to emphasize his/her mental state. The absurdity of the idea that the writer did not previously realize what he/she now knows is an indication of the gravity of the epiphany the writer has experienced.

14. **Answer:** A) Inevitable

    Rationale: The writer not only views figurative death as inevitable but views it as such to the extent that it has already happened without him/her even that he/she is "already dead."

15. **Answer:** D) Poignant

    Rationale: The writer expresses disbelief about not having realized that he/she was not living. Regret and sadness are implied through this belief as the writer has led a life unfulfilled.

Questions 16-20

16. **Answer:** B) Technology has made modern warfare faster and more precise but also more vulnerable.

    Rationale: The passage emphasizes how technology has fundamentally changed warfare, highlighting both its benefits (speed, precision) and its challenges (vulnerabilities).

17. **Answer:** B) Satellite-guided missiles

    Rationale: The passage specifically mentions satellite-guided missiles as an example of how technology has expanded warfare into new domains.

18. **Answer:** C) Securing digital networks will become increasingly critical.

    Rationale: The passage discusses the vulnerabilities introduced by technology, especially in cyber warfare, and suggests that securing networks will be a priority in defense strategies.

19. **Answer:** B) Decisions made by AI systems without human input.

    Rationale: The term "autonomous decision-making" is used in the context of AI's role in assisting or making decisions during military missions.

20. **Answer:** C) Cyber warfare can cause significant disruptions without physical conflict.

    Rationale: The passage discusses how cyberattacks can disrupt communication systems and disable infrastructure, showing the significant impact they can have without involving physical force.

Questions 21-25

21. **Answer:** B) Airpower has consistently redefined the nature of warfare.

    Rationale: The passage discusses how advancements in military aviation have repeatedly transformed warfare, making option B the best summary of the main idea.

22. **Answer:** B) P-51 Mustang

    Rationale: The passage highlights the P-51 Mustang as an example of advanced fighters during World War II, distinguishing it from earlier or modern aircraft.

23. **Answer:** B) They have made airpower more efficient and less risky for pilots.

    Rationale: The passage states that drones are essential tools in modern warfare, implying their efficiency and the safety they provide by removing pilots from direct danger.

24. **Answer:** B) The continuous effort to improve airpower and aviation technology.

    Rationale: The passage describes advancements in military aviation as a reflection of humanity's determination to innovate and dominate aerial technology.

25. **Answer:** B) Stealth technology and jet propulsion became priorities.

    Rationale: The passage highlights the Cold War as a period of rapid advancements in jet propulsion and stealth technology, suggesting their importance in military aviation development.

TEST 10. PHYSICAL SCIENCE PRACTICE TEST ANSWERS

1. **Answer:** C. Fossil records

   Rationale: The earth is composed of four distinct concentric spherical layers of material. At the center of the earth is the inner core, composed of a solid metallic sphere of nickel and iron. The next layer is the outer core which surrounds the inner core and is composed of molten liquid nickel and iron. The third concentric layer is the mantle. The mantle surrounds the outer core. The mantle is composed primarily of rocky silicate mineral compounds. At the deepest regions of the mantle, this material is extremely hot and behaves like a molten liquid. The mantle material cools as it approaches the surface of the earth, becoming more like a solid substance with the consistency of putty. Although this region is technically a rocky solid, it can flow or ooze when subjected to pressure. The outermost layer is the earth's crust, a thin layer of solid rocky mineral compounds that literally floats on top of the upper mantle layer. At the crust layer, the surface material interacts with the surface environment of the earth to produce a wide variety of solid rocky mineral compounds that are not found in the deeper mantle. Although fossils are found only on or within the earth's crust, scientists do not use this fact to distinguish the crust from the other layers of the earth. If the earth sustained no life, it would have no fossils, but it would still have a crust layer.

2. **Answer:** D. Lithosphere

   Rationale: The lithosphere is the solid outer layer of the earth. It includes the earth's crust and the upper layer of the underlying mantle.

3. **Answer:** B. Charged particles from the sun interacting with the earth's magnetic field

   Rationale: The sun continuously emits vast amounts of charged particles into space. This is known as the solar wind. When these particles intercept the earth, they are directed and concentrated by the earth's magnetic field to the north and south poles. As this concentration of charged particles collides with air molecules, it emits light of varying colors, resulting in a beautiful dynamic display of curtains of changing colors across the extreme northern and southern latitudes. At the North Pole regions, this phenomenon is called the Aurora Borealis. At the southern pole regions, it is known as the Aurora Australis.

4. **Answer:** A. Basalt

   Rationale: Sedimentary rock is formed by accumulating particulate material that is gradually deposited in layers and compressed, fusing and transforming the material together over very long periods of time. Shale, coal and sandstone are examples of sedimentary rock. Basalt is formed right when it cools from a molten state, such as molten lava from a volcano. Rock formed in this fashion is known as igneous rock.

5. **Answer:** C. 1,000 times more energetically powerful

   Rationale: The Richter scale is logarithmic—for every increase of one Richter unit number, the measured intensity of ground motion or earthshaking that results from an earthquake increases by a power of 10. The energy required to increase an earthquake's ground shaking motion by a factor of 10 requires approximately 31.6 times as much energy compared to an earthquake that is one Richter unit smaller. Therefore, since the difference between 7 and 5 is 2, the Richter scale 7 earthquake generates 10^2 times (100 times) more intense ground shaking compared to the Richter scale 5 earthquake. This corresponds to an increased energy of $31.6 * 31.6 \approx 1000$ or approximately 1,000 times more energy required to generate a Richter scale 7 earthquake compared to a Richter scale 5 earthquake.

6. **Answer:** B. Stratosphere

   Rationale: Modern aircraft cruise in the Stratosphere, at about 35,000 feet above the earth.

7. **Answer:** B. Stratus

   Rationale: Depending on the altitude at which you encounter fog, it may be almost any type of cloud, particularly at very high elevations near mountain peaks. Stratus clouds are the only type of cloud that can form at ground level at the lowest; those elevations are, by definition, the elevations at or near sea level.

8. **Answer:** C. It has 17 protons

   Rationale: This is the definition of an atomic number. The number of protons found in the nucleus is the same as the charge number of the nucleus since protons are positively charged. The atomic number uniquely identifies a chemical element. In an uncharged atom, the atomic number is equal to the number of electrons, since each electron has a negative charge.

9. **Answer:** B. Electric current: Watt

Rationale: Current is measured in amperes (amps)—memorize it!

10. **Answer:** C. 5 miles

Rationale: A meter is approximately 3.28 feet. An 8K race refers to a race covering a total distance of 8,000 meters. This converts to a distance in feet of 8,000 times 3.28, which is 26,240 feet. A mile is 5,280 feet. 26,240 divided by 5,280 is equal to about 4.97. Therefore, an 8K race would be 4.97 miles, or about 5 miles.

11. **Answer:** B. A relatively tiny rock and mineral core with an enormous thick outer layer of gas

Rationale: Neptune and the other "gas giants"—Jupiter, Saturn, and Uranus—are planets with rocks on the inside and enormous volumes of gas on the outside. The inner four planets—Mercury, Venus, Earth, and Mars—have negligible "atmospheres" over their solid surfaces. Answer D describes the composition of several moons of the gas giants (and possibly many other objects, including large comets and asteroids, within our solar system).

12. **Answer:** C. Mars and Jupiter

Rationale: The Asteroid Belt is located between Mars and Jupiter. It consists of numerous irregularly shaped bodies called asteroids. Astronomers believe the Asteroid Belt represents material that would have formed a planet between Mars and Jupiter if the gravitational forces of Jupiter did not have such a disruptive effect on the planet-forming processes. The total mass of the Asteroid Belt is about 4% that of Earth's moon. The largest four asteroids are named Ceres, Vesta, Pallas and Hygiene.

13. **Answer:** D. Phylum, class, order, family, genus, species

Rationale: The biological classifications, in descending order, are:

i. Life
ii. Domain
iii. Kingdom
iv. Phylum
v. Class
vi. Order
vii. Family
viii. Genus
ix. Species

14. **Answer:** D. II only

Rationale: All animals can move actively on their own at some stage of their life cycle. Cell walls are distinctive biological envelopes of cells found outside of the inner cell membrane. They are features of bacterial, plant and fungal cells, but are not characteristic of animal cells. Although almost all members of the Animal kingdom have a nervous system, sponges (which are members of the phylum Porifera of the animal kingdom) do not have nervous systems. All vertebrates are animals, but not all animals are vertebrates. There are more species of animals without vertebrae than with vertebrae.

15. **Answer:** C. Sponges

Rationale: Sponges are not fungi. Sponges are animals of the phylum Porifera, or "pore bearers." They are multicellular organisms that have bodies full of pores and channels, allowing water to circulate through them. Sponges do not have nervous, digestive or circulatory systems; instead, they rely on maintaining a constant water flow through their bodies to obtain food and oxygen and to remove waste. Yeast is a broad common term for two classes of organisms within the fungi kingdom—the class Ascomycetes and the class Basidiomycetes. Most molds are also members of the Fungi kingdom (the exception is slime molds, which are members of the Protista kingdom). All species of mushrooms are members of the fungi kingdom.

16. **Answer:** D. Time

Rationale: Taxonomy trees represent changes over time. Generally, the past is represented at the bottom of the tree, and the present is represented at the top of the tree.

17. **Answer:** A. To change an organism so that it is more likely to produce viable offspring

Rationale: Biological evolution is the process where mutations in an organism's genetic code result in an increased probability that the code will be passed to future generations of the organism. This process allows organisms to adapt to changes in the environment and to compete more successfully for survival.

18. **Answer:** A. Selection

    Rationale: This two-word term is a key concept in understanding evolution. The word design is in quotation marks because that word implies intention, and evolution is, by definition, a neutral process that does not require any outside intelligence to occur.

19. **Answer:** B. Comets

    Rationale: Asteroids orbit the sun in approximately the same way planets do. Comets, however, can have elongated elliptical orbits that bring them in the vicinity of the sun only occasionally. This is why Halley's Comet is visible only every 75 years. The best way to visualize the difference is to think of comets moving quickly toward the sun and then being thrown way back into deep space in a slingshot- like motion.

20. **Answer:** B. Solid, liquid and gas/vapor

    Rationale: Matter can exist in one of four phases: solid, liquid, gas and plasma. In the solid phase, water molecules are locked into a fixed position and cannot move relative to surrounding water molecules—this is the solid form of water (ice). As the temperature of the water molecules increases to the melting point of water, the water molecules become energetic enough to begin to move freely with respect to other water molecules, but are still attracted to each other, so they remain in physical contact—this is the liquid phase of water. The gas phase of water occurs when the water temperature reaches the boiling point. The water molecules are so energetic that they cannot be constrained by their attraction to each other. They become freely moving individual molecules that expand away from other water molecules and fill the entire volume of any container or, if uncontained, mix with the other gas molecules in the atmosphere. The plasma phase does not exist for water molecules. In the plasma phase, the temperatures are so high that the water molecules are broken down into individual hydrogen and oxygen atoms. The electrons of the individual atoms have so much thermal energy that they break free from their atomic nuclei. This results in a phase of matter consisting of a mixture of very hot atomic nuclei and free electrons.

    Evaporation is the term for a phase change of a substance from a liquid to a gas. This is also known as boiling. Condensation is the term for a phase change of a substance from a gas to a liquid. Sublimation is the name for a phase change of a substance from a solid directly to a gas. This process can be observed in the laboratory as a sample of frozen carbon dioxide (or "dry ice") turns directly to vapor at room temperature.

    In chemistry, the terms solute, solvent and solution refer to volumes of liquids that have other chemicals dissolved in the liquids. The liquid is the solvent. The dissolved chemicals are the solutes. The combination of the liquid and the dissolved chemicals are called the solution.

21. **Answer:** B. Mesozoic

    Rationale: Dinosaurs roamed the earth during the Mesozoic era, which consists of the Triassic, Jurassic and Cretaceous periods.

22. **Answer:** A. When a population is separated by geographic change

    Rationale: Geographic speciation occurs when a subpopulation of a species is geographically separated from the other members of the species. This subpopulation is no longer able to interbreed with the other species members. Over time, this subpopulation will evolve independently from the original species and can eventually evolve into an entirely new species.

23. **Answer:** C. Igneous

    Rationale: The answer choices are the three main rock types. Igneous rock is formed through the cooling and solidification of magma or lava. Memorize it! It may be interesting (or not) for you to know that igneous rocks are classified according to mode of occurrence, texture, mineralogy, chemical composition and geometry.

24. **Answer:** C. Sometimes A, sometimes B, sometimes both A and B

    Rationale: Fungi can reproduce sexually or asexually because they can produce both haploid and diploid cells. This allows them to adjust their reproduction to conditions in their environment. When conditions are generally stable, they reproduce quickly asexually. Fungi can increase their genetic variation through sexual reproduction when conditions are changing, which might help them survive.

Almost all fungi reproduce asexually by producing spores, which are haploid cells produced by mitosis, and so they are genetically identical to their parent cells. Fungi spores can develop into new haploid cells without being fertilized. Sexual reproduction involves the mating of two haploids. During mating, two haploid parent cells fuse, forming a diploid spore that is genetically different from its parents. As it germinates, it can undergo meiosis, forming haploid cells.

25. **Answer:** A. Parietal

Rationale: The parietal lobe is one of the four major lobes of the cerebral cortex in the brain. It integrates sensory information from all parts of the body. This region of the cerebral cortex is called the primary somatosensory cortex.

26. **Answer:** C. $w = m \times g$

Rationale: It's important to remember that mass is defined as the amount of matter that exists in an object. Matter possesses inertia, so it is a measure of an object's resistance to movement. On the other hand, weight is defined as the product of the object's mass and the gravitational acceleration being applied to that object. Therefore, $w = m \times g$.

27. **Answer:** B. 6 miles/h^2

Rationale: Acceleration is defined as the change in velocity over time:

$$A = \Delta V / \Delta T$$

In this question, the change in velocity has been from 43 miles/h to 64 miles/h for a change of 21 miles/h. The change in time is 3.5 h. Therefore, you can calculate the acceleration simply by substituting the numerical values:

$$A = \frac{\Delta V}{\Delta T} = \frac{\frac{21\text{miles}}{\text{h}}}{3.5\text{ h}} = 6\text{ miles/h}^2$$

28. **Answer:** A. A 10-pound object dropped from 70 ft

Rationale: Potential energy (PE) is measured by the following equation:

$$PE = M \times G \times H$$

where M is the mass, G is the gravitational acceleration, and H is the height. Gravitational acceleration is simply the acceleration of an object caused by the force of gravity on the earth; the conventional standard value is 32.174 ft/s^2 (9.8 m/s^2).

To answer this question correctly, it is important that you convert all units so that they correspond. First, convert each of the three scenarios, remembering the following:

$$\begin{aligned} 1\,foot &= 12\,inches \\ 1\,yard &= 3\,feet \\ 1\,ton &= 2{,}000\,pounds \end{aligned}$$

Now, convert each of the three scenarios given:

- 10 pounds dropped 70 ft
- 58 pounds dropped 4 yards = 58 pounds dropped 12 ft
- tons dropped 300 inches = 20 pounds dropped 25 ft

Since the gravitational constant is the same for all objects and all masses, the PE can be compared by multiplying the two terms:

$$10 \times 70 = 700$$

$$58 \times 12 = 696$$

$$20 \times 25 = 500$$

Answer choice A is the largest of the three numbers (700). Now you must multiply it by the gravitational constant to calculate the potential energy:

$$PE = M \times G \times H = 700 \times 32.174\frac{ft}{s^2} = \text{about } 22{,}520 \text{ joules.}$$

In this case, the "winner" involves the lightest object dropped the greatest distance.

29. **Answer:** B. Joule

    Rationale: Work, by definition, only occurs whenever a force has been applied to an object, and that object moves some distance. Remember the product of newtons (force) and meters (distance), determines joules.

30. **Answer:** C. Greater distance the object must travel

    Rationale: A 500-pound steel ball must be placed onto its pedestal three feet above the ground. Your maximum applied force is not adequate to lift the ball vertically. But if you arrange an inclined plane with a sloping surface of about 5 degrees, the top of the ramp ends at the top of the pedestal. It will now be possible to roll the steel ball up the ramp to place it on the pedestal. The ramp can then be removed, and your sculpture is complete and ready to be admired. The ball had to be moved 40 feet up the ramp instead of a vertical height of 3 feet, but you were able to move it using the simple machine, which allowed you to achieve the goal of lifting the massive sphere.

31. **Answer:** D. Move the fulcrum in the direction of the load to be lifted

    Rationale: Increasing the force generated by your body at will is not feasible, so answer choice A is not a viable option. Balancing the lever on the fulcrum may not necessarily help, since it depends on where it was originally, so answer choice B may not be helpful for you to achieve your goal. Increasing the height of the fulcrum simply increases the distance that the load must be moved, so answer choice C is unlikely to be helpful. Only answer choice D (moving the location of the fulcrum so that it is closer to the load) will necessarily be helpful. Experience on a teeter-totter tells us that this is true.

32. **Answer:** A. Gear

    Rationale: This is the definition of the mechanical advantage of a gear. For example, in turning a smaller gear, it means you will turn it more times to complete one rotation of a larger gear, but it translates into an increase in the torque that will be applied due to the larger radius of the larger gear. In the same way, if you turn a larger gear, you will have to apply more torque, but it will result in a faster rotation of the smaller gear, and that might be the desirable result of using the gear system.

33. **Answer:** B. Output Force : Input Force

    Rationale: Mechanical advantage is simply the ratio of the output force to the applied input force. Again, a simple machine increases either the amount or the direction of an applied force. Inevitably, the simple machine increases the output force; therefore, the mechanical advantage must be greater than "one." For example, a simple inclined plane allowed the Egyptians to raise very heavy blocks of stone high enough to place them on top of the previous level of stone blocks. The Egyptians did not have electrical or chemical power to move enormous stones of the pyramids. Only by using huge earthen inclined planes could the Egyptians increase the magnitude of the applied force generated by humans and animals to move those stones and construct the magnificent pyramids over 3,000 years ago.

34. **Answer:** D. Neither A nor B

    Rationale: A system of pulleys can change the magnitude of the applied force when used in the form of a "block-and-tackle." However, the simplest form of a pulley (with a single fixed rotating disk) simply changes the direction of the applied force without increasing the output force. It is possible to have a pulley system that moves the load in the same direction as the direction of the applied force; for example, you could have a block-and-tackle on a sailing ship that lowers the sails by pulling down on the rope. Therefore, neither answer choices A nor B are necessarily always true.

35. **Answer:** D. None of the above

    Rationale: A wedge is a simple machine that is designed to amplify the applied force. The mechanical advantage is defined as the ratio of the length of the slope divided by the height of the wedge. The more gradual the slope, the longer the incline. The longer incline will increase the mechanical advantage. Similarly, the smaller the height of the wedge, the greater the mechanical advantage since the ratio is dividing by a smaller number

$$Force_{output} = Force_{applied} \times \text{length of the incline/height of the incline}$$

If we apply a force of 10 newtons, to the face of the wedge with an incline length of 10 cm, and an incline height of 1 cm, what is the corresponding output force of the wedge, The use of the wedge amplifies the input force by 10 times! The output force would be 100 newtons.

36. **Answer:** B. Reduces the coefficient of friction

Rationale: The reason cars have wheels is to reduce the friction coefficient between the ground and the wheel, i.e., to reduce the friction force experienced when performing work. This makes the wheel one of the greatest inventions/discoveries in the history of humankind. The wheel replaces the ground friction coefficient with a friction coefficient at the axle of the wheel. If the axle is well-greased and/or has good bearings, then the friction coefficient is significantly reduced, meaning that the force required to pull or push an object becomes greatly reduced.

37. **Answer:** A. O moves at 3 ft/h and weighs 1 ton

Rationale: The momentum of an object is equal to its mass multiplied by its velocity; in other words, momentum is an object's tendency to keep moving. To answer this question correctly, it is important that you do two things: (1) convert all units so that they correspond, and (2) notice that the question is asking which scenario has the least momentum, not the most.

First, "convert" each of the four scenarios using the following conversion factors:

$$\begin{aligned} hour &= 60\ minutes \\ 1\ hour &= 3{,}600\ seconds \\ 1\ minute &= 60\ seconds \\ 1\ ton &= 2{,}000\ pounds \end{aligned}$$

Now "convert" each of the four scenarios given:

A. 3ft/h that weighs 1 ton has momentum as follows:

$$3\text{ft}/3600\text{s} \times 2{,}000 \text{ pounds}$$

$$0.00083 \text{ ft/s} \times 2{,}000 \text{ pounds or}$$

$$1.76 \text{ pounds} - \text{ft/s}$$

B. 5,000 ft/h that weighs 2 pounds has momentum as follows:

$$5{,}000 \text{ ft}/3{,}600 \text{ s} \times 2 \text{ pounds or}$$

$$1.39 \text{ ft/s} \times 2 \text{ pounds or}$$

$$\text{pounds} - \text{ft/s}$$

C. 1,000 ft/min that weighs 25 pounds has momentum as follows:

$$1{,}000 \text{ ft}/60 \text{ s} \times 25 \text{ pounds or}$$

$$16.67 \text{ ft/s} \times 25 \text{ pounds or}$$

$$417 \text{ pounds} - \text{ft/s}$$

D. 300,000 ft/s that weighs 0.001 pounds has momentum as follows:

$$300{,}000 \text{ ft/s} \times 0.001 \text{ pounds or}$$

$$300 \text{ ft} - \text{pounds/s}$$

Since all answers are in pounds-ft/s, the scenario with the least momentum is 1.76 pounds-ft/s. In this case, the smallest momentum is a very heavy object that is moving very slowly.

38. **Answer:** D. All the above

Rationale: Inclined planes, levers, and wedges are all examples of simple machines. A simple machine does not "create" force; instead, it enables applied force to be used in a useful way.

39. **Answer:** A. It increases the torque produced

Rationale: When you are in a lower gear on a bicycle, your legs move rapidly to maintain an adequate speed. However, it's also easier to climb hills in a lower gear, because you're able to apply more torque to the bicycle wheels. Remember that when going uphill, you must exert more force against gravity. As you know intuitively, climbing a hill in a higher gear may not be possible, because you can't move the pedals enough to apply the required torque to the wheels.

40. **Answer:** D. All the examples listed ARE wedges

Rationale: The basic shape of the three objects listed are the same. The axe is probably the most "dramatic" example of how powerful a wedge can be in amplifying applied power. Using only the power that can be generated by a human body, a person can cut through the trunk of a tree. This is done by applying muscle power to a very thin edge, and the reduction in surface area increases the effectiveness of the applied force.

41. **Answer:** C. 25 newtons

Rationale: Since force is the product of mass times acceleration, acceleration is the change in velocity divided by the time. For this problem, acceleration is $15/3$. The force is $5 \times 15/3$, or 25 newtons of force.

42. **Answer:** D. 8 newtons

Rationale: Since force is the product of mass times acceleration, acceleration is the change in velocity divided by the time. For this problem, acceleration is $12/4.5$. The force is $3 \times 12/4.5$, or 8 newtons of force.

43. **Answer:** A. 23,275 joules

Rationale: Since the potential energy in Earth's gravity is the product of mass times gravitational acceleration times the height, you need the gravitational acceleration value, which is $9.8\ m/s^2$. The product becomes $25 \times 95 \times 9.8$, which is 23,275 joules of potential energy.

44. **Answer:** D. 19,110 joules

Rationale: Since the potential energy in Earth's gravity is the product of mass times gravitational acceleration times the height, you need the gravitational acceleration value, which is $9.8\ m/s^2$. The kinetic energy that you apply must be enough to equal the potential energy at the peak of the trajectory. The product becomes $1.3 \times 1500 \times 9.8$, which is 19,110 joules of energy.

45. **Answer:** C. 345,679 joules

Rationale: The kinetic energy formula is one half the product of the mass times the square of the velocity. The conversion of 80 km/h to meters per second means multiplying by one thousand and dividing by 3600 (the number of seconds per hour). That value is 22.2222, which must be squared and divided by two, or $\sim 246.91\ (m/s)^2$. Multiplying by 1400, the answer becomes 345,679 joules.

46. **Answer:** A. Ionized

Rationale: Elements usually have the same number of electrons as protons. In the case where the opposite is true (if an atom has different number of protons and electrons), it is described as being "ionized." When an atom is ionized, it acquires a negative or positive charge by gaining or losing electrons.

47. **Answer:** C. A magnetic field is generated

Rationale: An inductor is a component in an electric circuit that can store voltage as a magnetic field. As electrons pass through an inductor (or coil), a magnetic field is generated.

48. **Answer:** C. The direction electrons move

Rationale: Alternating current (AC) and direct current (DC) are both forms of electricity in that they both are based on the movement of electrons through a circuit; however, the electrons move through the circuit differently. With DC, the flow of electrons is constant and unidirectional. The best example of a direct current source is a battery. With AC, the flow of electrons moves back and forth in an alternating "wave."

49. **Answer:** A. Low resistance, flexible, common

Rationale: Electrical resistance in a wire is a direct function of the wire's material and its thickness—the more conductive the wire's material, and the larger the diameter of the wire, the lower the resistance. Copper has very low resistance and is most often used as a material for wire/cable. Other materials have low resistance and flexibility properties, but are not so commonly used, because they are rare and/or more expensive. Copper is relatively common and affordable.

50. **Answer:** D. Reducing the rate that electrons can flow

Rationale: Resistors are simply components that are made to reduce the rate at which electrons can flow; therefore, resistors change the voltage and current within a circuit.

51. **Answer:** C. Current

Rationale: If electrons move through a wire made of conductive material, the concept can be modeled as being similar to water moving through a hose. A longer hose limits flow like a longer wire, and a larger diameter hose allows flow similar to a larger diameter wire. When the "hose" is a wire, that means it has resistance. Friction opposes and limits the motion/flow. Resistance is "friction" at an atomic level, and the energy lost from electrons and atoms "rubbing against each other" is converted to heat. The term for measuring resistance is ohms. Note: The terminology is incorrect to describe the "flow of current," since that is the "flow of the flow of charges." Current is more correctly thought of as the "quantity of charge motion."

52. **Answer:** A. Voltage

Rationale: The amount of current (amperes) moving through a circuit is determined by the resistance in the circuit and the voltage that is applied. The voltage is measured in volts, and it is a measure of how much each electron's kinetic energy increases. Higher voltage means each electrical charge experiences a greater attraction in the circuit. Voltage propels current through resistance, so the higher the voltage, the more current will pass through a given resistance.

Voltage is sometimes referred to as *electromotive force* or *EMF*. It is unrelated to how much electricity there is, just as the pressure in a hose does not tell you how much water is present. EMF propels current through the circuit. Without EMF, electrical charge still exists in each conductor, but it will have no motion.

53. **Answer:** B. 0.10 amps

Rationale: This is simply a problem that requires knowledge of Ohm's laws. If you want to find the number of amperes (amps), or current (I) flowing through a circuit, use this formula:

$$I = V/R$$
$$I = 12/120$$
$$I = 0.1 \text{ amps}$$

54. **Answer:** B. Amp (ampere)

Rationale: This is the definition of an ampere, and it is critically important to know.

55. **Answer:** A. Volt (voltage)

Rationale: The amount of current moving through a circuit is determined by the applied voltage or EMF. When there is water (electrons/current) in the hose (circuit), it is always full, but it may come out with a pressure that squirts it 5 feet, or it might come out with a pressure that squirts it 50 feet—depending on the pressure. Volts are a measure of the force moving the current through a circuit. Without applied voltage, electrical charge is in the circuit, but it is not mobile; therefore, there is no charge motion (current) in the circuit.

56. **Answer:** C. 40 milliamperes

Rationale: Ohm's law tells you that current (I) is the ratio of V/R. Since $V = 6$ volts and $R = 150$ ohms, the current is $6/150$ amperes. That ratio is 0.040 amperes, or 40 milliamperes.

57. **Answer:** B. 525 volts

Rationale: Ohm's law tells you that voltage (V) is the product of $I \times R$. Since $I = 10.5$ amperes and $R = 50$ ohms, the current is 10.5×50 volts. That product is 525 volts.

58. **Answer:** A. 0.80 ohms

Rationale: Ohm's law tells you that resistance (R) is the ratio of V/I. Since V = 6 volts and I = 7.5 amperes, the resistance must be 6/7.5 ohms. That ratio is 0.80 ohms.

59. **Answer:** D. Electron

Rationale: Neutrons and protons are part of the nuclear structure and are fixed in the structure of the conductor. Electrons are the mobile charges that are in the orbitals of the conductor molecules. Voltage applied to a conductor creates a force that moves those charges through the conductor.

60. **Answer:** B. Voltage

Rationale: Electromotive force (EMF) is a term that is the same as voltage. When EMF is applied, the electrons are the mobile charges that are attracted to a positive potential and repelled by a negative potential. Voltage applied to a conductor moves those charges through the conductor.

Final Thoughts

Congratulations on reaching the end of this comprehensive AFOQT Study Guide! If you have made it this far, you can feel confident in your preparation. This book has been meticulously designed to cover all aspects of the AFOQT test, providing you with the knowledge and practice needed to excel. You have invested time and effort into your studies, and that dedication is commendable. As you approach your test date, be confident that you are well-prepared.

Also, remember that this is just one step in your academic and career journey. Your performance on this test does not define your worth or potential. Approach the test with a positive mindset and trust in your preparation.

Finally, thank you for choosing Spire Study System's AFOQT Study Guide. Your support and trust in this book mean a lot to us. If you have found this guide helpful and comprehensive, please leave a positive review on Amazon. Your feedback helps other students find the right resources and supports the continued creation of quality study materials.

Best of luck with your AFOQT test, and may your hard work lead you to success in your next endeavors!

Andrew T. Patton
Chief Editor
Spire Study System
Email: MyBookFeedback@outlook.com

Made in the USA
Middletown, DE
04 May 2025

74873277R00190